Advances in Geosciences

Volume 16: Atmospheric Science (AS)

ADVANCES IN GEOSCIENCES

Editor-in-Chief: Wing-Huen Ip *(National Central University, Taiwan)*

A 5-Volume Set

Volume 1: Solid Earth (SE)
ISBN-10 981-256-985-5

Volume 2: Solar Terrestrial (ST)
ISBN-10 981-256-984-7

Volume 3: Planetary Science (PS)
ISBN-10 981-256-983-9

Volume 4: Hydrological Science (HS)
ISBN-10 981-256-982-0

Volume 5: Oceans and Atmospheres (OA)
ISBN-10 981-256-981-2

A 4-Volume Set

Volume 6: Hydrological Science (HS)
ISBN-13 978-981-270-985-1
ISBN-10 981-270-985-1

Volume 7: Planetary Science (PS)
ISBN-13 978-981-270-986-8
ISBN-10 981-270-986-X

Volume 8: Solar Terrestrial (ST)
ISBN-13 978-981-270-987-5
ISBN-10 981-270-987-8

Volume 9: Solid Earth (SE), Ocean Science (OS) & Atmospheric Science (AS)
ISBN-13 978-981-270-988-2
ISBN-10 981-270-988-6

A 6-Volume Set

Volume 10: Atmospheric Science (AS)
ISBN-13 978-981-283-611-3
ISBN-10 981-283-611-X

Volume 11: Hydrological Science (HS)
ISBN-13 978-981-283-613-7
ISBN-10 981-283-613-6

Volume 12: Ocean Science (OS)
ISBN-13 978-981-283-615-1
ISBN-10 981-283-615-2

Volume 13: Solid Earth (SE)
ISBN-13 978-981-283-617-5
ISBN-10 981-283-617-9

Volume 14: Solar Terrestrial (ST)
ISBN-13 978-981-283-619-9
ISBN-10 981-283-619-5

Volume 15: Planetary Science (PS)
ISBN-13 978-981-283-621-2
ISBN-10 981-283-621-7

A 6-Volume Set

Volume 16: Atmospheric Science (AS)
ISBN-13 978-981-283-809-4
ISBN-10 981-283-809-0

Volume 17: Hydrological Science (HS)
ISBN-13 978-981-283-811-7
ISBN-10 981-283-811-2

Volume 18: Ocean Science (OS)
ISBN-13 978-981-283-813-1
ISBN-10 981-283-813-9

Volume 19: Planetary Science (PS)
ISBN-13 978-981-283-815-5
ISBN-10 981-283-815-5

Volume 20: Solid Earth (SE)
ISBN-13 978-981-283-817-9
ISBN-10 981-283-817-1

Volume 21: Solar Terrestrial (ST)
ISBN-13 978-981-283-819-3
ISBN-10 981-283-819-8

Advances in

Geosciences

Volume 16: Atmospheric Science (AS)

Editor-in-Chief

Wing-Huen Ip

National Central University, Taiwan

Volume Editor-in-Chief

Jai Ho Oh

Pukyong National University, South Korea

World Scientific

NEW JERSEY • LONDON • SINGAPORE • BEIJING • SHANGHAI • HONG KONG • TAIPEI • CHENNAI

Published by

World Scientific Publishing Co. Pte. Ltd.
5 Toh Tuck Link, Singapore 596224
USA office: 27 Warren Street, Suite 401-402, Hackensack, NJ 07601
UK office: 57 Shelton Street, Covent Garden, London WC2H 9HE

British Library Cataloguing-in-Publication Data
A catalogue record for this book is available from the British Library.

ADVANCES IN GEOSCIENCES
A 6-Volume Set
Volume 16: Atmospheric Science (AS)

ISBN-13 978-981-283-808-7 (Set)
ISBN-10 981-283-808-2 (Set)
ISBN-13 978-981-283-809-4 (Vol. 16)
ISBN-10 981-283-809-0 (Vol. 16)

Typeset by Stallion Press
Email: enquiries@stallionpress.com

Printed in Singapore.

EDITORS

REVIEWERS

The Editors of Volume 16 (Atmospheric Science) would like to acknowledge the following referees who have helped review the manuscripts published in this volume:

K. H. Chang
Anmin Duan
Xueliang Gue
J. G. Jhun
Ashwani Kulkarani
J. Y. Lee
Li Maoshan
Sanjay Malik
Mani Murail
Albert Rugumayo
K. H. Seo
U. S. Singh
N. Sontakke
Young Suno
Ramesh Yadav
Gang Chen
Gennady Glenko
Jinhai He
C. H. Kim
D. K. Lee
N. H. Lin
S Mahapatra
Detlef Muller
R. Park
A. N. V. Satyanarayan
G. P. Singh
S. O. Shoban
Manoj Kumar Srivastava
Morid Taqib
Tianjun Zhou

PREFACE

The present volume set of Advances in Geosciences (ADGEO) contains papers from the Busan annual meeting in 2008 and some from the Singapore annual meeting in 2009. As Editor-in-Chief, I must apologize to the AOGS members and authors for this delay. Since 2006 we have published 20 volumes in total. This publication project has been supported by the AOGS Council, World Scientific Publication Company (WSPC), the team of hard working editors and the broad membership and participants of AOGS. As with the main purpose of the Society, ADGEO is meant to promote information exchange and to forge scientific cooperation in the area of Earth science and environmental study. As witnessed by the negotiation efforts at the United Nations Climate Change Conference in Copenhagen in December 2009, all these issues have become more and more important and vital in the Asia-Pacific region. It is not a matter of exaggeration in saying that the solution to global warming, if there is one, has to come from the emerging economies and developing countries covered by AOGS. By design, ADGEO has its fundamental role to play. In practical terms, it is actually a difficult task because of many factors involved in deciding the quality of manuscripts, editorial and review processes, publication procedure, scientific impacts, readership, policy of the AOGS Council, and last but not least, marketing from the point of view of the publisher. Any small mishap in this long chain of interactive steps could lead to a major discontinuity. We have encountered such a situation with the publication of the Busan manuscripts. It is only with the cooperation of the authors, the ADGEO editorial team, WSPC, and the AOGS Secretariat Office, that we are able to produce these volumes, albeit a long delay. With this lesson learned, we hope to consolidate the ADGEO management and editorial system so that it would become an essential publication in our understanding of Earth and space science and information tool books in the battle against climate change. Finally, I would like to take this opportunity to thank the Volume Editors-in-Chief who are the driving force in making ADGEO possible: A. Bhardwaj (Planetary Science), M. Duldig (Solar and Terrestrial Science), J.P. Gan (Ocean Science), J.H. Oh (Atmospheric Science), N.S. Park (Hydrological Science), K. Satake and C.H. Lo (Solid Earth). They have to work very hard to ensure both the quantity and quality of the published papers in ADGEO. Of equal importance, the support from WSPC is essential and its foresight in identifying the academic and social values of Earth science and environmental study to be sustained and articulated by ADGEO is very much appreciated.

Wing-Huen Ip
Editor-in-Chief

PREFACE TO AS VOLUME

This special issue of Atmospheric Science (AS) contains the scientific papers of both the Busan and Singapore conferences. Since public interest in atmospheric research is increasing as concerns grow about mean monsoon conditions, climate change, air pollution and its effect on monsoon variability, climate change due to anthropogenic and natural processes, visibility, and public health. Scientists from around the world were invited to present their original research, and to submit full manuscripts for peer review and publication in AOGS volumes. The submitted papers cover a large area of atmospheric fields like the understanding of monsoon and its variability, the study of extreme events, the effect of Lunar Synodic influence on weather and climate, Dipole variability in Indian monsoon, cyclone/typhoon study, suspended solid particles, etc. Other papers on air pollution cover a wide range of new information, including ambient and indoor air pollution monitoring, air quality modeling, aerosols optical thickness, and integrated air quality monitoring. I hope that this special issue will be very useful for young researchers, as they will get a wide range of research articles in a single volume.

Having encountered many difficulties in determining the quality of manuscripts, editorial and review processes, publication procedure, scientific impacts, readership, policy of the AOGS Council, and viewpoints of the publisher, I am happy to complete the editing of all the manuscripts. All this was only possible with the cooperation of the authors, the ADGEO AS and ADGEO editorial teams, WSPC, and the AOGS Secretariat Office. Finally, I would like to take this opportunity to thank the Volume Editors in AS who are the driving forces in making ADGEO possible: G. P. Singh (Banaras Hindu University), C. Wu (National Taiwan University), and K.-J. Ha (Pusan National University). They have worked very hard to ensure both the quantity and quality of the published papers in the AS volume of ADGEO.

Jai Ho Oh
AS Volume Editor-in-Chief

CONTENTS

Editors v

Reviewers vii

Preface ix

Preface to AS Volume xi

Relationship Between Activity of West Pacific Subtropical High and Diabatic Heating During Heavy Rain in South China in June 2005 1

Wang Lijuan, Guan Zhaoyong and Yan Chengyu

Lunar-Synodic Component in the East Asian Winter Monsoon 13

Chi-Hua Wu and Wen-Shung Kau

August Rainfall in Korea and its Association with Circulations 23

Kyung-Ja Ha and Kyung-Sook Yun

Changma and Shifting Peak in Summer Rainfall of Korea: ENSO Influence 35

Kyung-Ja Ha, Sun-Seon Lee, P. N. Vinayachandran and Jong-Ghap Jhun

Prediction of Regional Drought Over Korea Using an Analog Method 47

Su-Bin Oh, Do-Woo Kim, Ji-Sun Lee, Hi-Ryong Byun and Ki-Seon Choi

Decadal Changes in the Hadley Circulation 61
Yongyun Hu and Chen Zhou

Simulation of Record Rainfall Event Over Mumbai on 26 July, 2005 75
G. P. Singh, Jai-Ho Oh and H. K. Chaudhary

The Leading Mode of Variability in the Indian Monsoon Region in the Absence of the ENSO Variability in the NCEP CFS 83
Deepthi Achuthavarier and V. Krishnamurthy

Longitudinal Oscillations of the South Asian High and the Subtropical Western Pacific High During Boreal Summer 93
Ling Zhang and Xiefei Zhi

Singular Vector and ENSO Predictability in a Hybrid Coupled Model 109
Xiaobing Zhou and Youmin Tang

Study of Lightning Activity Over Indian Subcontinent 121
H. S. Chaudhari, M. R. Ranalkar, Y. V. Kumkar, Jai Ho Oh and G. P. Singh

Improving Neural Network Models for Forecasting Seasonal Precipitation in Southwestern Iran: The Evaluation of Oceanic–Atmospheric Indices 133
Mohammad J. Nazemosadat, Peyman Setoodeh and Ali A. Safavi

Land Surface Energy Budget Over the Tibetan Plateau Based on Satellite Remote Sensing Data 147
Yuichiro Oku and Hirohiko Ishikawa

Long Term Variations in Temperature in Association with Convective Available Potential Energy in the Upper Troposphere Using Radiosonde Data Over Delhi (28.3°N, 77.1°E) and Kolkata (22.3°N, 88.2°E), India 155

S. K. Dhaka, R. Sapra, V. Panwar, M. Kaur, A. Goel, T. K. Mandal and A. R. Jain

Impact of Climate Change on the East Asia Droughts 165

Do-Woo Kim, Ki-Seon Choi, Ji-Sun Lee and Hi-Ryong Byun

Relationship between Summer Tropical Cyclone Genesis Frequency and Winter Aleutian Low Oscillation 179

Ki-Seon Choi, Do-Woo Kim, Su-Bin Oh, Ji-Sun Lee and Hi-Ryong Byun

Geomorphic Influences on the Variability of Coastal Erosion and Deposition on Ambae Island, Vanuatu, Caused by Cyclone Funa in January 2008 193

James P. Terry

Stratospheric Ozone Climatology and its Variability Over Ankara 203

Deniz Demirhan Bari, Ceyhan Kahya, Sema Topcu and Selahattin Incecik

Near Surface Atmospheric Methane Concentration at Suburban Environment of Guwahati by Differential Absorption Lidar System 215

Manoj Saikia, Minakshi Devi and Ananda Kumar Barbara

Source Apportionment Analysis of Measured Fine Particulate Matter in a Semi-Arid Urban Airshed in Corpus Christi, U.S.A 227

Saritha Karnae and Kuruvilla John

Multi-Scale Organization of Water Vapor Over Low and Mid-Tropical Africa 241

Ondego Joel Botai, Venkataraman Sivakumar, Willem Ludwig Combrinck and Cornelis Johannes De Wet Rautenbach

Aerosol Measurements Over South Africa Using Satellite, Sun-Photometer and Lidar 253

Venkataraman Sivakumar, Melaku Tesfaye, Wondimu Alemu, Ameeth Sharma, Christoph Bollig and Gizaw Mengistu

Mean Ozone and Water Vapor Height Profiles for Southern Hemisphere Region Using Radiosonde/Ozonesonde and Haloe Satelite Data 263

Venkataraman Sivakumar, Desalegne Tefera, Gizaw Mengistu and Ondego Joel Botai

Ground Based Lidar Observations of Anomalies in Middle Atmospheric Temperature Profiles Over a Tropical Station Gadanki (13.5°N, 79.2°E) 271

C. Nageswara Raju, M. Krishnaiah and Y. Bhavani Kumar

Statistical Distribution Models for Urban Air Quality Management 285

A. Deepa and S. M. Shiva Nagendra

First Ozone Sounding Results Over the UAE 299

Tariq Majeed, Mazhar Iqbal, Onuod Al-Marzouqi, David W. Tarasick, Jonathan Davies, Samuel J. Oltmans, Anne M. Thompson, Abdullah Mandoos Mohammad Al-Abri, S. Khalid Zaidi and Phil Rogers

Magnetic Measurements of Atmospheric Dust Deposition in Soils 311

Aleš Kapička, Eduard Petrovský, Hana Grison, Vilém Podrázský and Pavel Křížek

Impact of Vegetation on the Indian Summer Monsoon: Model Sensitivity Experiments 321

Trilochan Pattanaik, H. S. Chaudhari, Jai Ho Oh, Ashish Dhakate and G. P. Singh

Lidar Observations of Stratospheric Aerosol Over Gadanki 331

C. Nageswara Raju, M. Krishnaiah and Y. Bhavani Kumar

Advances in Geosciences
Vol. 16: Atmospheric Science (2008)
Eds. Jai Ho Oh *et al.*

RELATIONSHIP BETWEEN ACTIVITY OF WEST PACIFIC SUBTROPICAL HIGH AND DIABATIC HEATING DURING HEAVY RAIN IN SOUTH CHINA IN JUNE 2005

WANG LIJUAN
Key Laboratory of Meteorological Disaster of Jiangsu Province,
Nanjing University of Information Science & Technology,
Nanjing 210044, China
wljfw@163.com

GUAN ZHAOYONG
Nanjing University of Information Science & Technology, Nanjing 210044, China

YAN CHENGYU
Qinhuangdao Meteorological Bureau, Qinhuangdao, 066000, China

In terms of NCEP/NCAR daily reanalysis data, the position variation of the west Pacific subtropical high (WPSH) in June 2005 and its relation to diabatic heating are analyzed based on the complete vertical vorticity equation. The results show that the position variation of WPSH is associated with the diabatic heating in subtropical areas. In comparison with the climatology, there is strong heating on the north side of WPSH and relatively weak ITCZ convection on the south. Each of westward extension of WPSH corresponds to a significantly enhanced heating to the west of WPSH. In mid-troposphere, the vertical variation of heating on the north (south) of WPSH during 12–24 June, 2005, is basically greater(less) than the climatology, which is unfavorable to the northward movement of WPSH. In the mid and late of June 2005, the vertical variation of heating over the eastern coast of the Arabian Sea and the Bay of Bengal (to the west of WPSH) is largely higher than the climatology, which is in favor of the increase of anti-cyclonic vorticity on the west of WPSH, inducing westward extension of WPSH. As a result, the heating on the north and south, and to the west of WPSH work together to make WPSH extend more southward and westward in June 2005, which is in favorable to rainbelts maintaining in South China.

1. Introduction

In June 2005, large-scale intense rainfall events followed in succession in the southern Jiangnan area (north of 25°N, east of 110°E) and much of South

China with the consequence that parts of Guangxi, Guangdong, Fujian, Jiangxi, Hunan and Zhejiang were hit by torrential rains, with exceptionally heavy rains measured on a local basis. These events resulted in flooding and landslides, making for considerable casualties and loss of property. In particular, during June 17–26 a persistent strong rainfall happened to South China and the east of Jiangnan region, characterized by extensive cover and high severity of 100∼200 mm in general and the highest being 400∼1100 mm on a local basis. Overall, the rainfall measurements were 100∼300 mm higher compared to normal for South China areas. In June 2005, the west Pacific subtropical high (WPSH thereafter) was southward and westward of mean, allowing the confluence in the South China Sea of easterly flows from the south side of the WPSH with the SW warm, moist flows from the equatorial India Ocean and the Bay of Bengal, both branches of airflows transporting along the west side of the WPSH plenty of vapor into South China and the Jiangnan area. South China resided in the center of stronger vapor convergence leading to the occurrence and maintenance of torrential rains [1,2].

Many meteorologists suggest that the flood/drought in South China has close relationship with the seasonal shift of WPSH [3–6]. In particular, WPSH west-east location and north-south movement are responsible for East Asian monsoon establishment, precipitation over Yangtze River valley and South China. As a primary member of the East Asian monsoon, WPSH associates closely to diabatic heating and its west-east and north-south shifts, to a large extent, is due to the spatial patterns of diabatic heating [7]. Wu *et al.* [8–10] also pointed out that the latent heating of condensation induced by the East Asian monsoon is a key role affecting WPSH locations and intensity. Thereby in this paper, through comparisons with long-term mean status, the position variations of WPSH during heavy rain over South China in June 2005 are mainly studied, and the complete vertical vorticity equation is used to explain how the diabatic heating affects the activity of WPSH. The current study can provide a better understanding of the physical cause of flood in South China due to the shift of WPSH location and give some factual bases and ideas to predict the flood/drought.

2. Datasets and Methods

The daily NCEP/NCAR reanalysis datasets with $2.5^\circ \times 2.5^\circ$ horizontal resolution and WPSH ridge line Index data from CMA.

2.1. *Heating source calculation*

Atmospheric heating source Q_1 in the paper is calculated as follows [11]:

$$Q_1 = c_p \left(\frac{\partial \bar{T}}{\partial t} + \vec{V} \cdot \nabla \bar{T} + \bar{\omega} \left(\frac{p}{p_0} \right)^{\kappa} \frac{\partial \bar{\theta}}{\partial p} \right), \tag{1}$$

where $\kappa = R/c_p$, R and c_p are the constants for dry air gas and specific heat at constant pressure respectively and θ is the potential temperature. The three terms in the right brackets describe local variation, horizontal advection and vertical transport separately. Q_1 denotes the heating rate per unit time and mass.

Equation (1) can be rewritten as.

$$Q_1 = Q_R + L(\bar{c} - \bar{e}) - \frac{\partial}{\partial p} \overline{(S'\omega')}. \tag{2}$$

Equation (2) is integrated from p_t (100 hPa) to p_s (surface pressure) in the whole layer and reaches the following form of

$$\langle Q_1 \rangle = \langle Q_R \rangle + LP + SH, \tag{3}$$

where $\langle \rangle = \frac{1}{g} \int_{p_t}^{p_s} ()\, dp$. It can be seen that the atmospheric apparent heating source $\langle Q_1 \rangle$ is consisted of radiative heating (cooling)$\langle Q_R \rangle$, latent heating from precipitation LP and surface sensible heating SH.

2.2. *Discussion on the complete vertical vorticity equation*

The complete vertical vorticity Eq. [12] can be written as

$$\frac{\partial \zeta}{\partial t} + \vec{V} \cdot \nabla \zeta + \beta v = (1-\kappa)(f+\zeta)\frac{\omega}{P} - (f+\zeta)\frac{Q}{\theta} + \frac{1}{\alpha}\frac{d}{dt}\left[\frac{P_E}{\theta_Z} - C_d\right]$$
$$+ \frac{1}{\theta_z} F_\zeta \cdot \nabla \theta_z + \frac{f+\zeta}{\theta_z}\frac{\partial Q}{\partial z} - \frac{1}{\theta_z}\frac{\partial v}{\partial z}\frac{\partial Q}{\partial x} + \frac{1}{\theta_z}\frac{\partial u}{\partial z}\frac{\partial Q}{\partial y}, \quad \theta_z \neq 0, \tag{4}$$

where $\theta_z = \partial\theta/\partial z$, P_E is Ertel geopotential vorticity, C_d is the parameter representing atmospheric thermal structure, and others are meteorological symbols in common use. Respectively, the first four terms on the right-hand side represent the vertical motion, heating source, internal variation in the atmospheric thermodynamics and effect of frictional dissipation on vorticity; the last three terms on the right-hand side are contributions from spatial inhomogeneous heating.

Taking no account of the effects of internal variation in the atmospheric thermodynamics, heating source and frictional dissipation but only the apparent heating source Q_1, Eq. (4) can thus be rewritten as

$$\frac{\partial\zeta}{\partial t} + \vec{V}\cdot\nabla\zeta + \beta v = (f+\zeta)(1-\kappa)\frac{\omega}{P} + \frac{f+\zeta}{\theta_z}\frac{\partial Q_1}{\partial z} - \frac{1}{\theta_z}\frac{\partial v}{\partial z}\frac{\partial Q_1}{\partial x} - \frac{1}{\theta_z}\frac{\partial u}{\partial z}\frac{\partial Q_1}{\partial y}. \tag{5}$$

From a scale analysis[8], the vertical motion and horizontal inhomogeneous heating induced by the atmospheric apparent heating source are at $10^{-11}\sim10^{-12}$ order of magnitudes, which is one order of magnitude or more smaller than the forcing by apparent heating source vertical variation (10^{-10}). Then, Eq. (5) is changed to

$$\frac{\partial\zeta}{\partial t} = \frac{f+\zeta}{\theta_z}\frac{\partial Q_1}{\partial z} - \beta v - \vec{V}\cdot\nabla\zeta. \tag{6}$$

3. Behavior of WPSH in June 2005

Figure 1 shows the 5880 gpm contours of WPSH at the 500 hPa level on June 17–26, 2005 and in climatological June. It is obvious that WPSH in 2005 distributed along a west-east zone, spanning about 20 latitudes, with

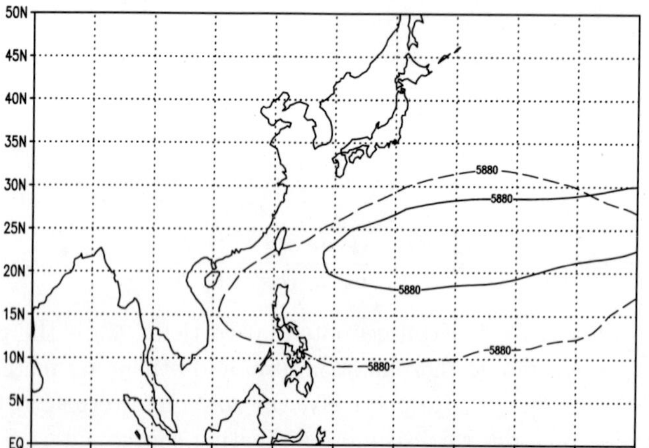

Fig. 1. WPSH characteristics as represented by 5880 gpm contours at 500 hPa during June 17–26, 2005 (the dashed line) and of climatological mean in June (the solid line), unit: gpm.

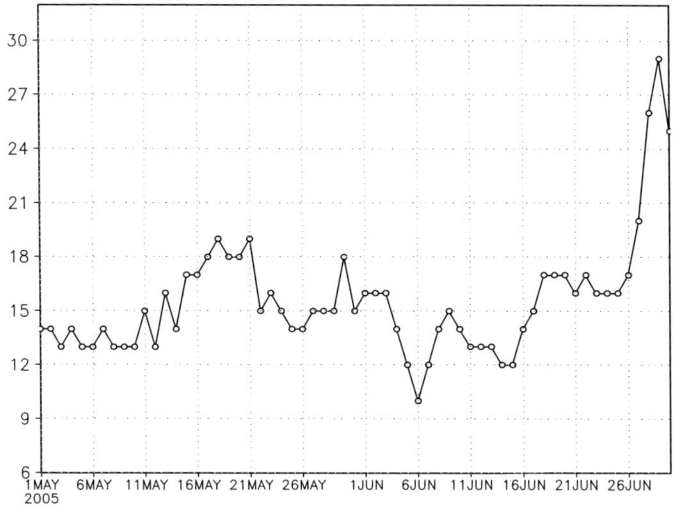

Fig. 2. The evolution of ridge of subtropical high along 120°E during May–June, 2005.

the 120°E ridge staying around 16°N and west-extending ridge point around 110°E. Comparing with the climatological mean, the WPSH ridge in 2005 shifted southwards about 6 latitudes and the 5880 gpm contour extended westwards by about 20 longitudes. Confluence of mid-latitude northwesterly flows with southwesterly on the north side of WPSH controlled much of South China areas. As a primary cause, WPSH anomalies induced flood over South China.

Figure 2 depicts the evolution of WPSH ridge line along 120°E during May–June 2005, where we can see that the ridge line took on a more southern position compared to normal. In June 3–6 the ridge retreated from 16°N to the south of 12°N, followed by its gradual northward lifting to 15°N and re-withdrawal from June 9, residing around 12°N on June 15, and another northward shift began from June 16 till day 25, with the ridge line maintained constantly at 16∼17°N, which was responsible for the rain band staying long in South China, with strong rainfall relatively concentrated and lasting for a long period. The WPSH made another northward jumping, leading to the beginning of the period of Plum rainfall in the Yangtze River valley, with the ridge line crossing 24°N on June 28. When the ridge line was south of 20°N, warm, moist airflows carried by WPSH met weak cold air moving southward from the mid-latitude just in South China, forming a front zone and thereby generating a stable rain band.

4. Relationship between Spatial Distribution of Diabatic Heating and Location of WPSH

There are complex interactions between WPSH and convective heating on the north/south sides and east/west sides. By scale analysis and numerical simulations based on the complete vertical vorticity tendency equation, Wu *et al.* [8–10] confirmed that condensation latent heating had an important role to play in determining the summertime location and intensity of the Eastern Hemispheric WPSH. Thus, this section will focus on the effects of convection-induced diabatic heating on WPSH in June 2005 and the difference from the climatological mean.

Figure 3 shows that WPSH shifts along a north-south direction in June. As for the climatological mean (Fig. 3(a)), there is, through the year, a strong heating center south of WPSH that is stronger in evidence than the north side of WPSH. That is due to continuous convective development at ITCZ south of WPSH. WPSH moves northward by season, and during different periods, there is various heating distribution in WPSH ridge and the north of it. In the first ten days of June, WPSH withdraws eastwards and locates east of 130°E after the onset summer monsoon, and convection develops on the west of WPSH (120–130°E) due to weak western ridge and a corresponding weak heating appears over the WPSH

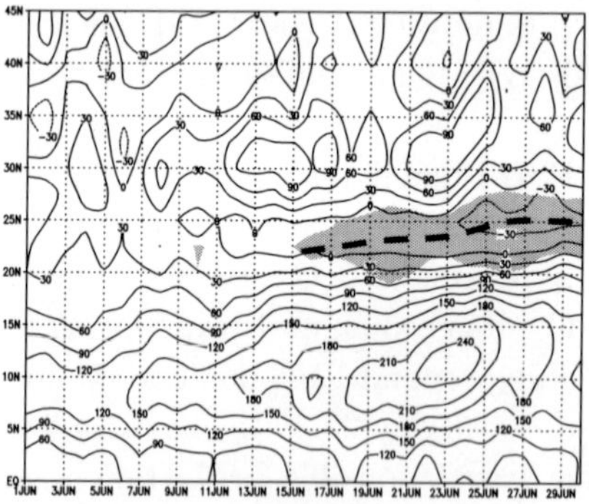

(a)

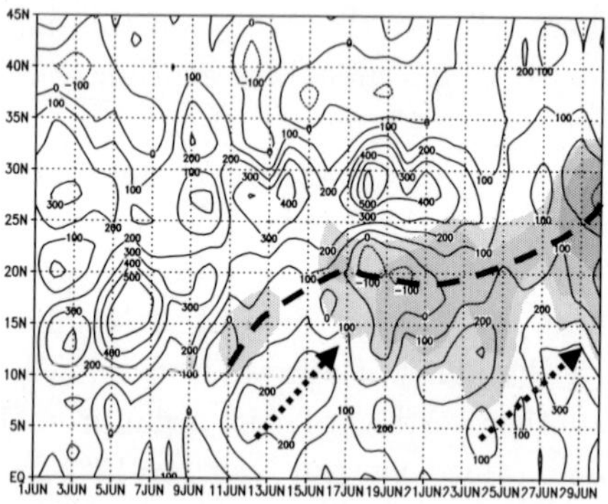

(b)

Fig. 3. Time–latitude cross section of atmospheric apparent heating source $\langle Q_1 \rangle$ of climatology along 120–150°E (a) and in 2005 along 110–140°E (b) (unit: W/m^2). Areas with 500 hPa geopotential height larger than 5880 gpm are shaded, and thick dashed lines represent mean locations of the ridge. Dotted lines with arrows denote the axis of atmospheric heating source extremum corresponding to the northward moving of WPSH.

region at 120–150°E. In the middle ten days of June, along with the northward movement, WPSH locates on the cooling area where radiative cooling and upward sensible heating are dominant. There is the heating areas of subtropical monsoon rainband to the north of WPSH, where $\langle Q_1 \rangle$ significantly increased. It is noted that the contour of 5880 gpm only appears during this period, which suggests a strongest WPSH that time. The area encircled by the 5880 gpm contour is corresponding to descending cloudless region, which is thus surrounded by zero heating contour.

Different from the climatology, the WPSH ridge steadily maintained at the south of 15°N during the early June, 2005 (Fig. 3(b)), and there was strong heating on the north of WPSH. WPSH jumped northwards around 11 June and its ridge line maintained at about 20°N from 16 to 26 June, which was favorable to the rainbelt staying in South China. At the same time, $\langle Q_1 \rangle$ increased remarkably on the north of WPSH. WPSH moved northward again on 26 June. It is noted that the heating on the south of WPSH extended northward as WPSH shifted northward(as shown by arrows in Fig. 3(b).

Figure 4 shows the east-west shift of WPSH in June. Climatological southern boundary of WPSH advances continuously from 15°N to 20°N, with convection developing on the south of WPSH. While there is weak heating in the WPSH area at 10–25°N. There are high-value heating centers to the west of WPSH (110–120°E) followed by the western extensions

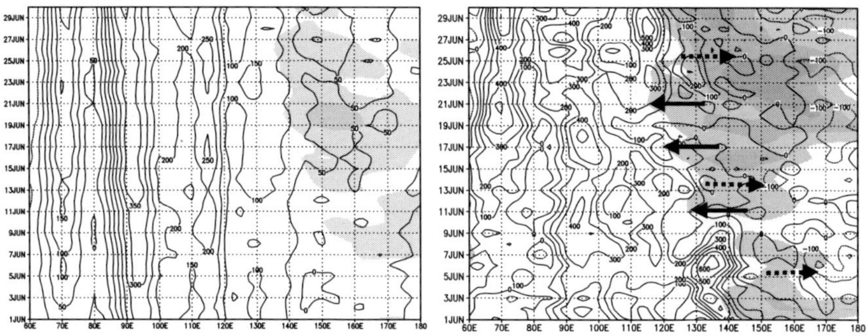

(a) (b)

Fig. 4. Time–longitude cross section of apparent heating source $\langle Q_1 \rangle$ of climatology (a) and 2005 (b) along 10–25° N (the 500 hPa area lager 5880 gpm is shaded). Solid (dotted) thick lines with arrows describe the westward extension (eastward retreat) of WPSH.

of WPSH (Fig. 4(a)). In comparison with the climatology, WPSH was characterized by westward extension in June 2005 (Fig. 4(b)). In the first ten days of June, the western ridge point of WPSH located at the east of 140°E; On 11 June, WPSH extended westward to 130°E (as shown by solid arrows in Fig. 4(b)), and then moved quickly eastward (as shown by dotted arrows in Fig. 4(b)) and again extended to 120°E around 16 June. WPSH extended westward thirdly to the west of 120°E on 21 June. It is noted that there were high-value heating centers to the west of WPSH (the Bay of Bengal area) company with the western extensions of WPSH. While the high-value heating centers on the west of WPSH (the east of 110°E) were corresponding to the retreats of WPSH. In general, there was relatively weak heating in the WPSH area surrounded by the 5880 contour and the WPSH center was usually a cooling area.

5. Effects of Vertical Variation of Diabatic Heating on WPSH Position

5.1. *Effects of vertical variation of diabatic heating on the north and south of WPSH on its southward and northward movements*

On the south/north side of WPSH (south: 110–140°E, 5–15°N; north: 110–140°E, 20–30°N), easterly (westerly) is prevailed ($v \approx 0$), and vorticity advection along the x axis is quite weak. The calculation results confirm this fact (figure not shown). So following the Eq. (6), we obtain

$$\frac{\partial\zeta}{\partial t} \approx \frac{f+\zeta}{\theta_z}\frac{\partial Q_1}{\partial z}. \tag{7}$$

That is, the local change of vorticity depends mainly on the vertical variation of diabatic heating on the south/north side of WPSH.

Figure 5(a) and 5(b) shows day-to-day evolution of $\frac{f+\zeta}{\theta_z}\frac{\partial Q_1}{\partial z}$ on the north and south sides of WPSH at 500 hPa for the climatological mean and June 2005, respectively. The north heating had greater vertical variation than the climatology during 12–24 June, 2005, i.e. $(\frac{f+\zeta}{\theta_z}\frac{\partial Q_1}{\partial z})_{2005} > (\frac{f+\zeta}{\theta_z}\frac{\partial Q_1}{\partial z})_{\text{mean}} > 0$, so that $(\frac{\partial\zeta}{\partial t})_{2005} > 0$. the increase of cyclonic vorticity on the north of WPSH was unfavorable for the northward movement of WPSH; whereas, the south heating had smaller vertical variation than the climatology in all time, namely $(\frac{f+\zeta}{\theta_z}\frac{\partial Q_1}{\partial z})_{2005} < 0 < (\frac{f+\zeta}{\theta_z}\frac{\partial Q_1}{\partial z})_{\text{mean}}$, so that

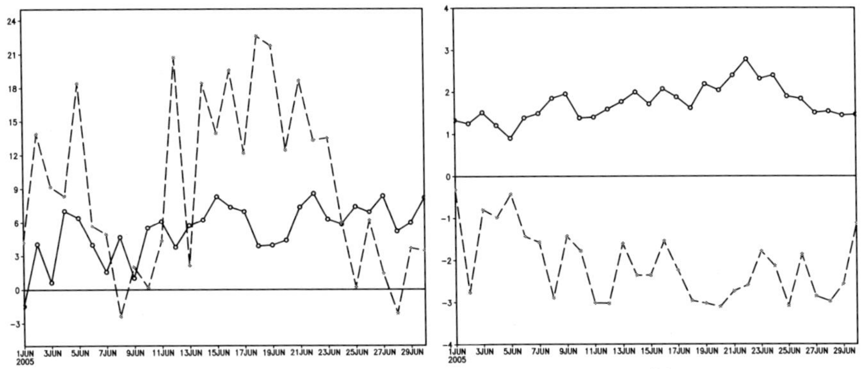

(a) (b)

Fig. 5. Day-to-day evolution of $\frac{f+\zeta}{\theta_z}\frac{\partial Q_1}{\partial z}$ on the north (a, 110–140° E, 20–30° N) and south (b,110–140° E, 5–15° N) sides of WPSH at 500 hPa for the climatological mean (the solid line) and June 2005 (the dashed line). unit: $10^{-11}\mathrm{s}^{-2}$.

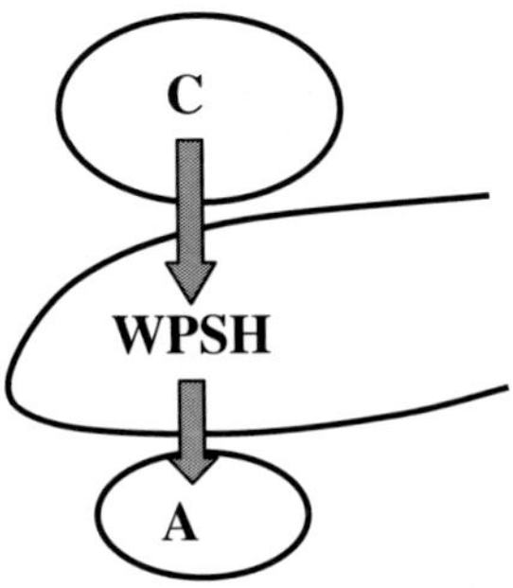

Fig. 6. Effects of the diabatic heating on the north (south) side of WPSH on the movement of WPSH.

$(\frac{\partial\zeta}{\partial t})_{2005} < 0$ was favorable for the increase of anti-cyclonic vorticity on the south side of WPSH and southward movement. With the effects of both north and south diabatic heating, the WPSH extended more southwards in June 2005 (Fig. 6).

5.2. *Effect of vertical variation of diabatic heating to the west of WPSH on its westward movement*

In this section, we will discuss the effect of vertical variation of diabatic heating to the west of WPSH on WPSH's position.

During the heavy rain in South China, there were two strong heating centers to the west of WPSH (60–100°E, 15–25°N), and one was over the east coast of Arabian Sea and the other was over the east of Bay of Bengal. What are the effects of the strong heating centers on WPSH's position? It will be discussed in the following section.

From Eq. (6), βv and $-\vec{V} \cdot \nabla\zeta$ are much smaller than $\frac{f+\zeta}{\theta_z}\frac{\partial Q_1}{\partial z}$ to the west of WPSH around heavy rain period based on calculation (figure not shown). So the local variation of vorticity depends mainly on the vertical change of daibatic heating. Eq (6) is changed to

$$\frac{\partial\zeta}{\partial t} \propto \frac{f+\zeta}{\theta_z}\frac{\partial Q_1}{\partial z}. \tag{8}$$

If $\frac{f+\zeta}{\theta_z}\frac{\partial Q_1}{\partial z} > 0$, $\frac{\partial\zeta}{\partial t} > 0$. When the heating source is apart from WPSH, the cyclonic vorticity around the heating source is favorable for the generation of anti-cyclonic circulation to the east of the heating source (on the west of WPSH), which induces the westward movement of WPSH (Fig. 7(a)).

When the heating source is close to WPSH, the cyclonic vorticity around the heating source is favorable for the eastward retreat of WPSH (Fig. 7(b)).

Figure 8 shows the day-to-day evolution of $\frac{f+\zeta}{\theta_z}\frac{\partial Q_1}{\partial z}$ over 60–100°E, 15–25°N area (to the west of WPSH) at 500 hPa for the climatological mean and June 2005. On the whole, the west heating from 11 to 30 June had larger vertical variation than climatology, namely, $(\frac{f+\zeta}{\theta_z}\frac{\partial Q_1}{\partial z})_{2005} > 0 > (\frac{f+\zeta}{\theta_z}\frac{\partial Q_1}{\partial z})_{\text{mean}}$, which was favorable for the increase anti-cyclonic vorticity

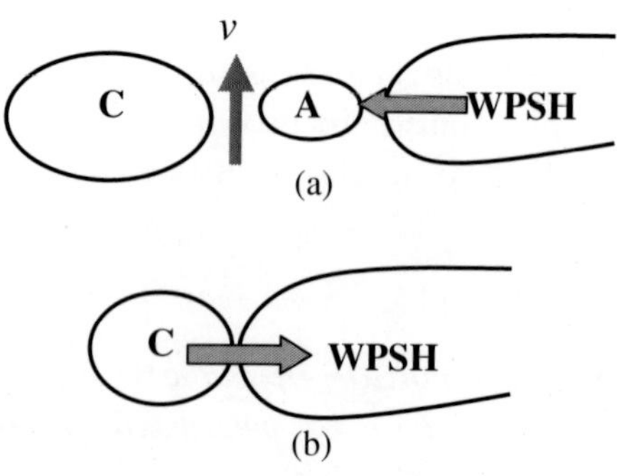

Fig. 7. Effects of the diabatic heating to the west side (a) and on the west side (b) of WPSH on the movement of WPSH.

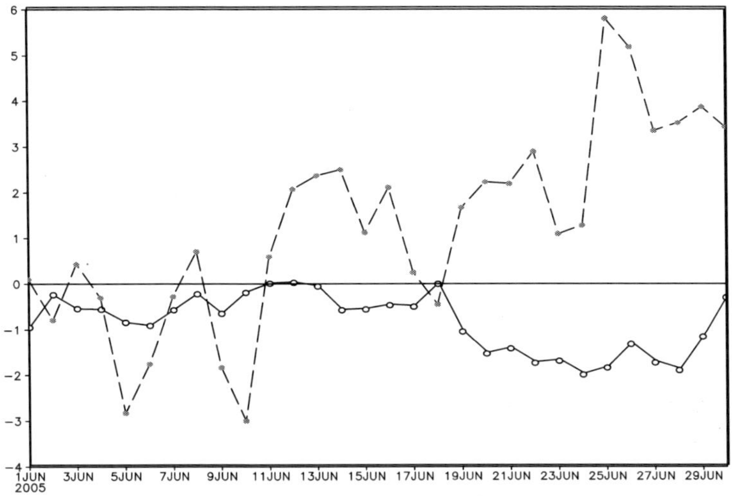

Fig. 8. Day-to-day evolution of $\frac{f+\zeta}{\theta_z}\frac{\partial Q_1}{\partial z}$ to the west (60–100°E, 15–25°N) side of WPSH at 500 hPa for the climatological mean (the solid line) and June 2005 (the dashed line). unit: $10^{-11}\mathrm{s}^{-2}$.

to the east of the heating source (on the west of WPSH), leading to the westward movement of WPSH.

6. Conclusions

(1) The position variation of WPSH is associated with the diabatic heating in subtropical areas. In comparison with the climatology, there is strong heating on the north side of WPSH and relatively weak ITCZ convection on the south in June 2005.
(2) The heating to the west of WPSH is generally greater in June 2005 and each of westward extension of WPSH corresponds to a significantly enhanced heating to the west of WPSH.
(3) In mid-troposphere, the vertical variation of heating on the north (south) of WPSH during 12–24 June, 2005 is basically greater (less) than the climatology, which is unfavorable to the northward movement of WPSH.
(4) In the mid and late of June 2005, the vertical variation of heating over the eastern coast of the Arabian Sea and the Bay of Bengal (to the west of WPSH) is largely higher than the climatology, which is in favor of the increase of anti-cyclonic vorticity on the west of WPSH, inducing westward extension of WPSH.

Acknowledgments

The authors would like to thank NCEP/NCAR and CMA for providing the data. This work is jointly supported by the National Basic Research Program of China (No. 2004CB418302; No. 2004CB418301) and Key Lab for Meteorological Disaster Experiment, NUIST (KLME060204).

References

1. W. Lijuan, G. Zhaoyong and H. Jinhai, *J. Nanjing Institute Meteorol.* **30** (2007) 145–152.
2. X. Wenbing, L. Jiangnan and Y. Cai, *J. Trop. Meteorol.* **23** (2007) 90–97.
3. S. Tingfei and L. Huibang, *J. Tropical Meteorol.* **19** (2003) 17–26.
4. Y. Dandan, Z. Ren and H. Mei, *J. Tropical Meteorol.* **23** (2007) 78–84.
5. W. Guoxiong and C. Jifan, *Beijing* (Science Press, 2002), pp. 36–47.
6. L. Riyu, *J. Meteorol. Soc. Jap.* **79** (2001) 771–783.
7. H. Ronghui and L. Weijing, *Chinese J. Atmos. Sci.* (1998), pp. 107–116.
8. W. Guoxiong, L. Yimin and L. Ping, *Acta Meteorologica Sinica* **57** (1999) 257–263.
9. L. Yimin, L. Hui, L. Ping, *et al.*, *Acta Meteorologica Sinica* **57** (1999) 385–396.
10. L. Yimin, W. Guoxiong, L. Hui, *et al.*, *Acta Meteorologica Sinica* **57** (1999) 525–538.
11. M. Yanai, S. Esbensen, and J. H. Chu, *J. Atmos. Sci.* **30** (1973) 611–627.
12. W. Guoxiong and L. Huanzhu, *Acta Meteorologica Sinica* **57** (1999) 1–15.

Advances in Geosciences
Vol. 16: Atmospheric Science (2008)
Eds. Jai Ho Oh *et al.*

LUNAR-SYNODIC COMPONENT IN THE EAST ASIAN WINTER MONSOON

CHI-HUA WU* and WEN-SHUNG KAU
Department of Atmospheric Sciences,
National Taiwan University, Taipei, Taiwan
**chhwu@ntu.edu.tw*

This study is an extension of the lunar-synodic influence on the weather and climate. Using data from satellite observations and model reanalysis in a lunar month during East Asian winter monsoon (EAWM), the atmospheric condition surrounding the semi-permanent anticyclone (SPA) is investigated. The SPA in the EAWM exists at lower-tropospheric atmosphere center near Taiwan. Between the first and the last quarter of the moon, the northward movement of the SPA ridge is assessed, to be the cause of the enhancement of windward coastal precipitation in East and Southeast Asia. Corresponding to the SPA variation across the full moon, the geopotential height over the SPA region (17.5–27.5°N; 115–135°E) increased at lower-tropospheric atmosphere and decreased at higher-tropospheric atmosphere. Associated with the squishing atmosphere, the middle- and low-level clouds with thick optical thickness (MLTC) increased significantly, which was highly related to the increased precipitation (there was 1/4 MLTC enhancement in the SPA region calculated by three-fifths of the lunar months between 1985 and 2006). It is suggested that the lunar-synodic cycle might modulate the EAWM precipitation when the atmospheric conditions change and favor development of thick clouds.

1. Introduction

The moon's phase was often imagined to a great deal of myth and folklore in human cultures since ancient times. In the past, the impact of lunar cycle on human behavior was reported broadly in the psychological study, e.g., accidents [Puharich, 1973]. Persinger [1989] suggested that the Earth's own internally-generated magnetic field relating to the lunar-synodic cycle could be a speculative chain for the extrasensory perception (ESP) performance. In meteorology, the statistical linkage between the moon and weather was also mentioned early in the last century [Meech, 1901]. Meteorologists further attempted to detect lunar component in the weather data [e.g., Adderley and Bowen, 1962; Hanson *et al.*, 1987; Roy,

2006]. A concise summary of lunar influence on climate can be found in Camuffo [2001].

The lunar influence on the Earth's atmosphere exists in different timescales, including the lunar-diurnal tidal force [Hamilton, 1984]; the lunar-monthly influence [Bradley *et al.*, 1962]; and in long-term timescales, the maximum lunar declination periodicity with 18.61-year period was assessed to influence on the El Niño / Southern Oscillation (ENSO) [Cerveny and Shaffer, 2001]. In the early 1960s, the lunar-synodic cycle (with ~29.53-day period) was almost simultaneously indicated to have influence on heavy precipitation in Australia [Adderley and Bowen, 1962] and in United States [Bradley *et al.*, 1962]. The authors made similar discovery but in different regions that extreme precipitation events occurred less/more frequently a few days before/after a full moon (Bowen's signal). Adderley and Bowen [1962] suggested that the Moon perhaps modulates the amount of meteor dust collected by the Earth. Lethbridge [1970] further indicated that the disturbance of magnetosphere could be the reason. By focusing on the spatial progression over the United States in the phase of lunar-precipitation relationship, Hanson *et al.* [1987] suggested that the lunar-synodic cycle modulates the precipitation in the form of long-wave circulation in the troposphere.

The solar-lunar mechanism was also considered for the influence in a lunar month [e.g., Hejkrlik, 2007]. Brier and Bradley [1964] divided the precipitation data into years with high and low solar activities, and found that most of the variations in the total lunar-synodic precipitation cycle come from years with less than average solar activity. Besides, scientists also concerned the lunar-synodic influence on other climate variables, including surface and air temperature [Balling and Cerveny, 1995]; thunderstorm frequency [Markson, 1971]; and ice nuclei [Bigg, 1963], etc. Roy [2006] recently studied the lunar-phase impact on the Indian monsoon rainfall, and found that the maximum in the frequency and amount of precipitation in the interior of India occurred a few days after a full moon. The observational evidence may be actually unable to confirm the expected global-scale impact as mentioned in Camuffo [2001]. Roy's study offered a thinking connecting the Bowen's signal to the monsoon regions. To characterize the lunar-synodic component in the monsoon system for assessing its possible impact is the purpose of this study, which would be meaningful for the prediction in the monsoon regions.

Cevolani and Bortolotti [1987] studied a possible seasonal-lunar influence on precipitation data in the Alps and the Po valley, and pointed out that the lunar influence was better recognizable in cold months because of lower vertical velocities in the atmosphere than in warm months. Indeed, the cause of the Asian monsoonal precipitation (generally light in average) in winter is relatively simple than in summer. The lunar cycle has been shown to affect heavy precipitation more than mean precipitation values as focused in the Bowen's signal. In this study, the lunar-synodic component on the meteorology is investigated in the season with relatively light precipitation, which may be different to the Bowen's signal.

2. Data and Methodology

2.1. *Data*

Data used in this study cover a period of 22 years from 1985 to 2006. The data with 1-day temporal resolution include: (a) Winds and geopotential height from Version 2 of the National Centers for Environmental Prediction reanalysis (NCEP-R2) [Kanamitsu *et al.*, 2002]; and (b) Cloud amount (coverage, or more correctly hydrometeor cover, since the radar is sensitive to both cloud and precipitation size particles) from the International Satellite Cloud Climatology Project (ISCCP) D1 data set [Rossow and Dueñas, 2004]. The precipitation data with pentad temporal resolution is also used from Climate Prediction Center Merged Analysis of Precipitation (CMAP). It is linearly interpolated to the daily pentad temporal resolution. All the data have a spatial resolution of 2.5° latitude-longitude.

2.2. *The semi-permanent anticyclone and lunar-synodic month*

Figure 1 shows the winter-mean 850 hPa streamline and the standard deviation (shading) of 7–30-day band-pass filtered cloud amounts of middle- and low-level thick clouds (MLTC) during wintertime from December 1985 to February 2006. The Tibetan Plateau is roughly denoted by dashed contour line at 1500 m height in the figure. In the ISCCP cloud categories, the MLTC is classified as the optical thickness greater than 3.6 for cloud top lower than the 440 hPa level. As shown in Fig. 1, the MLTC clouds over the semi-permanent anticyclone (SPA) center near Taiwan had the most variation in the sub-monthly timescale (corresponding to a lunar month), when compared to other monsoon regions (e.g., Bay of Bengal and Japan). The 7–30-day band-pass filtered variance of MLTC occupied more that

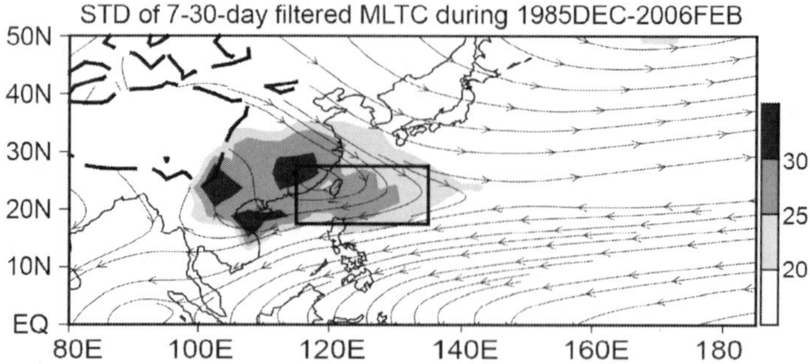

Fig. 1. Winter-mean (December 1985 to February 2006) 850 hPa streamline and the standard deviation (shading) of 7–30-day band-pass filtered cloud amounts of middle- and low-level thick clouds (MLTC, unit: %). The Tibetan Plateau and the semi-permanent anticyclone (SPA) region (17.5–27.5° N, 115–135° E) are denoted by 1500 m dashed contour line and the box respectively.

40% of total variance in the SPA region (17.5–27.5°N; 115–135°E, the box denoted in Fig. 1).

Associated with the SPA characteristics, Chen and Chen [2003] indicated that the northeasterly monsoon flow over Taiwan generally reaches the maximum intensity in winter. The ridge axis develops into a SPA near Taiwan (∼23.5°N, 121°E) in early January. In late January, the west Pacific subtropical high axis at the 850 hPa level extends westward and weakens the SPA. In the upper levels, the westerly jet axis reaches its southernmost position in January when the monsoon cold surges are most frequent [Yeh and Chen, 1984].

The lunar-synodic cycle with 29.53-day period is exactly at ∼63° heliocentric latitude. Traditionally, a lunar month (from New Moon to New Moon) was taking what we see from Earth as a reference, e.g., in the Chinese lunar calendar made by the ancient Chinese thousand years ago. For the data with 1-day temporal resolution, we first define 63 full moons referred by the lunar calendar publicly in Taiwan (Chinese lunar calendar) in the analyzed winters (1985–2006). All of the full moons are then defined as the 15th day and assigned the 14 days before and after the full moon. In Fig. 2, the cumulative distribution of lunation according to the domain-mean MLTC in the SPA region is shown, calculated by the 63 lunar months. As the calculation, a threshold value was given when the amount of MLTC is greater than long-term average (43.4%, averaged in 63 lunar-synodic months). The 100% in the figure means that all the MLTC amounts (63 cases) in the same day relative to lunation are greater than the

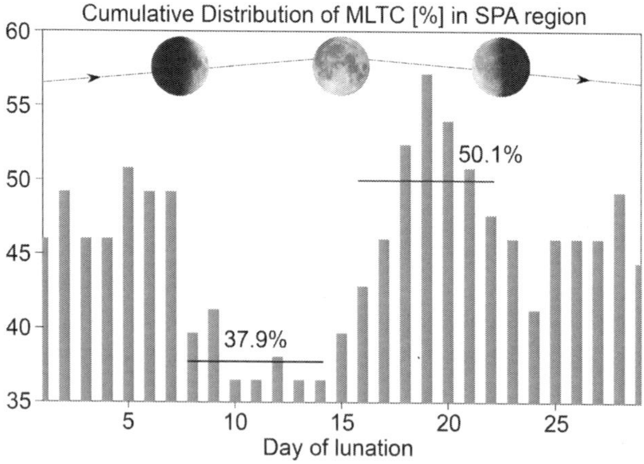

Fig. 2. The cumulative distribution of MLTC clouds (unit: %) in lunation. See detail in the text.

threshold. As shown in Fig. 2, the 7-day average of the MLTC cumulative distribution prior to the full moon is ~38% (24 of 63 days), which is much less than the 7-day average right after the full moon (~51%, 32 of 63 days). Such the lunar-cloud relationship across the full moon is similar to the lunar-precipitation relationship of the Bowen's signal.

3. Results

3.1. *Clouds and circulation*

For clearly demonstrating the change across the full moon, we choose 38 lunar-synodic cycles (60% of 63 lunar months) for the following analyses. A simple criterion for the 38 lunar months is that the 7-day-mean MLTC in the SPA region after the full moon is greater than before the full moon. Using data from 38 lunar months, Fig. 3 shows all the domain-mean MLTC in the SPA region (dots, 38×29 cases) relative to lunation. The standard deviations (bars) calculated by 38 cases per day relative to lunation is plotted center by average (solid line). The composite MLTC of total lunar months (63 months) is also shown in the figure (dashed line). It is expected that the enhancement of MLTC across the full moon is well defined through 38-lunar-month composite (at the 95 confidence level) better than total-lunar-month composite (Fig. 3). The weekly-mean MLTC in the SPA region composited by 38 months increased 1/4 (varied from ~40% prior to the

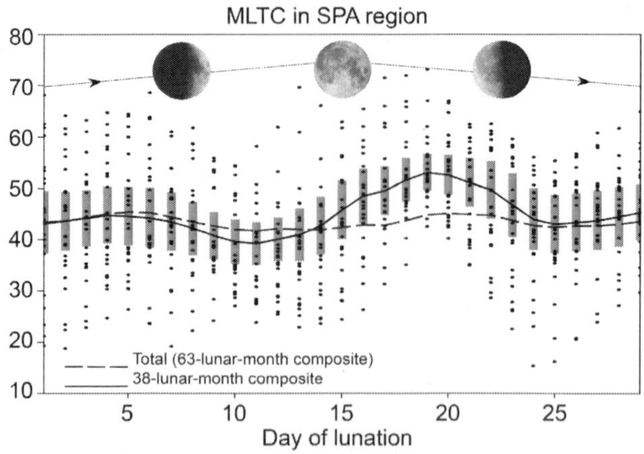

Fig. 3. All the domain-mean MLTC in the SPA region (dots, 38 × 29 cases) relative to lunation using data from 38 lunar months (See criterion in the text). The standard deviations (bars) calculated by 38 cases per day relative to lunation is plotted center by average (solid line). The composite MLTC of total lunar months (63 months) is also shown in the figure (dashed line).

full moon to ~50% after the full moon). The MLTC seems to be a good indicator for the lunar-synodic component of the winter monsoon.

For the relationship between the lunar components of MLTC and circulation in space, Fig. 4 shows the spatial distributions of MLTC and 850 hPa circulation before, during and after the full moon composited by 38 lunar months (dashed lines denote the SPA ridge). A few days prior to the full moon (day 11, upper panel), the major clouds located along southeast part of China and north part of Taiwan, and the SPA was centered over Taiwan. During the full moon (day 15, middle panel), the MLTC significantly increased corresponding to the change of SPA. A few days after the full moon (day 19, lower panel), the SPA seemed to be combined to the Asian high-pressure system at lower-tropospheric atmosphere. Associated with the SPA variation across the full moon a large amount of the MLTC expanded southeastward to Philippine Sea and southward to South China Sea, respectively.

It is important to illustrate the atmospheric condition associated with the cloud evolution in lunation. Using data composited by 38 lunar months, Fig. 5 shows the vertical distributions of infrared cloud amounts and geopotential height anomaly (remove the averaged lunar month) in the SPA region. As shown in Fig. 5, the geopotential height varied across the full moon with opposite signals in the high- and low-level atmospheres.

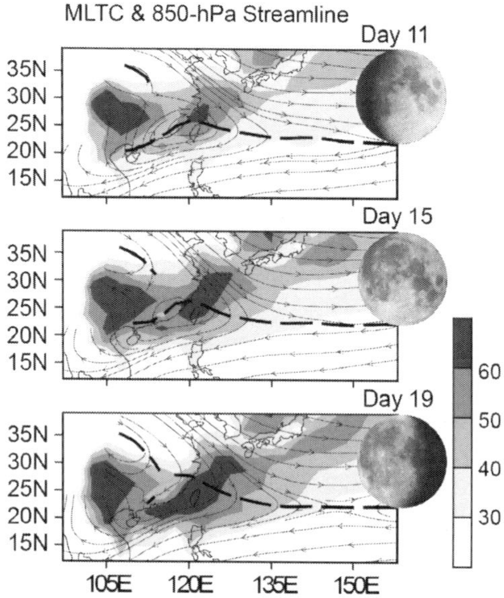

Fig. 4. The spatial distributions of composite MLTC (unit: %) and 850 hPa circulation (by 38 lunar months) before, during and after the full moon. The thick dashed lines denote the SPA ridge.

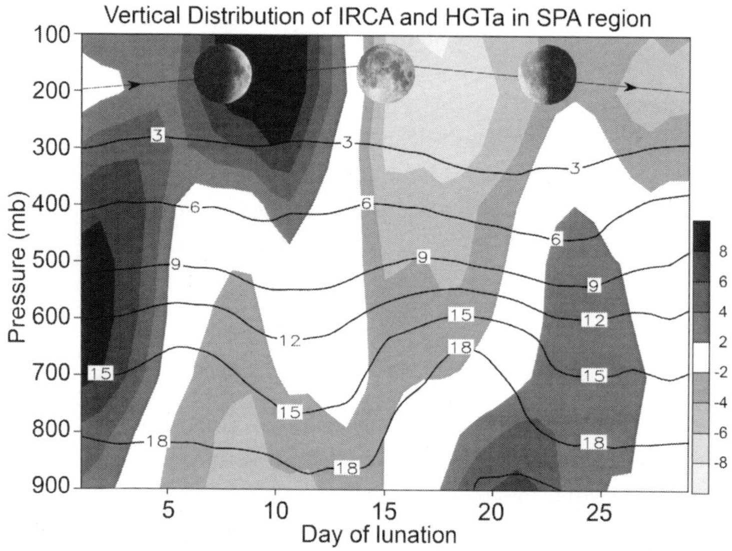

Fig. 5. The vertical distributions of infrared cloud amounts (unit: %) and geopotential height anomaly (remove the averaged lunar month, unit: m) in the SPA region, composited by 38 lunar months.

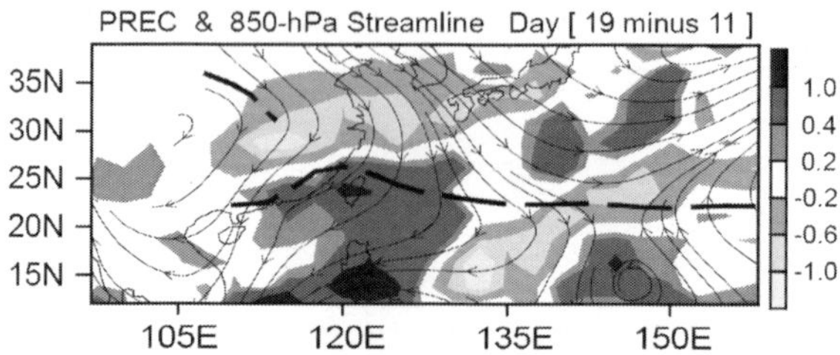

Fig. 6. The change of precipitation (unit: mm/day) and 850 hPa streamline from day 11 to day 19 relative to lunation. The thick dashed line denotes the SPA ridge at the full moon.

The atmospheric column, e.g., the depth between 200 and 850 hPa pressure levels was thicker before the full moon than after the full moon. Across the full moon, the geopotential height decreased ~13 m at 200 hPa level and increased ~5 m at 850 hPa level. The vertical air column squished almost 20 m. The clouds that varied majorly in middle- and low-levels might be associated with the squishing atmospheric condition.

3.2. *Precipitation*

In Fig. 6, the changes of precipitation and 850 hPa streamline from day 11 to day 19 relative to lunation are shown composited by 38 months. The thick-dashed line denotes the ridge of SPA at the full moon. As shown in Fig. 6, the precipitation near the SPA region across the full moon increased in the south part of the ridge and decreased in the north part. The enhanced precipitation may both correspond to the windward coastal region near Taiwan and Philippines and the southeastward extension of the MLTC. The evolution of the high-pressure system might be the cause of the changes of clouds and circulation.

4. Concluding Remarks

In this study, we focus on the association of clouds, precipitation and atmospheric conditions in lunation between the first and the last quarter of the moon. Using data composited by three-fifths of lunar months between 1985 and 2006, the lunar-synodic component of MLTC over SPA region is assessed to be a good indicator of the lunar-synodic component in the EAWM. Corresponding to the SPA variation across the full moon, the geopotential height over the SPA region increased at lower-tropospheric atmosphere and decreased at higher-tropospheric atmosphere. The geopotential-height difference between 200 and 850 hPa pressure levels over the SPA region decreased ~18 m, when the SPA ridge moved northward across the full moon. Associated with the squishing atmosphere, the 7-day mean MLTC amount over the SPA region after the full moon is found more than that before the full moon by ~10% (1/4 enhancement). It is suggested that the EAWM precipitation might be modulated by lunar-synodic cycle when the atmospheric conditions changed and favored development of thick clouds.

This paper reports preliminary results of the possible lunar impact on the EAWM. The result may call the attention of meteorologists e.g., for the value to the regional climate simulation of numerical models. Detailed analysis of the possible lunar-cycle impact on specific winter monsoon near Taiwan is attempted working in progress, e.g., the diurnal variation of precipitation affected by northeasterly monsoon.

Acknowledgments

This work was supported by National Research Council, Taiwan, under grant NSC 97-2111-M-002-009. The authors thank Ms. Hai-Wei Lin for modifying the figures, and are grateful for the following data used in this study: circulations from NOAA NCEP-R2 reanalysis; cloud data from ISCCP; and precipitation from CMAP.

References

1. L. W. Meech, *Mon. Weather Rev.* **29** (1901) 372.
2. D. A. Bradley, M. A. Woodbury and G. W. Brier, *Science* **137** (1962) 748.
3. E. E. Adderley and E. G. Bowen, *Science* **137** (1962) 749.
4. E. K. Bigg, *Nature* **197** (1963) 172.
5. G. W. Brier and D. A. Bradley, *Science* **21** (1964) 386.
6. M. D. Lethbridge, *J. Geophys. Res.* **75** (1970) 5149.
7. R. Markson, *Pure Appl. Geophys.* **84** (1971) 161.
8. A. Puharich, *Garden City* (Anchor Books, NY, 1973).
9. K. Hamilton, *Mon. Weather Rev.* **112** (1984) 1620.
10. F. W.-C. Yeh and G. T. J. Chen, *Quart. J. Meteorol.* **101** (1984) 9.
11. G. Cevolani and G. Bortolotti, *Nuovo Cimento* **10C** (1987) 593.
12. K. Hanson, G. A. Maul and W. McLeish, *J. Clim. Appl. Meteorol.* **26** (1987) 1358.
13. M. A. Persinger, *Research in Parapsychology 1988, Metuchen* (Scarecrow Press, NJ, 1989), p. 121.

14. R. C. Balling and R. S. Cerveny, *Science* **267** (1995) 1481.
15. D. Camuffo, *Earth, Moon and Planets* **85–86** (2001) 99.
16. R. S. Cerveny and J. A. Shaffer, *Geophys. Res. Lett.* **28** (2001) 25.
17. M. Kanamitsu, W. Ebisuzaki, J. Woollen, S.-K. Yang, J. J. Hnilo, M. Fiorino and G. L. Potter, *Bull. Am. Meteorol. Soc.* **83** (2002) 1631.
18. C.-S. Chen and Y.-L. Chen, *Mon. Weather Rev.* **131** (2003) 1323.
19. W. B. Rossow and E. N. Dueñas, *Bull. Am. Meteorol. Soc.* **85** (2004) 167.
20. S. S. Roy, *Geophys. Res. Lett.* **33** (2006) L01707, doi:10.1029/2005GL024771.
21. L. Hejkrlik, *Geo. Res. Abs.* **9** (2007) EUG2007-A-02520L.

Advances in Geosciences
Vol. 16: Atmospheric Science (2008)
Eds. Jai Ho Oh *et al.*

AUGUST RAINFALL IN KOREA AND ITS ASSOCIATION WITH CIRCULATIONS

KYUNG-JA HA
Division of Earth Environmental System,
Pusan National University, Busan 609-735, Korea
kjha@pusan.ac.kr

KYUNG-SOOK YUN
Division of Earth Environmental System,
Pusan National University, Busan 609-735, Korea

The recent characteristics of the summer rainfall in Korea have shown a significant change from the climatological features. In the present study, the changes in the Changma (July rainfall) and post-Changma (August rainfall) over the Korean Peninsula have been investigated using a long-term record based on the synoptic station data that includes Seoul, Busan, Daegu, Mokpo, and Gangneung. The association between the large-scale circulation and the changes in the rainfall has also been examined. The August rainfall has increased with time, while the July rainfall has not changed. The July rainfall has impacts of regional ocean variability, and the August rainfall has impacts of remote ocean variability on intensity, including the Indian Ocean and the tropical eastern Pacific. The August rainfall is strongly associated with the westward extension of the western North Pacific (WNP) high and position of the Bonin high. The August rainfall exhibits the extratropical Eurasian wave-like structure with a steady response during the summertime (June to August), while the July rainfall is closely related to the southwesterly monsoon flow during that month. Consequently, the August rainfall exhibits a clearer association with remote ocean variability and large-scale circulation.

1. Introduction

The summer monsoon rainfall in Korea is called Changma and its precipitation concentrated to July. However, the characteristics of summer rainfall in Korea have changed in the last half-century (e.g. Ho *et al.*, 2003[7], Wang *et al.*, 2007[14], and Kwon *et al.*, 2007[9]). In particular, the August rainfall has increased considerably (Cha *et al.*, 2007[2]). Ha and Ha (2006)[4] have also noted the increase of the "August mode" from the EOF analysis. The abrupt change in summer rainfall causes huge damage to human activities and property.

Although most prior studies have focused on the summertime (i.e. June–July–August), the rainfall in July and August has remarkably distinctive characteristics in their interannual variability (Cha *et al.*, 2007[2]). For example, the July rainfall is mainly affected by the East Asian summer monsoon front (Ha *et al.*, 2005[6]), while the August rainfall is modulated by complex atmospheric mechanisms (i.e. direct and indirect effects of typhoons, mesoscale complex systems, and thunderstorms). To understand the climate changes in the Korean summer rainfall, the similarities and discrepancies in the interannual variability of July and August rainfall should be investigated first. To do so, the changes in Changma (July rainfall) and Post-Changma (August rainfall) over the Korean Peninsula have been discussed, using long-term records based on the synoptic station data including those of Seoul, Busan, Daegu, Mokpo, and Gangneung. Furthermore, the characteristics of the sea surface temperature (SST) and circulation associated with the July/August rainfall anomalies are investigated in the present study.

2. Data and Method

2.1. *Data*

To ameliorate the long-term variability in Korean rainfall, we used the precipitation dataset obtained by the Korean Meteorological Administration (KMA) from 1912 to 2006. Five synoptic stations that made consistent observations from 1912 to the present were selected for this analysis. The five stations are represented in Table 1.

Composite and regression analyses for the circulations are carried out to find the association with a large-scale environment using the National Center for Environmental Prediction/National Center for Atmospheric Research (NCEP/NCAR) reanalysis data from 1948 to 2006 (59 years).

Table 1. The five synoptic stations selected in this study.

	Longitude (°E)	Latitude (°N)	Altitude (m)
Seoul	126.96	37.56	85.50
Busan	129.03	35.10	69.23
Daegu	128.62	35.87	57.64
Mokpo	126.38	34.80	37.88
Gangneung	128.88	37.73	25.91

The regression to the SST (obtained from HadISST) from the rainfall projection was obtained to investigate the spatial correlation distributions for the interested area.

2.2. *Methodology*

First, we try to detect the detection of climatic changes over the long-term record. We used the Pettitt (1980)[13] test, which is a change detection method in the median of the sequence of observation and a stout test of the change point resistant to outliers. Pettitt (1980)[13] derived the test statistics on whether there is change in the median of the sequence of observation on the basis of the rank of the observations. The Pettitt test uses a remarkably stable distribution and provides a reliable estimation of the change point resistant to outliers.

The Pettitt test procedure is as follows: First, the observations (X) are ranked from 1 to N (i.e. $X_1, \ldots, X_N$). Suppose R_i is the rank associated with the observation X_i, so we can calculate the sum of the ranks of the observations at each place (i.e. j) in the series as described below.

$$W_j = \sum_{i=1}^{j} R_i, \quad j = 1, 2, \ldots, N-1. \tag{1}$$

Then it is easier to rewrite U_j as a rank statistic.

$$U_j = 2W_j - j(N+1) \quad j = 1, 2, \ldots, N-1. \tag{2}$$

The value of j where the maximum in the absolute value of U_j occurs (that is, $K_{m,n}$) is the estimated change point in the sequence and is denoted by m $(n = N - m)$.

Now we conduct a statistical test to see whether the estimated change point m is significant using the sampling distribution of W_m. Thus, the significance probabilities associated with the value m are approximately given by

$$P = 2\exp\{-6W_m^2/(N^3 + N^2)\}. \tag{3}$$

The detailed extent is described by Pettitt (1979)[12]. The Pettitt test (Ha and Ha, 2006[4]) was applied for the detection of precipitation. For the July and August precipitation averaged over the five stations, the Pettitt test was performed with a significant level of 0.01. One significant change point was found in 1967 (1953) on the August (July) rainfall, respectively.

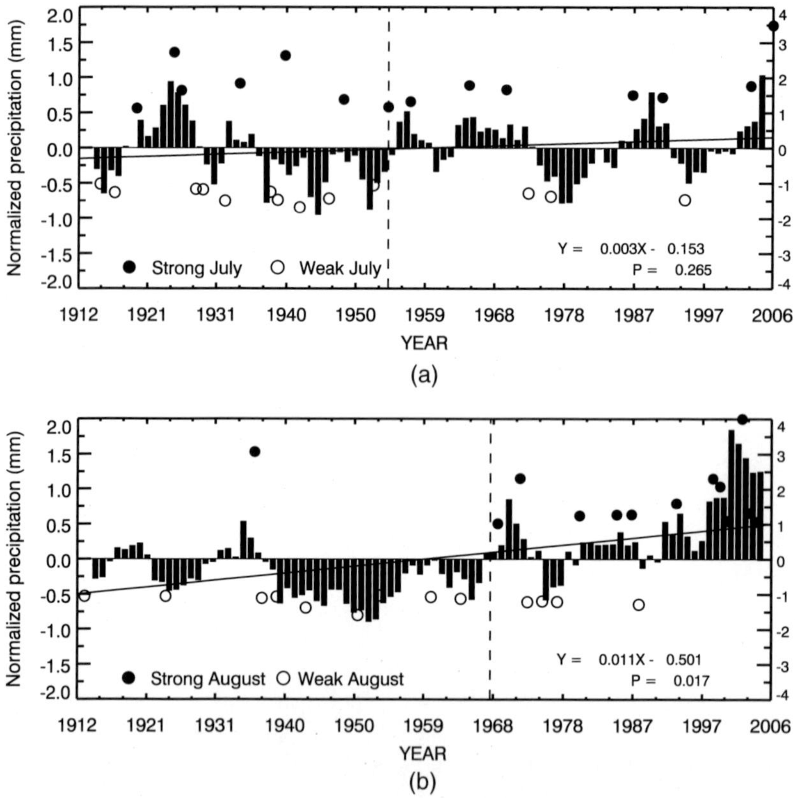

Fig. 1. Interannual variability of 5yr-running averaged (a) July and (b) August rainfall. The closed (open) circles indicate the rainfall that the normalized value is greater than 1.0 (less than −1.0) standard deviation (right axis), and the perpendicular dashed line denotes the change point of July/August rainfall. (a)–(b) The solid line indicates the linear trend of the series.

3. The Climate Change in July/August Rainfall

To investigate the long-term variability of the rainfall, the interannual variability of 5-year running averaged rainfall for July and August is compared as shown in Figs. 1(a) and 1(b). While the July rainfall does not show any significant change trend, the August rainfall displays a clear increasing trend after late 1960s. This upward trend significant at the 95% confidence level (i.e. $P < 0.05$) can be shown in the result from Mann-Kendall test (Mann, 1945[10]; Kendall, 1975[8]). The rainfall change in August is especially obvious after the change point (that is, 1967).

To locate the details of this change, extreme July and August rainfall years are displayed. The strong (weak) years are defined as having the rainfall anomalies greater than 1.0 (less than −1.0) standard deviation of normalized values, respectively for July and August. For July rainfall, the variability before the change point is somewhat larger than that after the change point (i.e. 1953). The numbers of the strong and weak years before the change point are approximately twice as many as that after the change point. This indicates that the July rainfall has had a relatively small fluctuation in recent years. On the other hand, the August rainfall exhibits distinct characteristics before and after its change point. Most of the strong August rainfall events occurred after 1967, while many weak August rainfall years appear before 1967.

The August rainfall shows a much clearer increasing trend compared to the July rainfall. The statistics also support the July and August precipitation results before and after the change points (Table 2). In comparison with July rainfall, August rainfall during latter period has a larger mean and standard deviation than during the first period. Consequently, after late 1960s, August rainfall undergoes increases with large deviations.

4. The Impacts of SST and Changes in Circulations

In this section, we try to investigate the SST and circulation changes in the July/August rainfall anomalies. First, to see the circulation change associated with the climate change in August rainfall, we have performed composite analysis for both the strong and weak August precipitation years (Fig. 2). The criterion for extreme August rainfall is the same as those shown in Fig. 1(a). The ten strong August rainfall years (1969, 1972, 1980, 1985, 1987, 1993, 1998, 1999, 2000, and 2002) are roughly shown in recent

Table 2. Statistics for the July and August precipitation before and after Pettitt's change point.

	Period	N	Mean	Median	SD
July	1912–1953	42	242.2	229.8	124.3
	1954–2006	53	270.8	260.2	103.6
August	1912–1967	56	189.8	188.5	78.1
	1968–2006	39	264.8	233.6	129.6

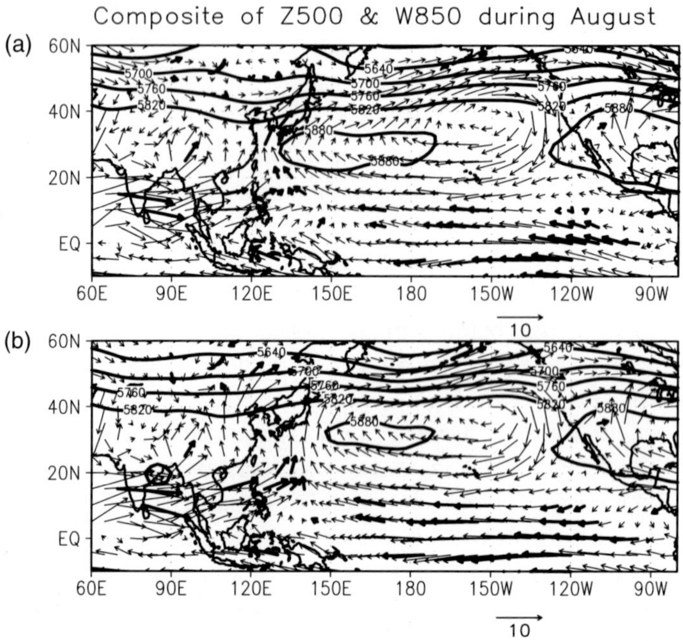

Fig. 2. Composite field of 500 hPa geopotential height (thick solid line) and 850 hPa wind (arrow) during August for (a) strong and (b) weak August rainfall years. The thick wind vector indicates the value significant at a 90% confidence level.

years, in comparison with the eight weak August rainfall years (1950, 1953, 1960, 1964, 1973, 1975, 1977, and 1988). In the strong August years, the 500 hPa geopotential height and 850 hpa wind during August show that the southerly wind from the western Pacific was expanded westward into the Korean Peninsula with expansion of the WNP high. Due to this westward expansion of the WNP high, low level winds are squeezed and the water vapor may transport into the Korean Peninsula (Zhou and Yu, 2005[15]). The westward extension of the high that is probably responsible for the August rainfall. It is known in the recent climate as the "Bonin high" in some studies (e.g. Enomoto *et al.*, 2003[3], and Ha and Lee, 2007[5]). The climate change in August rainfall is sensitive to the behavior of the Bonin high. Another possible factor is the weakening of tropical easterly winds. The weakened easterly winds in the central/eastern Pacific imply the El Niño effect on August rainfall during boreal summertime. It will be shown to be more evident in the regression of SST anomalies against the August rainfall in the Korean Peninsula (Fig. 3).

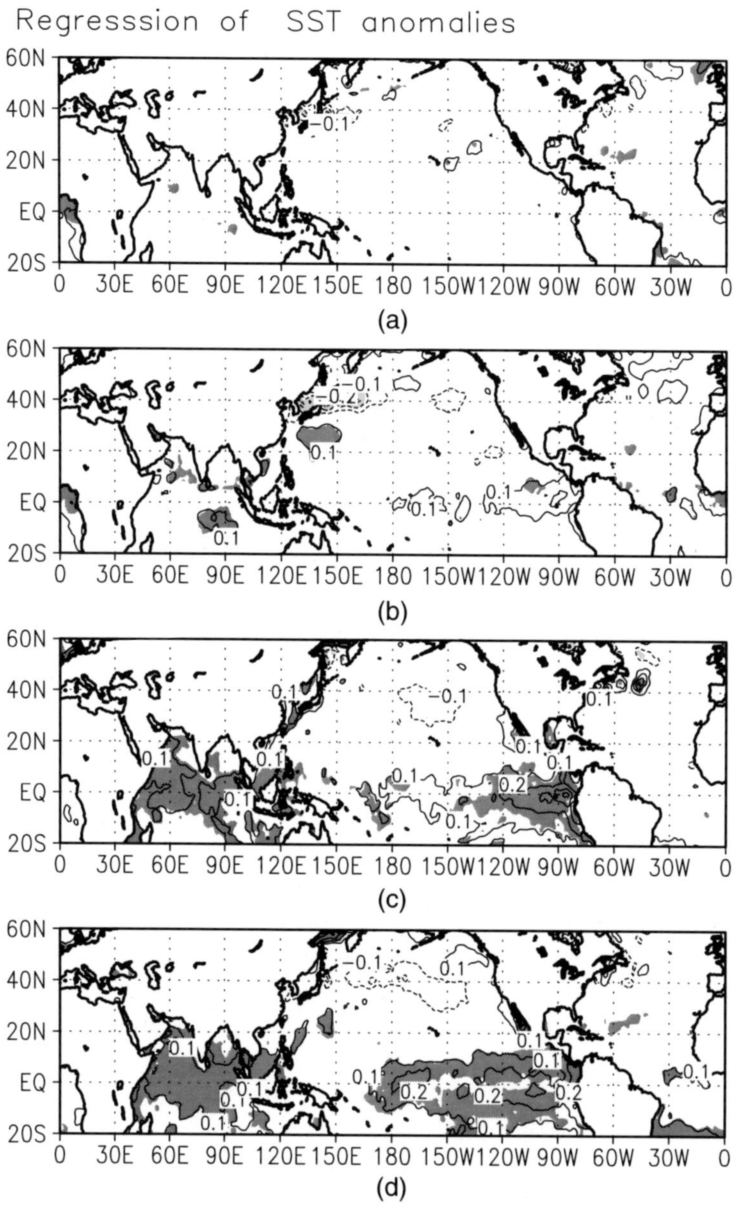

Fig. 3. Regression of the SST anomaly during (a) MAM and (b) JJA against the July rainfall anomalies and during (c) MAM and (d) JJA against the August rainfall anomalies. The shading indicates the values significant at a 95% confidence level.

The regression analysis of the SST anomaly against the July and August rainfall is performed from MAM to JJA. In the regressed SST field in lead times and concurrent time, the July rainfall exhibits the impacts of regional ocean variability (Figs. 3(a) and 3(b)). For example, during the summertime, warm anomalies in the WNP and cold anomalies over Korea and Japan appear. These may be associated with the East Asian summer monsoon rainfall front activity. The remote ocean (i.e. Indian Ocean and eastern Pacific) SST shows slightly weak warm anomalies. However, the August rainfall has impacts of remote ocean variability on intensity (Figs. 3(c) and 3(d)). A recent examination on global land monsoon rainfall variability showed similar SST anomaly structure and the result supports the remote Ocean impact on the August rainfall (Zhou *et al.*, 2008[16]).

In particular, the Indian Ocean warming has a persistent impact on the August rainfall. Considering the increasing trend in Indian Ocean (Alory *et al.*, 2007[1]), the warm anomalies in the long-term trend may play a vital role in modulating the circulation change associated with the August rainfall.

The WNP high with westward extension plays an important role in determining the intensity of the August rainfall. To look at the characteristics in large-scale circulation in detail, we have performed the regression analysis of the 500 hPa geopotential height anomalies against July and August rainfall anomalies (Fig. 4). To observe the persistence on the rainfall anomalies, the regression of the summertime (i.e. June–July–August) height anomalies is carried out in Figs. 4(b) and 4(d).

The July rainfall exhibits meridional wave structure, which is associated with the Pacific-Japan pattern (Nitta, 1983[11]). On the other hand, the August rainfall shows zonal wave-like pattern along the Eurasian continent. The structure also appears in the regressed 200 hPa geopotential height anomalies (not shown). This structure represents an equivalent barotropic structure with large amplitude. The equivalent barotropic pattern is modulated by the enhancement of Asian jet stream via stationary Rossby wave. This suggests that the interannual variability of August rainfall is related to the stationary Rossby wave along the Asian jet stream. During the summertime, the regressed field against the August rainfall also exhibits a similar wave-like pattern, while the one regressed onto July rainfall does not show any meaningful structure. This indicates that the August rainfall has a steady atmosphere response compared with the July rainfall.

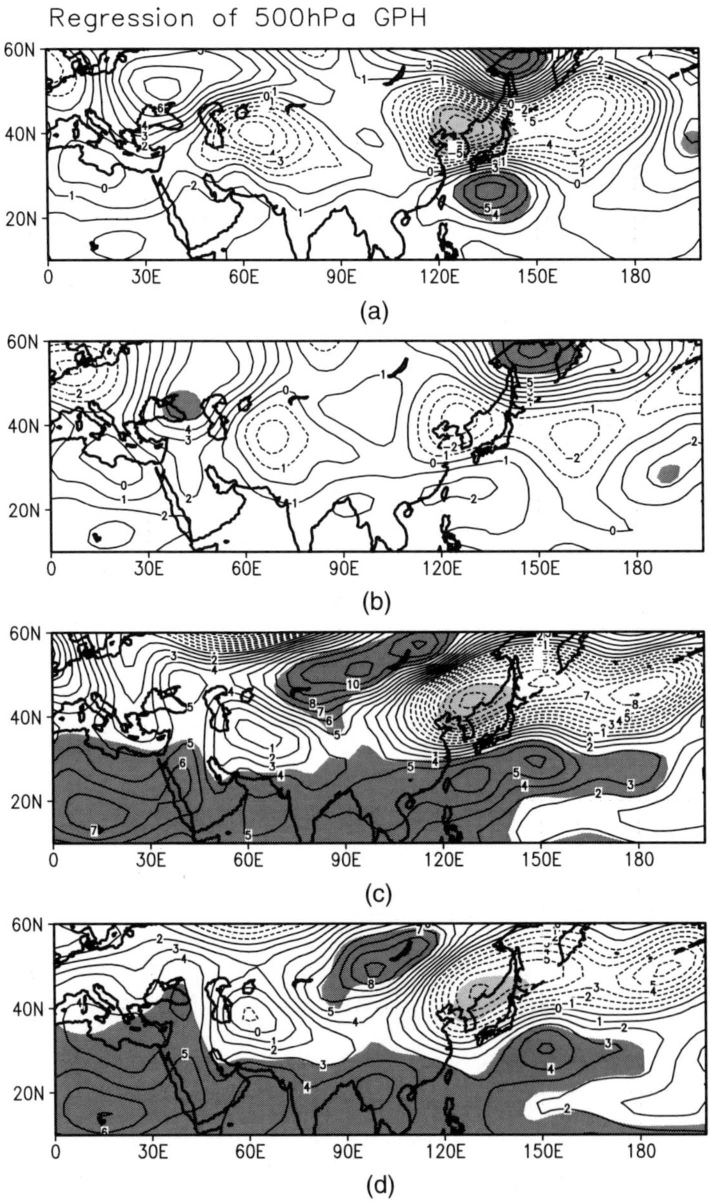

Fig. 4. Regression of the 500 hPa geopotential height during (a) July and (b) JJA against the July rainfall anomalies and during (c) August and (d) JJA against the August rainfall anomalies. The shading indicates the values significant at a 95% confidence level.

5. Summary and Conclusion

The August rainfall has increased with time, while the July rainfall did not show any significant rising trend. On the recent strengthening of August rainfall over the Korean Peninsula, the impacts of local and remote SST variability, and circulation were investigated in terms of the interdecadal variability of the July and August rainfall. The August rainfall is strongly related to the westward expansion of the subtropical WNP high and the presence of the Bonin high, which has an equivalent brotropic structure. In the regressed SST and circulation features, the July rainfall has impacts of regional ocean variability, and the August rainfall has impacts of remote ocean variability (i.e. eastern Pacific and Indian Ocean warming) on intensity. The August rainfall appears to have a more persistent wave-like structure for the regressed large-scale circulation than the July rainfall. This implies that the August rainfall has a more organized teleconnection than the July rainfall. The August rainfall also exhibits the extratropical Eurasian wave structure. The July rainfall is closely related to the PJ pattern, which is associated with the East Asian summer monsoon front movement. Consequently, the August rainfall appears in a different mechanism from the July rainfall. The August rainfall is more sensitive to the remote Ocean and large-scale circulation compared with the July rainfall. These results support that August rainfall is more useful for better seasonal predictions than July rainfall. These findings may contribute to the improvement in summer rainfall prediction.

Acknowledgments

This work was funded by the Korea Meteorological Administration Research and Development Program under Grant CATER 2008-4407 and the second stage of the Brain Korea 21 Project in 2008.

References

1. G. Alory, S. Wijffels and G. Meyers, Observed temperature trends in the Indian Ocean over 1960–1999 and associated mechanisms, *Geophys. Res. Lett.* **34** (2007) L02606, doi:10.1029/2006GL028044.
2. E. J. Cha, M. Kimoto, E. J. Lee and J. G. Jhun, The recent increase in the heavy rainfall events in August over the Korean Peninsula, *J. Korean Earth Science Society* **28** (2007) 585–597.
3. T. Enomoto, B. J. Hoskins and Y. Matsuda, The formation mechanism of the Bonin high in August, *Q.J.R. Meteorol. Soc.* **129** (2003) 157–178 .

4. K. J. Ha and E. Ha, Climatic change and interannual fluctuation in the long-term record of monthly precipitation for Seoul, *Int. J. Climatol.* **26** (2006) 607–618.
5. K. J. Ha and S.-S. Lee, On the interannual variability of the Bonin high associated with the East Asian summer monsoon rain, *Clim. Dyn.* **28** (2007) 67–83.
6. K. J. Ha, S. K. Park and K. Y. Kim, On interannual characteristics of climate prediction center merged analysis precipitation over the Korean peninsula during the summer monsoon season, *Int. J. Climatol.* **25** (2005) 99–116.
7. C. H. Ho, J. Y. Lee, M. H. Ahn and H. S. Lee, A sudden change in summer rainfall characteristics in Korea during the late 1970s, *Int. J. Climatol.* **23** (2003) 117–128.
8. M. G. Kendall, *Rank Correlation Measures* (Charles Griffin, London, 1975), p. 202.
9. M. H. Kwon, J.-G. Jhun and K.-J. Ha, Decadal change in east Asian summer monsoon circulation in the mid-1990s, *Geophys. Res. Lett.* **34** (2007) L21706, doi:10.1029/2007GL031977.
10. H. B. Mann, Nonparametric tests against trend, *Econometrica* **13** (1945) 245–259.
11. T. Nitta, Convective activities in the Tropical western Pacific and their impact on the northern hemisphere summer circulation, *J. Meteorol. Soc. Japan* **65** (1987) 373–390.
12. A. N. Pettitt, A non-parametric approach to the change-point problem, *Appl. Stat.* **28** (1979) 126–135.
13. A. N. Pettitt, Some results on estimating a change-point using nonparametric type statistics, *J. Statist. Comput. Simul.* **11** (1980) 261–272.
14. B. Wang, J.-G. Jhun and B.-K. Moon, Variability and singularity of Seoul, South Korea, rainy season (1778–2004), *J. Climate* **20** (2007) 2572–2580.
15. T.-J. Zhou and R.-C. Yu, Atmospheric water vapor transport associated with typical anomalous summer rainfall patterns in China, *J. Geophys. Res.* **110** (2005) D08104, doi:10.1029/2004JD005413.
16. T.-J. Zhou, R.-C. Yu, H. Li and B. Wang, Ocean forcing to changes in global monsoon precipitation over the recent half century, *J. Climate* **21** (2008) 3833–3852.

Advances in Geosciences
Vol. 16: Atmospheric Science (2008)
Eds. Jai Ho Oh *et al.*

CHANGMA AND SHIFTING PEAK IN SUMMER RAINFALL OF KOREA: ENSO INFLUENCE

KYUNG-JA HA* and SUN-SEON LEE
Division of Earth Environmental System,
Pusan National University,
Busan 609-735, Korea
**kjha@pusan.ac.kr*

P. N. VINAYACHANDRAN
Centre for Atmospheric and Oceanic Sciences,
Indian Institute of Science,
Bangalore & APEC Climate Center,
Busan 612-020, Korea

JONG-GHAP JHUN
School of Earth and Environmental Sciences,
Seoul National University,
Seoul 151-742, Korea

Since the mid-1970s, the maximum rainfall over Korea has been shifted from July to August. This has resulted in a significant increase in the August rainfall. Particularly, the relationship between August rainfall and ENSO changed from 1954–1975 (PI) to 1976–2002 (PII). The reason for this rainfall peak shift has been examined based on the change in teleconnection between August rainfall and ENSO. The interannual variability of August rainfall is significantly associated with SST variation over the eastern equatorial Pacific during PI period, while this SST anomaly does not exert an influence on August rainfall during the period PII. The low-level westerly and southerly (easterly and northerly) wind anomalies around the East China Sea related to strong (weak) August rainfall closely corresponded to El Niño (La Niña) during the PI. During the PII period, however, westerly-southerly wind anomalies around East China Sea in La Niña years were responsible for the high August rainfall over the East Asian region.

1. Introduction

The East Asian summer monsoon (EASM) rain bands associated with these regional features such as Mei-yu in China, Changma in Korea, and

Baiu in Japan appear at different phases of the monsoon cycle and exhibit variability at different time scales.[1,2] In early June, the monsoon trough is located over central China and Japan, over Korea in late June and over northern China in early July.[3]

Interannual and interdecadal variations in rainfall and the reasons for these have been an issue of considerable interest. The interdecadal variability in the summer climate (rainfall and temperature) in East Asia was found in the middle and the end of the 1970s in the variation of time coefficients of the second mode of singular value decomposition for summer rainfall and temperature in China.[4] Long-term precipitation records for Seoul were examined in search of characteristic climatic changes, and it was found that the "August mode" has enhanced during the modern era.[5] In addition, Wang *et al.*[6] showed that the occurrence of the rainy season summit for Korea tends to be delayed, and changed from the 37th pentad (30 June–4 July) during the period 1778–1807 to the 44th pentad (4–8 August) during the period 1975–2004.

In this study, long-term trends and characteristics of the July and August rainfall over the East Asia as well as the Korean Peninsula were investigated. Furthermore, in connection to the August rainfall increase, changes in low-level circulation associated with El Niño and southern oscillation (ENSO) were analyzed.

2. Data

The daily precipitation data at 12 weather stations measured by the Korea Meteorological Administration (KMA), spreading across Korea, from 1954 onward are used (hereafter referred to as "KMAR"). These weather stations are evenly distributed to represent an area mean value for Korea.[3] We have also used the CRU TS 2.1 Global Climate Dataset (hereafter referred to as "CRUR") to describe the spatial distribution. CRUR consists of 1224 monthly time series (1901–2002) of various climate variables, covering the global land surface (excluding Antarctica), at 0.5° resolution.[7] Hadley Center Global Sea-Ice and Sea Surface Temperature (HadISST) data set on a 1° latitude-longitude grid from 1871 is used.[8] For zonal and meridional winds at 850 hPa, the National Centers for Environmental Prediction-National Center for Atmospheric Research (NCEP-NCAR) the gridded (2.5° × 2.5° resolution) reanalysis data are used.[9] In precipitation data obtained from KMA, there is a missing period from 1950 to 1953 because of the Korean War. Thus, we analyzed the data from 1954.

3. Climatology and Shift of Rainfall Peak

From a statistical point of view, a change point is a location or time before which observations follow one distribution but follow another distribution afterwards.[10] The change point test was derived to determine whether a change in the median for the sequence of observation exists, based on the rank of the observations.[10] Pettitt's test[10] was applied to find a change point for Korean rainfall derived from CRUR for the period of 1954–2002. Based on the change point test, 1975 emerged as a significant point of change with a 95% confidence level, for both July and August rainfall. Therefore, the data sets were partitioned into two periods, 1954–1975 (hereafter, "PI") and 1976–2002 (hereafter, "PII").

To describe the fact that a shift of peak exists over the East Asian region, the spatial distributions of July and August rainfall during each period and its differences are shown in Fig. 1 using CRUR. July rainfall in the south and central parts of Korea is high during PI period. This pattern does not change, but the intensity decreases, particularly over the southern part of Korea. The rainfall over Shandong in China also decreased significantly. Meanwhile, August rainfall over Korea shows a more complex structure compared to July. Moreover, a large increase in August rainfall is seen over Korea and in the vicinity of Shanghai,

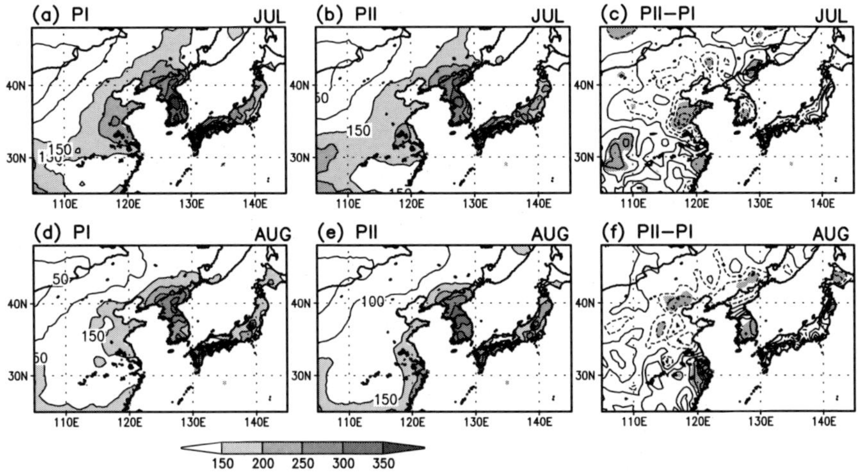

Fig. 1. The average rainfall over the East Asian region and its difference for PI and PII periods obtained from CRUR. July rainfall during (a) PI, (b) PII and (c) its difference. August rainfall during (d) PI, (e) PII and (f) its difference. The shadings in (c) and (f) represent the values significant at 90% confidence level.

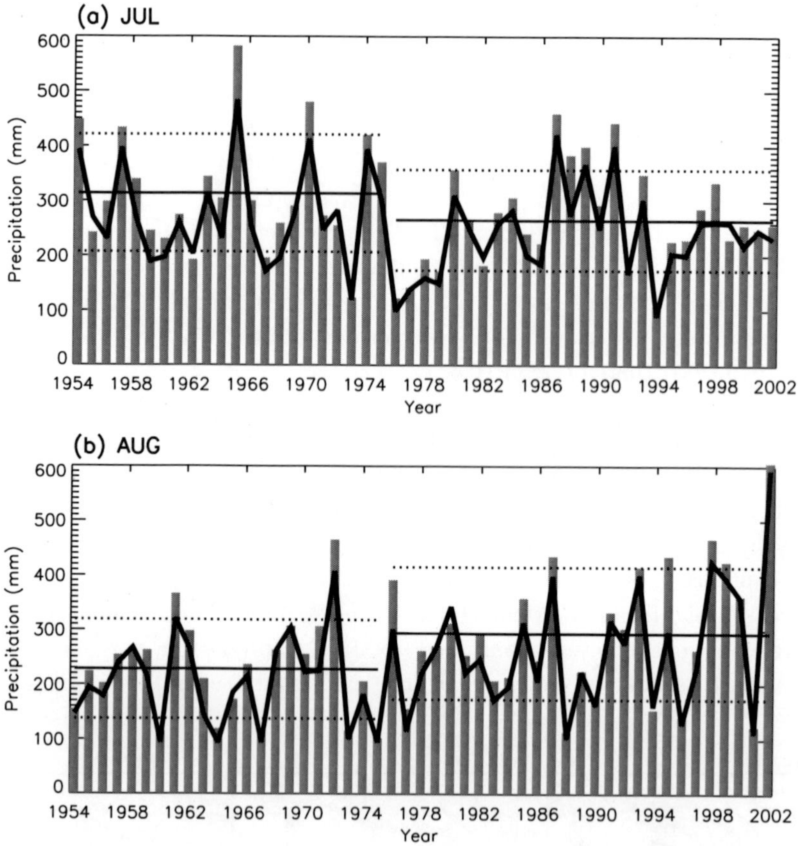

Fig. 2. The interannual variability of (a) July and (b) August rainfall obtained from CRUR (gray bar) and KMAR (black solid line) for 49 years (1954–2002). The solid lines indicate average rainfall for PI and PII and dotted lines indicate one standard deviation during PI and PII derived from CRUR.

China. The increase of August rainfall is also seen in annual cycle. The rainfall peak exists in July during PI, while that is seen in August during PII period.

Furthermore, we investigated the interannual variabilities of July and August rainfall from 1954 to 2002 (Fig. 2). The characteristics of rainfall change derived from KMAR shows similar pattern to that obtained from CRUR. The average and standard deviation of July rainfall decreased from the PI to PII period. For August rainfall, however, average and its standard deviation increased in the PII compared to the PI period.

4. Linkage with SST and Circulation Change in La Niña

4.1. *Relationship between SST and rainfall*

ENSO have an impact on the summer monsoon rainfall over the East Asia.[11,12] We examined the relationship between SST and Korean rainfall for both periods to understand whether SST exerts any control on the shift of Korean rainfall peak (Fig. 3). July rainfall has a strong negative relationship with SST around Korea and Japan from spring to summer during PI period. In the PII period, the SST over the Bay of Bengal and the East China Sea during the summer period is positively related to the

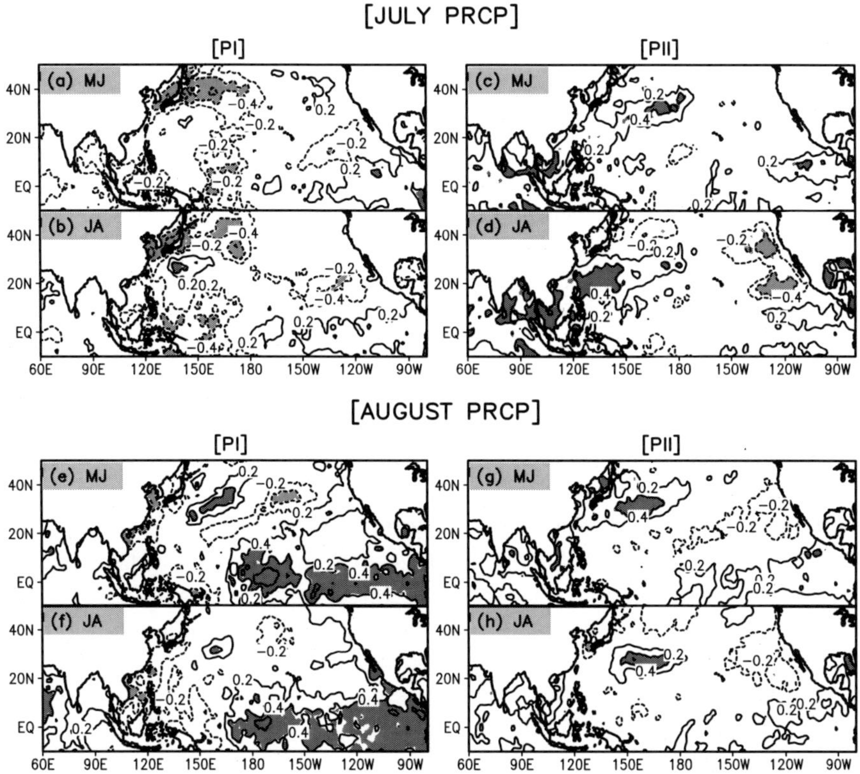

Fig. 3. The correlation coefficients between bi-monthly SST and rainfall. Left and right panels are the correlation coefficients during PI and PII, respectively. Shaded areas represent significance at 95% confidence level. Zero line is omitted. "MJ" in (a, c, e and g) means average SST from May to June. "JA" in (b, d, f and h) means average SST from July to August.

variability of July rainfall. The July rainfall has no teleconnection with El Niño during either the PI or the PII periods. On the other hand, August rainfall is simultaneously influenced by SST variation over the eastern Pacific during the PI period. An El Niño SST condition from May to August is highly connected with the strong August rainfall over Korea. A dramatic change in this relationship is seen during the PII period. No significant teleconnection between SST over the eastern equatorial Pacific and August rainfall has existed in recent years.

4.2. *Characteristics of circulation*

The large wind anomalies over the western North Pacific (WNP) can be caused by ENSO.[13–15] Therefore, we examined the low-level circulation associated with the rainfall intensity and ENSO. In order to explore the impact of ENSO via the circulation over the WNP on shift of rainfall peak, composite analysis was performed. Since August rainfall in Korea is related to SST over the eastern Pacific between May and August, El Niño (La Niña) years are defined when the averaged NINO3.4 anomaly index from May to August is above (below) 0.5 (−0.5). For rainfall, "strong" ("weak") years are defined when the normalized August rainfall is more (less) than 0.5 (−0.5) standard deviation.

During PI, August rainfall in El Niño years over Korea is stronger than in La Niña years. This is indicative of a positive relationship with the El Niño. Rainfall over the Yangtze River is also abundant. In comparison, during the PII, the difference between El Niño years and La Niña years is not significant (not shown). Associated with this rainfall feature, we examined the characteristics of low-level circulation anomalies. In the case of strong minus weak years, significant westerly and southerly wind anomalies can be found around the East China Sea and Korea (Fig. 4). The anticyclonic circulation becomes more dominant in strong years during PII period, and is related to the enhanced western Pacific high in the warm PDO period.[16] During PI, the low-level circulation difference between El Niño and La Niña years shows anomalous westerly and southerly winds over the East China Sea, similar to the strong minus weak rainfall year cases, although northwesterly anomalies are seen over some part of northern China. However, during the PII, these wind anomalies related to August rainfall were absent.

The distinctive characteristics of El Niño and La Niña years were further investigated using anomaly composite fields of El Niño and La Niña

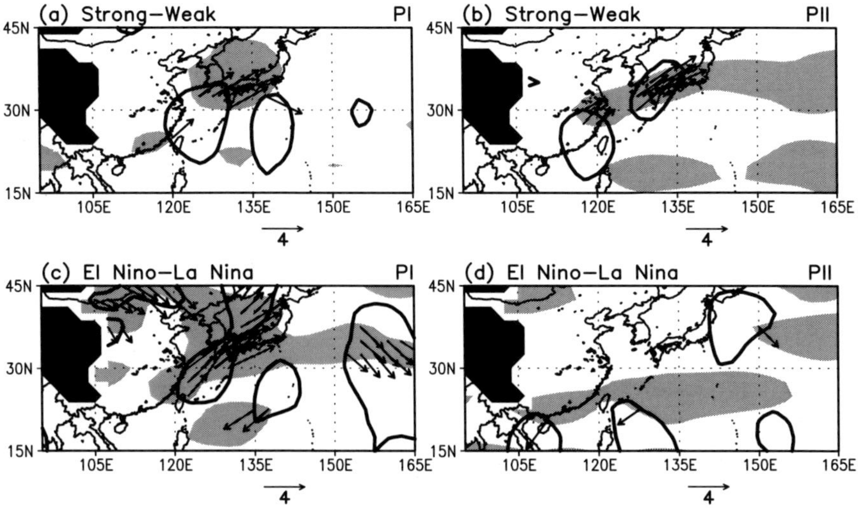

Fig. 4. The statistically significant differences (at the 90% confidence level) of winds (vectors) at their zonal (shading) and meridional (contours) components at 850 hPa between (a, b) strong and weak years, (c, d) El Niño and La Niña years during PI (left panels) and PII (right panels). The contour denotes the significance at 90% confidence level for the zonal wind. The shading indicates significant difference of meridional wind at 90% confidence level.

years for the two periods (Fig. 5). During the PI, low-level circulation and geopotential height at 850 hPa as well as SST in El Niño years have opposite anomaly phases to La Niña years. The strong anticyclonic circulation anomaly over the WNP as shown in 850 hPa wind vector and geopotential height fields can lead to the strong EASM and heavy rainfall over Korea. In La Niña years, however, the easterly and northerly anomalies may interrupt the transport of moisture into the East Asia region and make the weak EASM. During PII period, on the other hand, low-level circulation, geopotential height at 850 hPa and SST shows remarkable changes in La Niña years. Compared to PI, cold core of SST over the eastern Pacific moved westward and the warming over the Indo-Pacific region is notable. Moreover, the circulation around the East China Sea exhibits a reverse pattern for PI, with westerly and southerly wind anomalies. The anticyclonic circulation anomaly is also seen in 850 hPa geopotential height and this anomalous circulation pattern has the benefit for moisture transport into the East Asian region. This change in circulation is responsible for the change in the relationship between August rainfall and ENSO and strong rainfall in La Niña years during PII. These results suggest

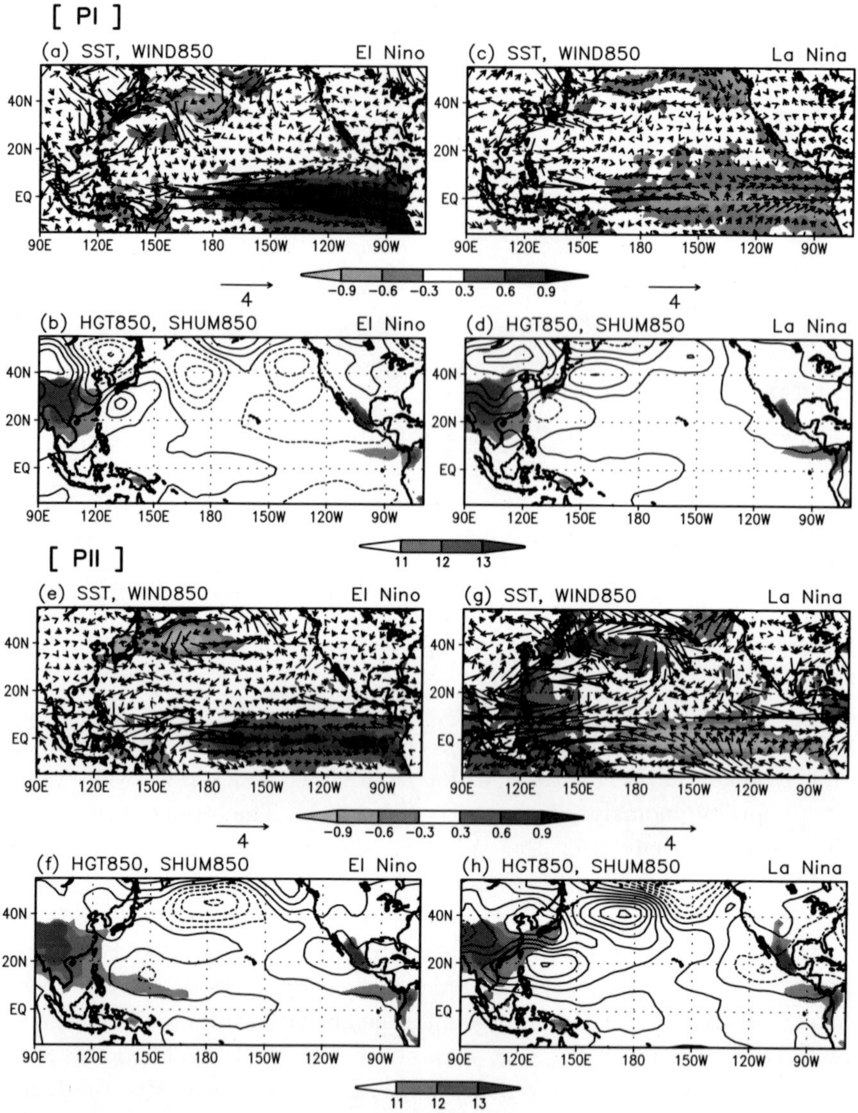

Fig. 5. Spatial pattern in El Niño and La Niña years. Anomalies of (a, c, e and g) SST (shading) and 850 hPa wind (vector), (b, d, f and h) geopotential height at 850 hPa (contour) in El Niño (left panels) and La Niña (right panels) years during PI and PII period. The specific humidity (shading) in (b, d, f and h) is average of El Niño and La Niña years, respectively.

that the manner in which ENSO affects the East Asian summer rainfall has changed since 1970s.

Recently, several studies suggested the relationship between the SST increase over the Indian Ocean and circulation over the WNP.[17,18] In relation to the circulation change in La Niña years during PII period, SST variability over the Indo-Pacific region can be an important factor because SST warming over that region is very conspicuous. Thus, the bi-monthly evolving pattern of SST in La Niña years during PII period is examined and the pattern of SST evolution is similar to the El Niño decaying summer as described by Wu *et al.*[17] (not shown). They showed that warming over the Indian Ocean in the ENSO decaying summer strengthens WNP anticyclone through the Kelvin wave-type low-level easterly response and anomalous Hadley circulation.

5. Summary and Conclusion

The rainfall over Korea has two peaks, one in July and other in August, in its annual cycle. Using KMAR and CRUR, changes in this annual cycle are investigated for the period 1954–2002. Until 1975, rainfall peak was seen in July and after 1975, the August rainfall exceeded that of July. In order to find the cause of shifting peak of Korean rainfall, the relationship between SST and rainfall for each period is examined. During PI period, warm SST anomaly over eastern Pacific is related to the intensity of August rainfall simultaneously. However, this positive relation is not seen during PII period.

ENSO can influence August rainfall over Korea via low-level circulation over the WNP. In the period of PI, the anomalous circulation pattern obtained from El Niño minus La Niña composite is comparable to that of strong minus weak years composite. In strong years and El Niño years during PI, anomalous westerly and southerly winds related to the strong August rainfall over Korea are dominant in the vicinity of Korea and the East China Sea. La Niña years have easterly and northerly wind anomalies around Korea and this is responsible for the weak August rainfall. During PII period, however, La Niña years during PII period had higher rainfall than PI period and this is associated with the circulation change over the East China Sea. The high August rainfall in La Niña years during PII is attributed to the westerly-southerly wind anomalies around East China Sea.

In relation to the circulation change in La Nina years during PII, warming over the Indo-Pacific region is considered to be a critical factor. Particularly, the evolution of SST in La Niña years during PII period is similar to the ENSO decaying summer pattern suggested by Wu *et al.*[17]. It needs further study to understand the possible mechanisms of circulation change in La Niña years during PII.

Acknowledgments

This work was supported by a grant from the Korean Ministry of Environment as "Ecotechnopia 21 project" and the second stage of the Brain Korea 21 Project.

References

1. H. Ueda, H. Yasunari, T. and R. Kawamura, Abrupt seasonal change of large-scale convection activity over the western Pacific in northern summer, *J. Meteorol. Soc. Japan* **73** (1995) 795–809.
2. K. M. Lau and S. Yang, Climatology and interannual variability of the Southeast Asian summer monsoon, *Adv. Atmos. Sci.* **14** (1997) 141–162.
3. C. H. Ho, J. Y. Lee, M. H. Ahn and H. S. Lee, A sudden change in summer rainfall characteristics in Korea during the late 1970s, *Int. J. Climatol.* **23** (2003) 117–128.
4. Z. Z. Hu, Interdecadal variability of summer climate over East Asia and its association with 500 hPa height and global sea surface temperature, *J. Geophys. Res.* **102** (1997) 19403–19412.
5. K. J. Ha and E. Ha, Climatic change and interannual fluctuations in the long-term record of monthly precipitation for Seoul, *Int. J. Climatol.* **26** (2006) 607–618.
6. B. Wang, J. G. Jhun and B. K. Moon, Variability and singularity of Seoul, South Korea, rainy season (1778–2004), *J. Climate* **20** (2007) 2572–2580.
7. T. D. Mitchell and P. D. Jones, An improved method of constructing a database of monthly climate observations and associated high-resolution grids, *Int. J. Climatol.* **25** (2005) 693–712.
8. N. A. Rayner, *et al.*, Global analyses of SST, sea ice, and night marine air temperature since the late 19th century, *J. Geophys. Res.* **108** (2003) 4407.
9. E. Kalnay, *et al.*, The NCEP/NCAR 40-year reanalysis project, *Bull. Am. Meteorol. Soc.* **77** (1996) 437–471.
10. A. N. Pettitt, A non-parametric approach to the change-point problem, *Appl. Stat.* **28** (1979) 126–135.
11. S. Shen and K. M. Lau, Biennial oscillation associated with the East Asian monsoon and tropical sea surface temperatures, *J. Meteorol. Soc. Japan* **73** (1995) 105–124.

12. Y. Wang, B. Wang and J. H. Oh, Impact of the preceding El Niño on the East Asian summer atmosphere circulation, *J. Meteorol. Soc. Jap.* **79** (2001) 575–588.
13. B. Wang, R. Wu and X. Fu, Pacific-East Asian teleconnection: How does ENSO affect East Asian climate? *J. Climate* **13** (2000) 1517–1536.
14. R. Wu and B. Wang, Interannual variability of summer monsoon onset over the western North Pacific and the underlying processes, *J. Climate* **13** (2000) 2483–2501.
15. E. J. Lee, J. G. Jhun and C. K. Park, Remote connection of the northeast Asian summer rainfall revealed by a newly defined monsoon index, *J. Climate* **18** (2005) 4381–4393.
16. Y. Zhang, J. M. Wallace and D. S. Battisti, ENSO-like interdecadal variability: 1900–1993, *J. Climate* **10** (1997) 1004–1020.
17. B. Wu, T. Zhou and T. Li, Seasonally evolving dominant interannual variability modes of East Asian Climate, *J. Climate* **22** (2009) 2992–3005.
18. S. P. Xie, K. Hu, J. Hafner, H. Tokinaga, Y. Du, G. Huang and T. Sampe, Indian Ocean capacitor effect on Indo-Western Pacific climate during the summer following El Niño, *J. Climate* **22** (2009) 730–747.

Advances in Geosciences
Vol. 16: Atmospheric Science (2008)
Eds. Jai Ho Oh *et al.*

PREDICTION OF REGIONAL DROUGHT OVER KOREA USING AN ANALOG METHOD

SU-BIN OH, DO-WOO KIM, JI-SUN LEE and HI-RYONG BYUN*
Department of Environmental Atmospheric Sciences,
Pukyong National University,
599-1 Daeyeon 3-dong, Nam-gu,
Busan 608-737, Republic of Korea
**hrbyun@pknu.ac.kr*

KI-SEON CHOI
Korea Meteorological Administration,
45 Gisangcheong-gil, Dongjak-gu,
Seoul 156-720, Republic of Korea

This study investigated a drought prediction method on the basis of similarity of spatiotemporal patterns of past droughts in Korea. The method was implemented in the following steps: First, drought areas in Korea were divided into four drought subregions by means of hierarchical clustering analysis. Second, spatiotemporal drought statistics for each subregion for the period from 1926 to 2008 were established. Temporal statistics involve the drought onset, end dates, duration, and regional drought intensities, measured by the Effective Drought Index (EDI). These statistics were collected over the four subregions, and spatial patterns of drought were examined. Third, the analogous drought events that had spatiotemporal patterns similar to those of the current or subject drought were selected. Fourth, the progress of the subject drought and the selected drought were compared. Finally, the progress pattern of the subject drought was predicted on the basis of the hypothesis that it will progress in a way similar to the selected analogous case. We applied this predicted method to several previous drought cases and evaluated the prediction efficiency. The results showed that this method was efficient in predicting droughts for about 1 year.

1. Introduction

Droughts have occurred since many centuries, and the damage caused by or influence of droughts has increased in recent years. However, because droughts progress slowly unlike other natural disasters and the damage caused by them is gradual, they give us enough time to determine ways in which they can be prevented. Thus, the damages can be reduced if a prediction and warning system is established. However, thus far, no theories have been fully recognized as useful for predicting droughts. Previous studies on droughts were limited to recognizing the signs of droughts before their occurrence; this was achieved by determining their causes through case analysis. As such studies were limited in their scopes, it was not easy to determine the causes of droughts. Hence, in some cases, antithetical theories as well as different causes or results have been suggested for the same droughts. For example, three case studies on droughts in Korea in 1994 (Park and Schubert, 1996; Guan and Yamagata, 2003; Yoo *et al.*, 2004) presented different conclusions on the causes of these droughts. Furthermore, a number of case studies on the drought that occurred in the USA in 1988 (Trenberth *et al.*, 1988; Palmer and Brabkovic, 1989; Qu *et al.*, 1994; Namias, 1991) pointed out different causes.

With such limited resources, researchers tried to predict droughts because of the urgency of doing so. However, because of insufficient studies having been conducted on the causes of droughts, statistical approaches were mainly applied for prediction. In such efforts, statistical approaches such as the probability distribution (Yevjevich, 1967), Markov chain model (Lohani and Loganathan, 1997; Canelliere *et al.*, 2007), neural network (Kim and Valdes, 2003; Morid *et al.*, 2006), low-order discrete autoregressive moving average models (Chung and Salas, 2000), seasonal autoregressive integrated moving average model (Mishra and Desai, 2005), and rotated empirical orthogonal function analysis (Kim *et al.*, 2005) were used. Nevertheless, such approaches were merely restricted to reporting the results rather than being actually applied to predict droughts.

This study also attempted to predict droughts using an analog method, in which a statistical approach was employed. The analog method is based on the assumption that the progress of a drought will be similar to that of the previous one if the climatic conditions at present are similar to those at the time at which the previous drought occurred. The analog method has been adopted mainly when it is required to predict droughts in spite of poor understanding of their causes. The analog method has been particularly applied to predict the path of typhoons and atmospheric circulation, precipitation, and temperature in the long term (Radinovic, 1975; Christensen *et al.*, 1981; Bergen and Harnack, 1982; Gutzler and Shukla, 1984). Furthermore, previous studies pointed out that the analog method was more useful for long-term predictions than for short-term predictions (Park and Lee, 2003; Xavier and Goswami, 2007). Hence, the analog method is expected to be a breakthrough to predict droughts in terms of the features of droughts that progress gradually in the long term.

Thus far, no studies were conducted using the analog approach to predict droughts, since it is very difficult to objectively quantify the previous data because of the obscure definition of the period or spatiality of droughts. This study partially attempts to resolve this difficulty by using the Effective Drought Index (EDI; Byun and Wilhite, 1999). Thus, this study substantially depends on the fact that the EDI is more precise and accurate than other drought indices (Kim *et al.*, 2009; Morid *et al.*, 2006). The EDI is useful for worldwide application, because it is independent of climatic characteristics of the locations (Byun and Wilhite, 1999). The previous studies concluded that the EDI is reasonable enough to assess the intensity and duration of a drought based on the comparison between the existing drought index and the EDI. The EDI calculation and the corresponding strengths are further described in Sec. 2.2.

The spatiality of drought was obtained by the clustering analysis. The clustering analysis is a method to divide subregions by similar characteristics. Most of the previous studies relied on precipitation to define climatological clusters (Moon, 1990; Lee and Park, 1999; Qian and Qin, 2008). However, these are not suitable for climatological drought clustering, because the precipitation data comes from a short period or a single season. Therefore, this study performed a clustering analysis relying on drought intensity using the EDI collected over the past 35 years. The spatiality of drought was used as fundamental data for the analog method.

However, as it is difficult to predict if cases identical to previous cases would occur again, the prediction of droughts using the analog method cannot be expected to be highly reliable in scientific aspects. This is the common problem of all analog methods that have been studied thus far. Nevertheless, the practical value of the analog method is considered to be significant, which is why it has been applied thus far.

2. Spatiotemporal Pattern of Drought

2.1. *Study area*

We used the precipitation data from 1925 to 2008. The operation of each observation station started in different years. There were only 6 observation

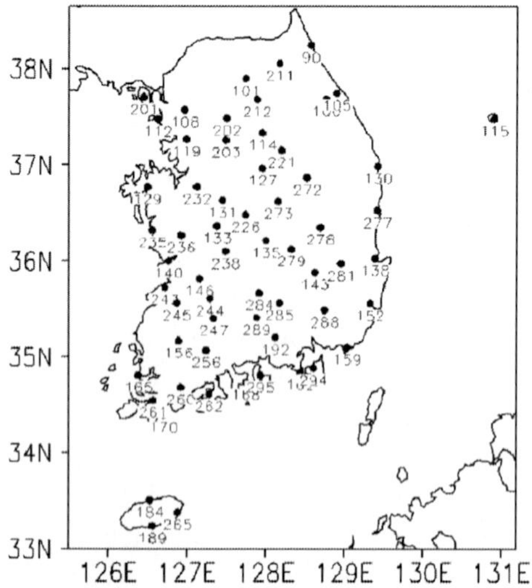

Fig. 1. The locations of precipitation stations used for the study.

stations in 1925 and the number of observation stations added up ever since. In 1973, a total of 61 observation stations were in operation. As shown in Fig. 1, the observation stations are distributed over nationwide.

2.2. *Effective drought index*

The EDI is used to measure an intensity of drought. The calculation process of the EDI is as follows:

$$EP_i = \sum_{n=1}^{i} \left[\left(\sum_{m=1}^{n} P_m \right) \Big/ n \right], \tag{1}$$

$$DEP = EP - MEP, \tag{2}$$

$$EDI = DEP/ST(DEP). \tag{3}$$

Equation (1) expresses Effective Precipitation (EP) expresses currently available water resources generated by past precipitation and reflects the depletion by runoff and evaporation with time. Here, P_m denotes the precipitation in m days before a particular day and i denotes the number of the days whose precipitation is summed for calculation of drought severity. Equation (2) is used to calculate the deviation of EP (DEP) from the climatological mean of EP (MEP) for each calendar day. Finally, Eq. (3) is used to calculate the standardized value of the DEP (EDI). Here, ST (DEP) denotes the standard deviation of the DEP of each day. For further details, refer to Byun and Wilhite (1999).

The main advantages of the EDI are as follows. (1) It gives a reasonable measure of the current level of water resources by considering daily precipitation accumulation with a weighting function with passing time. (2) It is effective to define the starting day, ending day, and duration of the drought because the EDI is expressed in the unit of days. (3) The EDI has universal applicability because it is independent of the climatic characteristics of a particular region. The feasibility of the EDI has been proved in previous studies (Byun and Lee, 2002; Yamaguchi and Shinoda, 2002; Morid *et al.*, 2006; Smakhtin and Hughes, 2007; Akhtari *et al.*, 2008).

This study calculated the EDI using the daily precipitation at 61 stations in Korea from 1925 to 2008 (84 years). The EDI values are listed in Table 1.

2.3. *Drought subregions*

Drought subregions were fixed in order to investigate the spatial pattern of droughts. By using the Statistical Package for Social Science (SPSS), hierarchical clustering analysis was carried out; the between-group linkage method with the Pearson correlation measure was used in this clustering procedure. For the clustering, time series from each station were calculated on the basis of a 35 years (1974–2008) monthly minimum EDI. As shown in Fig. 2(a), we used a clustering procedure for groups 61 to 1 and calculated the between-group correlations for each linkage. If the point where the coefficient between merged clusters increases markedly can be discerned, the clustering process can be stopped at this point. Thus, this study selected five as the appropriate number of clusters. Figure 2(b) shows the spatial

Table 1. The classification of the Effective Drought Index (EDI).

Effective Drought Index	Classification	Drought class (n)
EDI > 2.5	Extreme wet	0
$1.5 <$ EDI ≤ 2.5	Severe wet	0
$0.7 <$ EDI ≤ 1.5	Moderate wet	0
$0 <$ EDI ≤ 0.7	Weak wet (Normal)	0
$-0.7 <$ EDI ≤ 0	Weak drought (Normal)	1
$-1.5 <$ EDI ≤ -0.7	Moderate drought	2
$-2.5 <$ EDI ≤ -1.5	Severe drought	3
EDI ≤ -2.5	Extreme drought	4

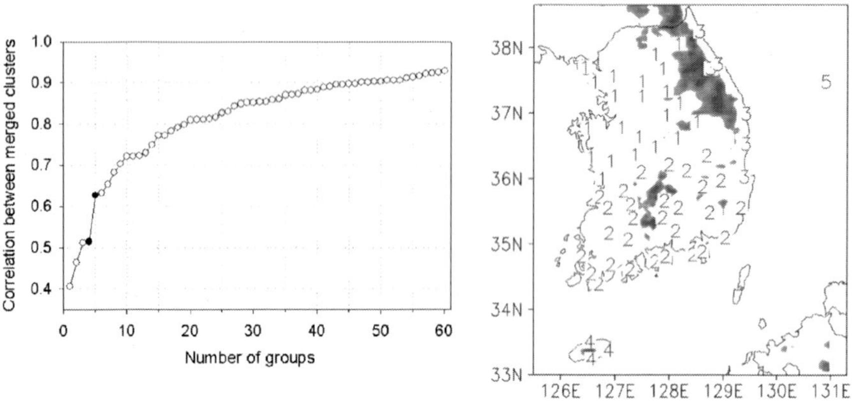

(a) (b)

Fig. 2. Hierarchical clustering analysis from the EDI for period 1974–2008. (a) Variation of Pearson correlation coefficient between merged clusters. The number of clusters is from 1 to 59. (b) Spatial distribution of five drought clusters: 1-Central part, 2-Southern part, 3-Eastern part, 4-Jeju Island and 5-Ulung Island. Topography higher than 500 m is shaded.

distribution of five clusters: central district (cluster 1; A), southern district (cluster 2; B), eastern district (cluster 3; C), Jeju Island (cluster 4; D), and Ulreung Island (cluster 5; E).

3. Experiments on the Drought Predictability using an Analog Method

3.1. *Construction of drought code*

The daily drought codes were generated for the intensity, duration, and spatial distribution of droughts acquired using the EDI. The temporal patterns of drought codes were obtained from the days of duration of the drought and the mean EDI at all stations. The spatial patterns were collected using the drought class on a cluster-by-cluster basis. The code type is ±Em Co AnBnCnDn, and the meaning of each type is described in Table 2. Cluster E was excluded from the analysis because the analysis of this cluster was difficult owing to presence of just one station in this cluster.

For example, the drought code on December 31, 2008, was −1.60 131 A2B3C2D1, which indicates that the mean EDI was −1.60, the duration of the droughts was 131 days, clusters A and C experienced moderate

Table 2. The explanation of each part of the daily drought code.

Code	Meaning
$\pm$Em	Mean EDI of total station
Co	Drought duration ($\pm$Em ≤ 0)
AnBnCnDn	Drought class of cluster A, B, C, D

drought, cluster B experienced severe drought, and cluster D experienced weak drought. As precipitation data on all stations in Korea exist from 1973, the drought indices for each cluster from January 1, 1926, to December 31, 2008, were acquired using the mean EDI of existing stations.

3.2. *Selection of subject cases*

Figure 3 shows the time series of the annual minimum EDI in four subregions (A, B, C, and D) and the deviation of annual precipitation averaged over 60 stations for the past 30 years (1979–2008). The years 1988, 1994, and 2001 were selected as the subject years because they showed an anomalously low EDI and precipitation at the same time. For 1988, cluster B showed the minimum EDI, -2.10, on December 30. In 1994, cluster C showed the minimum EDI, -2.35, on October 10; in 2001, cluster A, showed the minimum EDI, -2.14, on June 11. The days on which the minimum EDI below -1.0 just before the annual minimum value was observed were selected as the subjected days. The dates August 19, 1988, August 12, 1994,

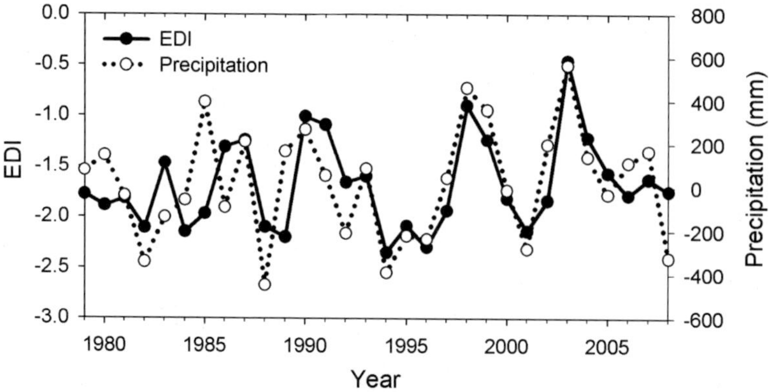

Fig. 3. The time series of the annual minimum EDI (solid line) and deviation of annual precipitation (dotted line) from 1979 to 2008 over 60 stations in Korea.

and May 10, 2001, were selected as the subject dates since they met the abovementioned criterion.

3.3. *Selection of analogous cases*

We selected only one analogous case, one that is most similar to the subject case among several drought cases, by considering the daily drought codes using the following steps:

Step 1. Select cases similar to the subject case by considering the spatial distribution and seasonality of droughts.

- 1.1. Sort out the cases that have the same spatial drought code (AnBnCnDn).
- 1.2. Choose the cases that have similar seasonality (±30 days from the subject case).

Step 2. If several similar cases are found in the same year, select only one case with the most similar intensity and duration.

- 2.1. Select the cases with the most similar mean EDI (Em).
- 2.2. Select the cases with the most similar duration of drought (Co).

Step 3. If similar cases are observed in several different years, select the year with the highest correlation coefficient by comparing the time series.

In accordance with the application of the analog method to the subject drought cases selected as described, August 19, 1988 (−0.93 023 A2B2C1D1) was similar to July 23, 1992 (−1.14 049 A2B2C1D1); August 12, 1994 (−0.96 076 A2B2C2D0) was similar to September 11, 1937 (−1.30 082 A2B2C2D0); and May 10, 2001 (−0.94 036 A2B1C2D1) was similar to June 7, 1996 (−0.95 034 A2B1C2D1).

3.4. *Results of experiment*

The subject cases and their analogous cases were examined to determine the similarity in their progresses. Figure 4 shows the comparison between the times series of the subject cases (black line) and analogous cases (gray line) in the subregion where the minimum EDI value was obtained for the subject drought case. For the case on August 19, 1988, both the subject case and the analogous case showed drought conditions for about 4 months

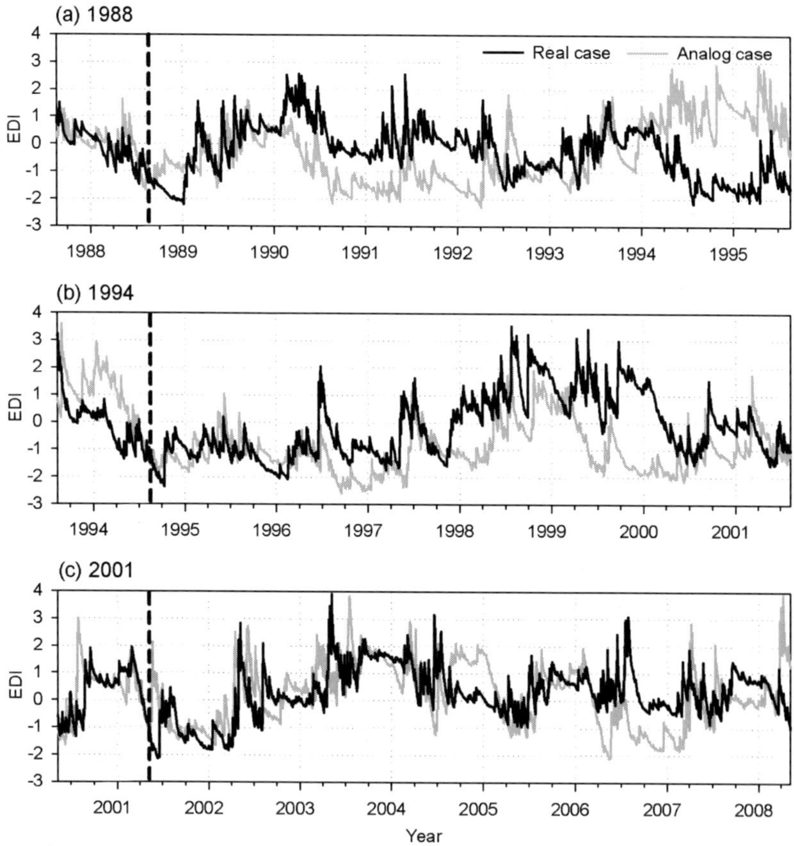

Fig. 4. Time series of daily EDI for subject drought cases (1988, 1994, and 2001; black line) and its analogous drought cases (gray line). Vertical dashed line denotes the start date of applying analog method.

(September–December 1988), although the intensity was stronger in the subject case. Two cases showed similar behavior for about 1 year (January 1989–January 1990). For the cases on August 12, 1994, and May 10, 2001, it was confirmed that the subject and analogous cases made similar progress for about 2 years (August 1994–May 1996) and 3 years (June 2001–June 2004), respectively. After the periods in which the analogous cases showed similar behavior, no more significant similarity was detected in all three cases because they showed different progresses for more than 6 months.

The root mean square error (RMSE) was calculated to examine the change in prediction accuracy with time. The RMSE is widely used to measure the difference between the actual case and the predicted case. It is calculated using the following formula:

$$RMSE = \sqrt{\frac{1}{n}\sum_{k=1}^{n}(p_k - o_k)^2}, \tag{4}$$

where o denotes the actually observed value, p denotes the predicted value, and n denotes the period of prediction test. The larger the value of the RMSE, the larger is the difference between a subject case and an analogous case.

This study investigated the change in the RMSE on the basis of the monthly minimum EDI with increasing prediction period in annual units (Fig. 5). For the cases on August 19, 1988, and August 12, 1994, the prediction errors increased rapidly after 1 year and 2 years, respectively. In contrast, the case on May 10, 2001, showed low errors compared with other cases, and the errors increased as the prediction period increased.

We extended the application period to 76 years (1926–2008). A drought event was defined as the period of consecutive negative EDIs with the minimum EDI of below −1. As a result, a total of 75 drought cases were selected, and the analog method was applied to them. In accordance with the mean RMSE of the 75 drought cases, it was found that the prediction errors increased rapidly when the prediction period increased to beyond 2 years (Fig. 6). In other words, the prediction using the analog method cannot be extended to beyond 2 years.

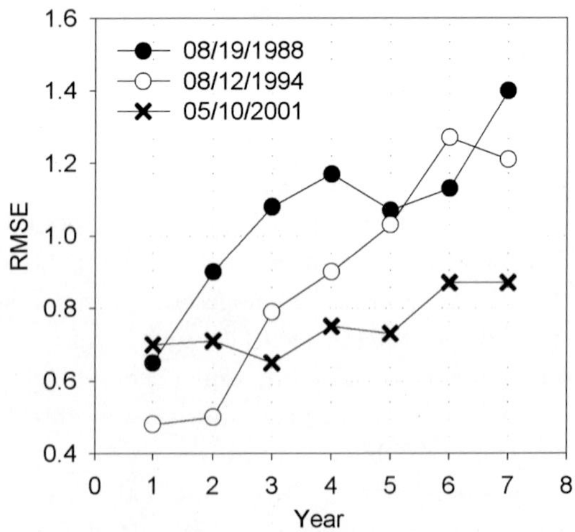

Fig. 5. Root mean square error (RMSE) of 7 years analog forecast.

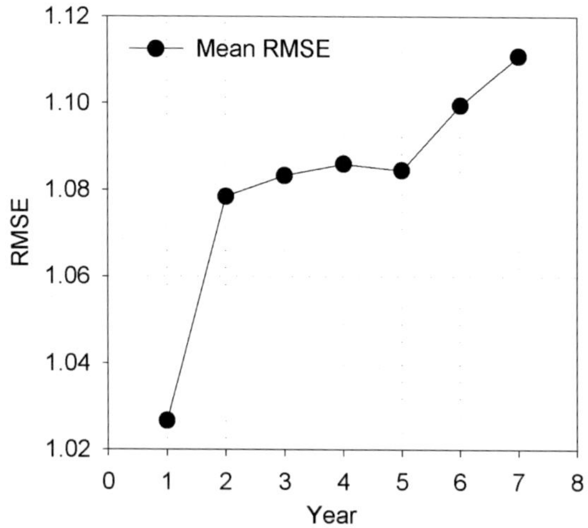

Fig. 6. Mean RMSE of 7 years analog forecast for 75 droughts from 1926 to 2008.

4. Discussion

This study investigated the prediction of droughts using the analog method. Although the analog method has some limitations as a statistical approach, it is useful to effectively predict droughts for about 1 year. We selected analogous cases on the basis of the similarity of variation in the EDI time series. However, the prediction results may be improved when the analog method is applied in consideration of several factors such synoptic-scale atmospheric circulation and climatic elements highly correlated with droughts. Further, the analog method could be successfully used to predict days on which the analogous and subject cases were not similar. The drought subregions and the daily drought codes presented in this study could probably be used as the fundamental data for drought pattern classification or case selection in future studies. This study is expected to prompt research on the general causes of droughts. If we classify drought cases in accordance with the spatiotemporal patterns of droughts based on daily drought codes and identify the general causes and features from all cases, we might obtain more valuable predictions of droughts.

Acknowledgments

This work was funded by the Korea Meteorological Administration Research and Development Program under Grant CATER 2006–2306.

References

1. R. Akhtari, S. Morid, M. H. Mahdian and V. Smakhin, Assessment of real interpolation methods for spatial analysis of SPI and EDI drought indices, *Int. J. Climatol.* **29** (2009) 135–145.
2. R. E. Bergen and R. P. Harnack, Long-range temperature prediction using a simple analog approach, *Mon. Weather Rev.* **110** (1982) 1083–1099.
3. H. R. Byun and D. H. Wilhite, Objective quantification of drought severity, *J. Climate* **12** (1999) 2747–2756.
4. D. K. Lee, *et al.*, Defining three rainy seasons and the hydrological summer monsoon in Korea using available water resources index. *J. Meteorol. Soc. Jap.* **80** (2002) 33–44.
5. A. Cancelliere, G. Di Mauro, B. Bonaccorso and G. Rossi, Drought forecasting using the standardized precipitation index, *J. Am. Water Resour. Assoc.* **21** (2007) 801–819.
6. R. A. Christensen, R. F. Eilbert, O. H. Lindgren and L. L. Rans, Successful hydrologic forecasting for California using an information theoretic model, *J. Appl. Meteorol.* **20** (1981) 706–713.
7. C. H. Chung and J. D. Salas, Drought occurrence probabilities and risks of dependent hydrological processes, *J. Hydrol. Eng. ASCE* **5** (2000) 259–268.
8. Z. Guan and T. Yamagata, The unusual summer of 1994 in East Asia: IOD teleconnections, *Geophys. Res. Lett.* **30** (2003) 1544–1547.
9. D. S. Gutzler and J. Shukla, Analogs in the wintertime 500 mb height field, *J. Atmos. Sci.* **41** (1984) 177–189.
10. D. W. Kim, H. R. Byun and K. S. Choi, Evaluation, modification, and application of the effective drought index to 200-year drought climatology of Seoul Korea, *J. Hydrol.* **378** (2009) 1–12.
11. S. Kim, C. K. Park and M. K. Kim, The regime shift of the northern hemispheric circulation responsible for the spring drought in Korea, Asia-Pacific, *J. Atmos. Sci.* **41** (2005) 571–585.
12. T. Kim and J. B. Valdes, Nonlinear model for drought forecasting based on a conjunction of wavelet transforms and neural networks, *J. Hydrol. Eng. ASCE* **8** (2003) 319–328.
13. V. K. Lohani and G. V. Loganathan, An early warning system for drought management using the Palmer drought index, *J. Am. Water Resour. Assoc.* **33** (1997) 1375–1386.
14. A. K. Mishra and V. R. Desai, Drought forecasting using stochastic models, *Stoch. Environ. Res. Risk Assess.* **19** (2005) 326–329.
15. Y. S. Moon, Division of precipitation regions in Korea through the cluster analysis, *J. Kor. Meteorol. Soc.* **26** (1990) 203–215.
16. S. Morid, V. Smakhtin and K. Bagherzadeh, Drought forecasting using artificial neural networks and time series of drought indices, *Int. J. Climatol.* **27** (2006) 2103–2111.
17. M. Moghaddasi, *et al.*, Comparison of seven meteorological indices for drought monitoring in Iran, *Int. J. Climatol.* **26** (2006) 971–985.
18. J. Namias, Spring and summer 1988 drought over the contiguous United States — Causes and prediction, *J. Climate* **4** (1991) 56–65.

19. T. N. Palmer and C. Brankovic, The 1988 United States drought linked to anomalous sea surface temperature, *Nature* **338** (1989) 54–57.
20. C. K. Park and S. D. Schubert, On the nature of the 1994 East Asian summer drought, *J. Climate* **10** (1997) 1056–1070.
21. J. G. Park and S. M. Lee, A regionalization of annual precipitation over South Korea, *J. Kor. Meteorol. Soc.* **29** (1993) 117–126.
22. W. H. Park and J. G. Lee, Long-range forecast using analog/anti-analog method with GDAPS forecasts, Asia-Pacific, *J. Atmos. Sci.* **39** (2003) 491–501.
23. W. H. Qian and A. Qin, Precipitation division and climate shift in China from 1960 to 2000, *Theor. Appl. Climatol.* **93** (2008) 1–17.
24. J. Qu, M. L. Sestak, A. R. Riebau, L. R. Smith and D. Ouren, A study of El Nino and Southern Oscillation (ENSO) impact on drought and wetness in the Western United States, *6th Conf. Clim. Var.*, Nashville Tennessee, 23–28 January 1994, pp. 101–104.
25. D. Radinovic, An analogue method for weather forecasting using the 500/1000 mb relative topography, *Mon. Weather Rev.* **103** (1975) 639–649.
26. V. U. Smakhtin and D. A. Hughes, Automated estimation and analyses of meteorological drought characteristics from monthly rainfall data, *Environ. Model. Software* **22** (2007) 880–890.
27. K. E. Trenberth, G. W. Branstator and P. A. Arkin, Origins of the 1988 North American drought, *Science* **242** (1988) 1640–1645.
28. P. K. Xavier and B. N. Goswami, Analog method for real time forecasting of summer monsoon subseasonal variability, *Mon. Weather Rev.* **135** (2007) 4149–4160.
29. Y. Yamaguchi and M. Shinoda, Soil moisture modeling based on multiyear observations in the Sahel, *J. Appl. Meteorol.* **41** (2002) 1140–1146.
30. V. Yevjevich, An objective approach to definitions and investigations of continental hydrologic droughts. Hydrology Papers Colorado State University, Fort Collins (1967).
31. S. H. Yoo, C. H. Ho, S. Yang, H. J. Choi and J. G. Jhun, Influences of tropical western and extratropical Pacific SST on East and Southeast Asian climate in the summers of 1993–1994, *J. Climate* **17** (2004) 2673–2687.

Advances in Geosciences
Vol. 16: Atmospheric Science (2008)
Eds. Jai Ho Oh *et al.*

DECADAL CHANGES IN THE HADLEY CIRCULATION*

YONGYUN HU† and CHEN ZHOU
Department of Atmospheric Sciences, Peking University
Beijing, 100871, China
†*yyhu@pku.edu.cn*

The Hadley circulation is one of the most important atmospheric circulations. The main goal of this paper is to investigate whether the Hadley circulation has changed in the past few decades as an integral part of global climate change. We focus on two key points: one is the horizontal scale of the Hadley circulation, and the other one is its strength. Using three meteorological reanalyses, we show that the Hadley circulation has a significant expansion of about 2 to 4.5 degrees of latitude in autumn for both hemispheres since 1979, and that two of the three reanalysis datasets show strengthening of the Hadley circulation in Northern-Hemisphere winter. Trends derived from general circulation model simulations are compared with observed changes in the Hadley circulation. It is found that most simulations show poleward expansion of the Hadley circulation, with weaker magnitudes. However, most simulations display weakening of the Hadley circulation, rather than strengthening.

1. Introduction

The Hadley circulation is one of the most important atmospheric circulations. It is a thermally driven meridional circulation, with warmer air rising in the tropics due to the release of latent heat and colder air sinking in the subtropics, generating an enclosed circulation in each hemisphere (Held and Hou, 1980). The Hadley circulation is fundamentally important to the global climate system. It does not only transport heat from the tropics to the subtropics and to high latitudes through extratropical eddies but also transports momentum flux to the subtropics (Lindzen, 1994). Both heat and momentum transports have important influences on subtropical jet streams, which consequently impact on waves and atmospheric circulations

*This work is supported by the NSF of China (40575031 and 40875042), the Ministry of Education of China (106002 and 20070001002), and the National Basic Research Program of China (2007CB411801).

at middle and high latitudes. Therefore, changes in the Hadley circulation have important impacts on global climate (Diaz and Bradley, 2004).

In the past few years, there have been growing interests in decadal changes in the Hadley circulation. These studies were concerned with two key problems: one is multi-decadal changes in intensity of the Hadley circulation, and the other one is its horizontal scales or width. Using outgoing longwave radiation (OLR) datasets from the Earth Radiation Budget Experiment (ERBE), Chen *et al.* (2002) and Wielicki *et al.* (2002) suggested that the Hadley circulation was intensified in the 1990s. Using reanalysis datasets, Quan *et al.* (2002), Hu *et al.* (2005) and Mitas and Clement (2005) showed that the winter cell of the Hadley circulation has large intensification in the past few decades. However, whether the intensification of the Hadley cell as suggested by some of these studies is real remains controversial (Trenberth, 2002; Held and Soden, 2006).

On the other hand, results from several recent studies suggest that the Hadley circulation has poleward expansion since 1979. Using microwave sounding unit (MSU) data, Fu *et al.* (2006) showed an enhanced warming in the mid-latitude troposphere. Such a change in tropospheric temperatures indicates a poleward shift in the maximum horizontal temperature gradient and thus implies a poleward shift of subtropical jet streams. Since the location of the subtropical jet stream marks the poleward edge of the Hadley circulation, Fu *et al.* (2006) suggest a broadening of the tropics. Using total ozone data, Hudson *et al.* (2006) showed that the subtropical upper tropospheric front in the Northern Hemisphere shifted poleward by about 2.75 degree latitudes over 1979–2003, which also implies a poleward shift of the tropospheric subtropical jet stream. Based on mass streamfunction calculations and OLR data analyses, Hu and Fu (2007) provided direct evidence that the Hadley circulation has expended poleward since 1979. Through an analysis of tropopause heights in the subtropics using radiosonde measurements and reanalysis data, Seidel and Randel (2007) found an expansion of the tropical belt for about 4.25 degree latitudes during 1979–2005.

Changes of the Hadley circulation in both width and strength have important implications for global climate changes, especially for subtropical regions where the Hadley circulation descends. A stronger Hadley circulation would lead to stronger downward motions in the subtropics, which consequently causes less precipitation in the subtropics. A broadening Hadley circulation would lead to poleward expansion of the subtropical dry zone in both hemispheres.

In this study, our main goal is to investigate whether the changes in the Hadley circulation are responses to global climate change due to increasing greenhouse gases. To investigate the problem, we compare the trends from reanalyses with that from general circulation model (GCM) simulations. We first use three meteorological reanalysis datasets to show changes in both the width and strength of the Hadley circulation in the recent three decades. Then, we examine changes in both width and intensity of the Hadley circulation in two kinds of GCM simulations: atmospheric GCM (AGCM) simulations forced by observed sea surface temperatures (SST) and coupled atmospheric and oceanic GCM (AOGCM) simulations forced by increasing greenhouse gases. Data and methods used in this study are described in Sec. 2. Results from reanalysis datasets are in Sec. 3. Results from GCM simulations are presented in Sec. 4. Discussion and conclusions are in Sec. 5.

2. Data and Methods

The Hadley circulation is characterized with the mean meridional mass streamfunction (MMS). MMS is calculated by vertically integrating monthly meridional winds in the conventional way (Holton, 1994). Clockwise circulation (the northern cell) is defined as positive, and anti-clockwise circulation (the southern cell) is defined as negative. The locations of poleward edges of the Hadley circulation are identified as the latitudes where MMS equals $0\,\mathrm{kg\,s^{-1}}$, which are obtained using linear interpolation. Poleward expansion of the Hadley cells is estimated by calculating linear trends of the edge latitudes. The three reanalysis datasets used in this study are from the National Center for Environmental Prediction/National Center for Atmospheric Research (Kalnay, *et al.*, 1996), the National Center for Environmental Prediction/Department of Energy (Kanamitsu, *et al.*, 2002) and the European Centre for Medium-Range Weather Forecasts (Uppala, *et al.*, 2005). For simplicity, they are denoted by NCEP/NCAR, NCEP/DOE, and ERA40, respectively. The ERA40 reanalysis used here is from January 1979 to August 2002 (24 years), and NCEP/NCAR and NCEP/DOE reanalyses are from January 1979 to December 2007 (29 years).

The two kinds of GCM simulation datasets are from the Atmospheric Model Intercomparison Project (AMIP) (Gates *et al.*, 1999) and the AOGCM simulations of the 20th century for the Fourth Assessment Report (AR4) of the Intergovernmental Panel on Climate Change (IPCC)

(Solomon *et al.*, 2007). As pointed out by Gates *et al.* (1999), AMIP was designed to simulate the atmosphere's evolution subject to the observed sequence of monthly averaged global sea surface temperature and sea-ice distributions from 1979 to 2000. Therefore, in AMIP simulations atmospheric compositions and the solar constant are fixed, while the AGCMs are forced by observed time-varying SST. AMIP includes 12 models: GISS_ER, IAP_FGOALS, IPSL_CM4, MIROC32_MEDRES, MPI_ECHAM5, CNRM-CM3, GFDL-CM21, MIROC32-HIRES, MRI-CGCM23-2A, NCAR-CCSM30, NCAR-PCM1, and UKMO-HADGEM1. The first 5 models have ensemble members of simulations, and the other 7 models have only single runs. To compare with these ensemble simulations, results from the 7 single runs are averaged and presented as ensemble numbers. For IPCC-AR4 simulations, we choose 10 of 23 model results, and all the 10 models have ensemble simulations. The models are: MPI-ECHAM5, UKMO-HADCM3, NCAR, GISS-EH, GISS-ER, GFDL-CM20, GFDL-CM21, NCAR-CCSM30, UKMO-HADGEM1, GISS-AOM. Though all simulation results are available from 1979, these simulations end at different years such as 1999 or 2000. For comparison, trends from all simulations are converted into trends over 20 years.

It is noticed that the poleward expansion in all the three reanalyses mainly occurs in their summer and autumn seasons for both hemispheres. In winter and spring seasons, poleward expansions are either very weak or slightly negative, as pointed out by Hu and Fu (2007). Therefore, our analysis focuses on autumn.

3. Results from Reanalysis Data

3.1. *Poleward expansion of the Hadley circulation*

Figure 1 shows the time series of poleward-edge latitudes of the Hadley circulation at the 500 hPa pressure surface in both boreal and austral autumn, derived from the three reanalyses. For the left panel (Figs. 1(a)–1(c)), all the three datasets show that the poleward edge of the northern Hadley circulation branch exhibits a systematic poleward expansion in September–October–November (SON) from 1979 onward. For NCEP/NCAR, the linear trend in poleward-edge latitudes is about 2.14° in latitude from 1979 to 2007, with a statistical significance above the 99% confidence level (student t-test value is 2.93). For NCEP/DOE and ERA40, the poleward expansions are 2.83° and 2.40° of latitude over 1980–2007

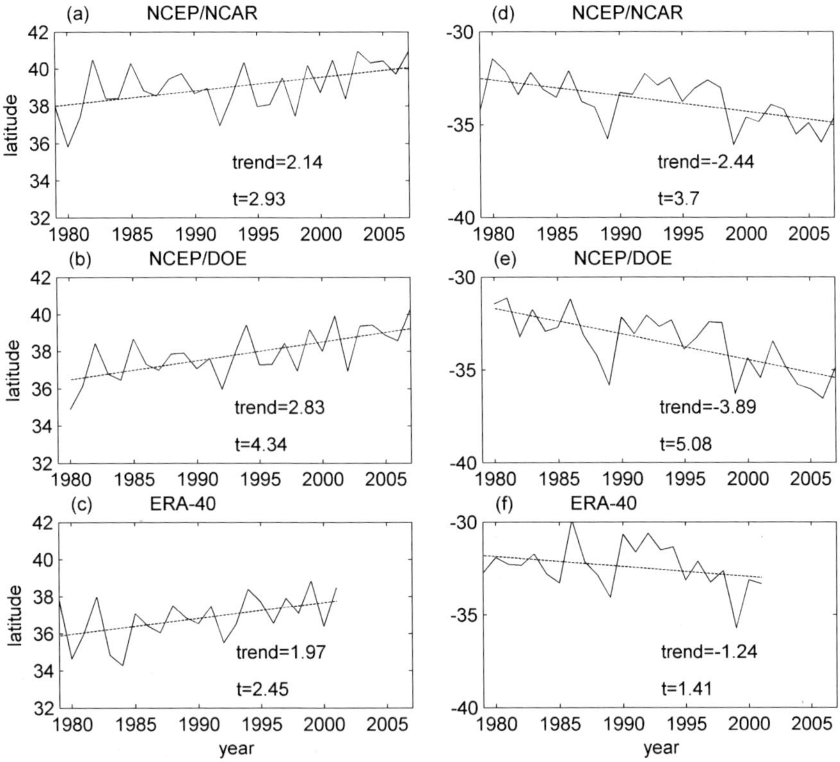

Fig. 1. Time series of poleward-edge latitudes of the Hadley circulation at 500 hPa. Plots on the left panel are for NH autumn (SON), derived from three reanalysis datasets. Plots on the right panel are for SH autumn (MAM). From top to bottom, the plots are for NCEP/NCAR, NCEP/DOE, and ERA40, respectively. Trends marked in the plots are the values for the years over which reanalysis data are available. Student t-test values are also marked in these plots. 1.7 approximately corresponds to the 90% confidence level.

and 1979–2001, respectively, with statistical significance all above the 98% confidence level. From Fig. 1, we can find that the poleward expansion of the northern Hadley circulation branch appears to be a systematic robust feature of the analyses, but not to be caused due to the Southern Oscillation/El Nino events (e.g., 1988, 1998, and so on). In addition, the significant poleward expansion is not limited over the period since 1979. It is also found in ERA40 and NCEP/NCAR reanalyses since the 1950s. Because reanalyses before 1979 are less reliable due to not including satellite observations, trends before 1979 are not shown.

Poleward expansion is also found for the southern Hadley-circulation branch. The three plots on the right panel (Figs. 1(d)–1(f)) show time series of poleward-edge latitudes of the southern Hadley-circulation branch at 500 hPa for March–April–May (MAM). Similar to that of the northern branch, the poleward edges of the southern branch in all the three datasets demonstrate systematic southward expansions. The largest southward expansion is found in NCEP/DOE, with magnitude close to 3.89 degree latitudes over the 29 years. All the trends are statistically significant.

3.2. *Intensification of the Hadley circulation*

In studying the decadal changes in the strength of the Hadley circulation, Mitas and Clement (2005) examined the maximum value of MMS. Here, we show height-latitude cross-sections of trends in MMS, which better demonstrate the spatial structures of changes in the Hadley circulation. Since the intensification of the Hadley circulation mainly exists in the branch of the winter hemisphere (Mitas and Clement, 2005), our analysis also focuses on the winter season for both hemispheres. Figures 2(a)–(c) show the trends in MMS for Northern-Hemisphere winter, derived from three reanalyses. Three plots display different spatial structures of trends in the tropics. For NCEP/NCAR, trends in the northern tropics are positive, except for relatively weak negative trends at the lowest layers. Maximum trends more than $2.5 \times 10^{10}\,\mathrm{kgs}^{-1}$ over 1979–2007 are located around 300 hPa. Because the streamfunction of the northern cell is defined positive, the positive trends indicate intensification of the northern branch. Comparison with the climatological maximum value of about $16 \times 10^{10}\,\mathrm{kgs}^{-1}$ for the northern branch in winter, the increase of the northern-branch strength is more than 15%. For NCEP/DOE, weak positive trends are found between 10° N and 30° N, however, strong negative trends are around the equator. Thus, it is difficult to determine whether the Hadley circulation is intensified or not. For ERA40, the northern tropics are dominated by positive trends, except for the top layers in the tropical troposphere. Comparison with the climatological maximum value of about $21 \times 10^{10}\,\mathrm{kgs}^{-1}$, the net increase is about 29%.

Plots on the right panel show trends in MMS for austral winter. For NCEP/NCAR, the southern tropics are dominated by weak negative trends. Since the southern branch is defined negative, the negative trends are indicative of intensification of the southern branch. Compared with that of the northern branch, the trends for the southern branch are much weaker. For NCEP/DOE, weak positive trends are found between the equator and 15° S, suggesting a weakening of the southern branch. For ERA40, the

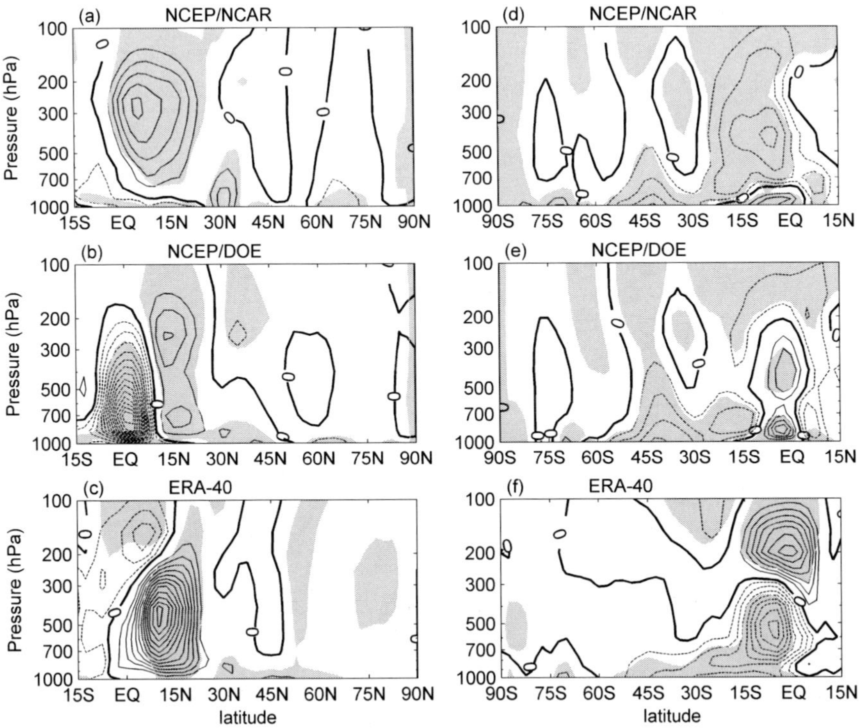

Fig. 2. Trends in zonal-mean mass streamfunction. Plots on the left are for NH winter (DJF), derived from three reanalysis datasets. Plots on the right are for SH winter (JJA). From top to bottom, the plots are for NCEP/NCAR, NCEP/DOE, and ERA40, respectively. Contour interval is 0.5×10^{10} kg s^{-1} per 29, 28, and 23 years, for NCEP/NCAR, NCEP/DOE, and ERA40, respectively. Shading areas are the regions where statistical significance is above the 90% confidence level.

upper tropical troposphere shows positive trends, while the lower tropical troposphere shows negative trends, suggesting that neither the maximum value of MMS nor the whole southern branch has significant changes.

4. Simulation Results

The radiative effect of increasing greenhouse gases causes warming for both SST and the troposphere (Solomon *et al.*, 2007). If the above decadal trends in the Hadley circulation are responses to global greenhouse warming, both SST and tropospheric warming would have contributions. The AMIP simulations provide a test bed to examine how the Hadley circulation responds to SST warming. The IPCC-AR4 AOGCM simulations include warming in both the troposphere and SST forced by increasing greenhouse gases. Here, we first analyze AGCM simulations. Then, we analyze the AOGCM simulations.

4.1. *Results from AMIP simulations*

AMIP simulations also demonstrate that poleward expansion of the Hadley circulation occurs in the autumn season for both hemispheres. Figure 3(a) shows trends in poleward-edge latitudes of the northern branch in SON. All ensemble simulations and the averaged results from 7 single runs show poleward expansion. Magnitude varies from 0.2° to 1.0° of latitude, with an average value of about 0.5° in latitude per 20 years. For the southern

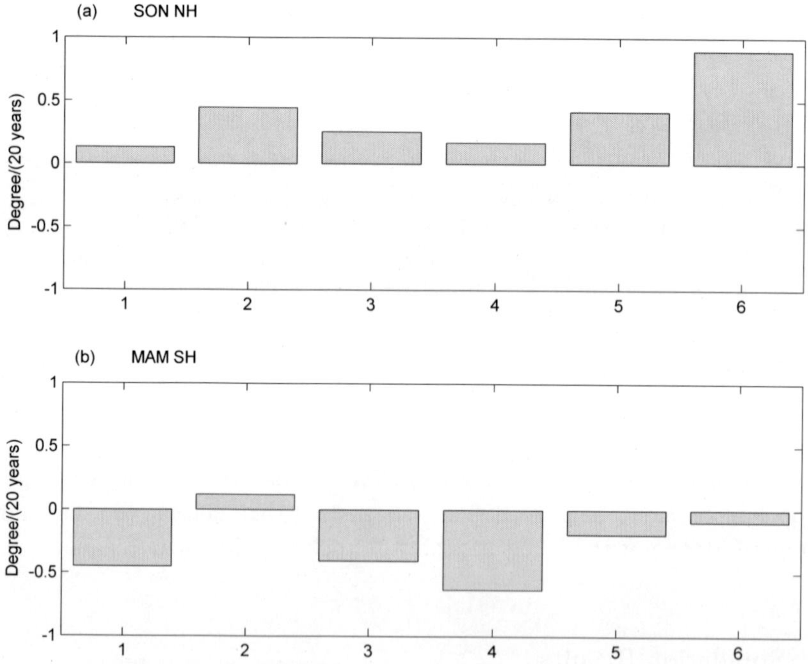

Fig. 3. Poleward expansion of poleward-edge latitudes of the Hadley circulation derived from AMIP simulations. In Fig. 3(a), positive (negative) trends indicate poleward (equatorward) shifts of the northern branch. In Fig. 3(b), positive (negative) values indicate equatorward (poleward) shifts of the poleward edge of the southern branch. The bars from left to right represent trends derived from GISS_ER, IAP_FGOALS, IPSL_CM4, MIROC32_MEDRES, and MPI_ECHAM5 ensemble simulations. The last bar represents the trend in averaged simulations from 7 single model runs.

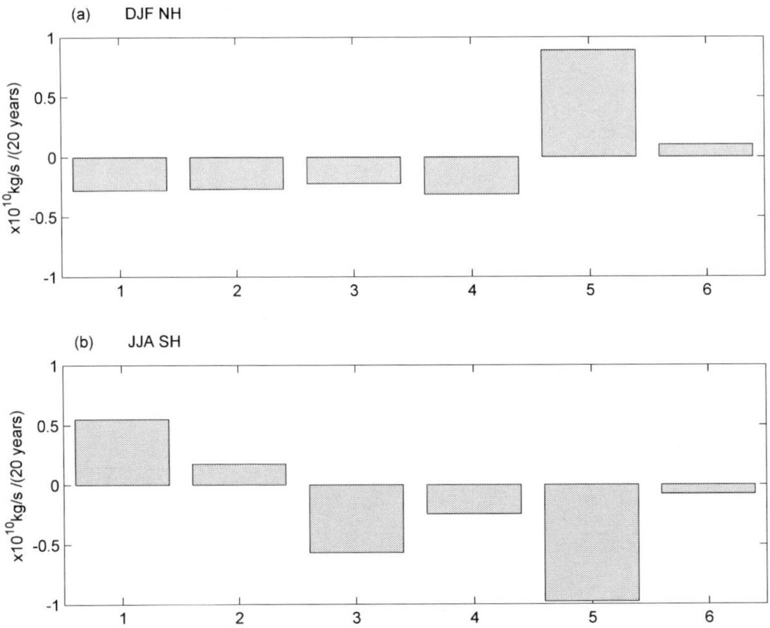

Fig. 4. Trends in the maximum zonal-mean mass streamfunction, derived from AMIP simulations. The bars are in the same order as that in Fig. 3. For both northern and southern branches, positive values indicate strengthening of the Hadley circulation, and negative values indicate weakening.

branch in austral autumn (MAM, Fig. 3(b)), the simulations also show poleward expansion, except for IAP_FGOALS. The averaged magnitude of poleward expansion is close to 0.4° in latitude over 1979–2000. The trends are statistically significant. Therefore, the results from AMIP simulations are consistent with the results in reanalyses, except for that the trends are much weaker than that from reanalyses.

Figure 4(a) shows trends in the maximum MMS of the northern Hadley-circulation branch in boreal winter. Four of the 5 models with ensemble simulations show weakening of the northern branch, and the averaged result from 7 single runs shows very weak strengthening. Note that these trends for either weakening or strengthening are statistically insignificant. For the southern branch in austral winter, 2 models of ensemble simulations show strengthening, whereas 3 others and the averaged result show weakening. Overall, AMIP simulations demonstrate weak weakening of the Hadley circulation, which does not support the result of intensification of the Hadley circulation in reanalyses.

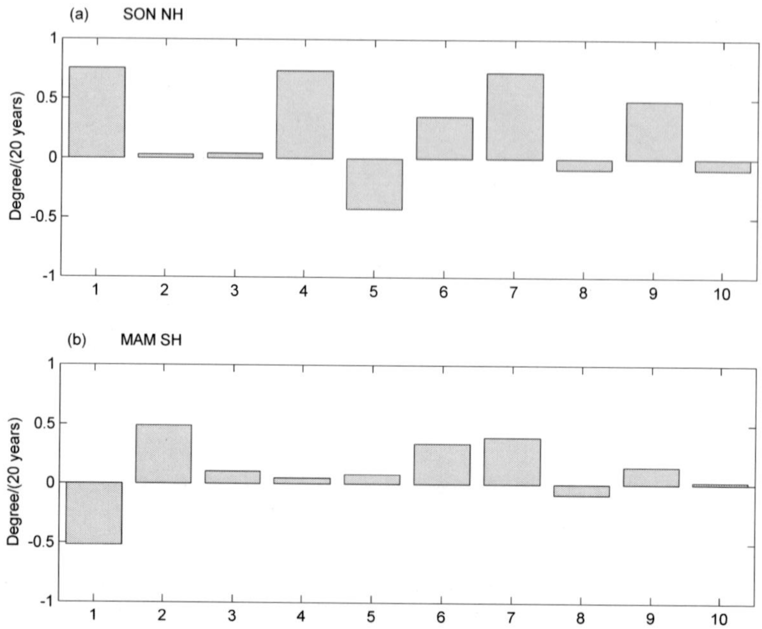

Fig. 5. Same as Fig. 3, except for IPCC-AR4 simulations. The bars from left to right correspond to MPI-ECHAM5, UKMO-HADCM3, NCAR, GISS-EH, GISS-ER, GFDL-CM20, GFDL-CM21, NCAR-CCSM30, UKMO-HADGEM1, GISS-AOM, respectively.

4.2. *Results from IPCC-AR4 simulations*

Figure 5(a) shows trends of poleward-edge latitudes of the northern branch in boreal autumn, derived from 10 model simulations. Seven of the models yield northward expansion, while the other 3 show equatorward shrinking. Averaged magnitude of poleward expansion among the 10 models is about 0.33^o in latitude, weaker than that derived from AMIP simulations. For the southern branch in austral autumn, 8 of 10 models show equatorward shrinking of the southern branch (Figure 5(b)), and 2 models show poleward expansion. However, our analyses show that the annual-mean yields weak poleward expansions of the southern branch, indicating that IPCC-AR4 simulations still generate poleward expansion of the southern branch, though the simulations could not capture the seasonality of poleward expansion in reanalyses. The annual-mean of the northern branch also shows poleward expansion.

Figure 6(a) shows trends in the maximum MMS values of the northern Hadley-circulation branch in boreal winter, derived from the same 10 model simulations. 8 of the 10 models display weakening, one model has almost no

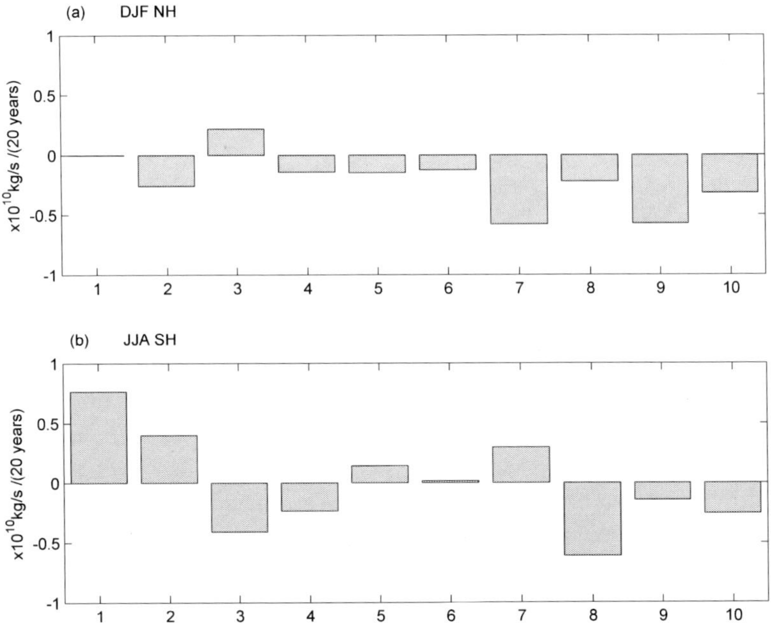

Fig. 6. Same as Fig. 4, except for IPCC-AR4 simulations. The bars are in the same order as that in Fig. 5.

changes, and only one model shows weakly strengthening. For the southern branch in austral winter (Fig. 6(b)), 5 models display weakly weakening, while another 5 show weak strengthening. The average of the 10 models yields a trend close to zero. However, for annual mean, the average of the 10 models shows weakly weakening. Overall, IPCC-AR4 simulations yield a weakened Hadley circulation, contradicting with that in reanalyses.

5. Discussion and Conclusions

Using three reanalysis datasets and two kinds of GCM simulations, we have examined decadal changes in both width and strength of the Hadley circulation. In general, both reanalyses and simulations show consistent results that the Hadley circulation has poleward expansion in both hemispheres since 1979. The results here, along with other observational results such as trends in total ozone (Hudson *et al.*, 2006), OLR trends (Hu and Fu, 2007), and trends in tropopause heights (Seidel and Randel, 2007), all suggest that the Hadley circulation has poleward expansion since 1979.

However, the signal of intensification of the Hadley circulation derived from NCEP/NCAR and ERA40 reanalyses is not supported by the results from NCEP/DOE reanalysis and GCM simulations. Moreover, most GCM simulations show weakening of the Hadley circulation. These inconsistent results of intensity changes in the Hadley circulation suggest that the intensification in NCEP/NCAR and ERA40 could be artificial due to problems of the reanalysis models and other reasons. Held and Soden (2006) pointed out that the ERA40 reanalysis model produced an unrealistically large positive trend in tropical precipitation, and that the artificially increased tropical precipitation would lead to stronger upward motions in the tropics and thus a strong Hadley circulation. Therefore, the strong intensification of the Hadley circulation in ERA40 may be artificial. Mitas and Clement (2006) suggested that the intensification of the Hadley circulation in NCEP/NCAR might be due to the weak cooling trends in the tropical upper troposphere, which cause stronger hydrostatic instability. Thus, upward tropical motions become stronger, which consequently leads to a strong Hadley circulation. They pointed out that the weak cooling trends are due to the update of sounding instruments for observations.

AMIP simulations reproduced poleward expansion of the Hadley circulation in both hemispheres and captured the seasonality very well. IPCC-AR4 AOGCM simulations for the 20th century yield poleward expansion of the northern branch in the same season (boreal autumn). They also reproduced poleward expansion of the southern branch based on the annual mean, though they do not capture the seasonality. The agreement between the reanalyses and simulations suggest that the observed broadening of the Hadley circulation might be a response to increasing greenhouse gases. However, it is important to point out that magnitudes of poleward expansion generated in GCM simulations are much weaker than that in reanalyses and other observational datasets. In particular, the magnitudes in IPCC-AR4 simulations, which include both SST and tropospheric warming, are even weaker than that in AMIP simulations. The difference between the AMIP-type and coupling GCM simulations is likely because the coupling models could not reproduce the observed SST warming. At current stage, it is not clear why GCMs are unable to reproduce magnitudes of poleward expansion comparable to observations.

One important question is how global greenhouse warming caused the broadening of the Hadley circulation. Hu and Fu (2007) argued that the observed poleward expansion is due to the weakening of baroclinic instability in the extratropics. It is because global warming is not uniform, with weaker warming in the tropics and stronger warming at higher latitudes (Fu *et al.*, 2006), which leads to a weakening of meridional temperature gradients in the extratropics. Consider that changes in meridional temperature gradients must cause changes in baroclinic wave activity. A weaker temperature gradient would cause weaker baroclinic wave activity, which allows angular momentum conservation extending further poleward (Held, 2000). Thus, the Hadley circulation becomes broader. This qualitative argument needs to be quantified in further studies. Moreover, whether the much weaker magnitudes of poleward expansion in GCM simulations are due to the lack of capability of GCMS in generating realistic meridional temperature gradients also needs to be confirmed.

References

1. J. Y. Chen, B. E. Carlson and A. D. Del Genio, *Science* **295** (2002) 838–841.
2. H. F. Diaz and B. Bradley, *The Hadley Circulation: Present, Past and Future* (Kluwer Academic Publishers, 2004).
3. Q. Fu, C. M. Johanson, J. M. Wallace and T. Reichler, *Science* **312** (2006) 1179.
4. Q. Fu, C. M. Johanson, S. G. Warren and D. J. Seidel, *Nature* **429** (2004) 55–58.
5. W. L. Gates, J. S. Boyle, and C. Covey, *et al.*, *Bull. Am. Meteorol. Soc.* **80** (1999) 29–55.
6. I. M. Held and A. Y. Hou, *J. Atmos. Sci.* **37** (1980) 515–533.
7. I. M. Held, *Proc. Program in Geophysical Fluid Dynamics* (2000), http://gfd.whoi.edu/proceedings/2000/PDFvol2000.html.
8. I. M. Held and B. J. Soden, *J. Climate* **19** (2006) 5686–5699.
9. J. R. Holton, *An Introduction to Dynamic Meteorology* (Academic Press, New York, 1994)
10. Y. Hu, K.-K. Tung and J. Liu, *J. Climate* **18** (2005) 2924–2936.
11. Y. Hu and Q. Fu, *Atmos. Chem. Phys.* **7** (2007) 5229–5236.
12. R. D. Hudson, M. F. Andrade, M. B. Follette and A. D. Frolov, *Atmos. Chem. Phys.* **6** (2006) 5183–5191.
13. E. Kalnay, M. Kanamitsua and R. Kistlera, *et al.*, *Bull. Am. Meteorol. Soc.* **77** (1996) 437–471.
14. M. Kanamitsu, W. Ebisuzaki and J. Woollen, *et al.*, *Bull. Am. Meteorol. Soc.* **83** (2002) 1631–1643.
15. R. S. Lindzen, *Ann. Rev. Fluid Mech.* **26** (1994) 353–378.
16. J. Lu, G. A. Vecchi and T. Reichler, *Geophys. Res. Lett.* **34** (2007) L06805, doi:10.1029/2006GL028443.
17. C. M. Mitas and A. Clement, *Geophys. Res. Lett.* **32** (2005) L03809, doi:10.1029/2004GL021765.

18. C. M. Mitas and A. Clement, *Geophys. Res. Lett.* **33** (2006) L01810, doi:10.1029/2005GL024406.
19. X. Quan, H. F. Diaz and M. P. Hoerling, *Conf. the Hadley circulation: Present, Past and Future*, Honolulu, Hawaii. 12–15 November 2002.
20. D. J. Seidel and R. J. Randel, *J. Geophys. Res.* **112** (2007) D20113.
21. D. J. Seidel, Q. Fu, W. J. Randel and T. J. Reichler, *Nature Geosci.* **1** (2008) 21–24.
22. S. Solomon *et al.*, *Climate Change 2007; The Physical Basis Cambridge* (Cambridge University Press, UK, 2007).
23. K. E. Trenberth, *Science* **296** (2002) 2095a.
24. S. M. Uppala, P. W. Kallberg and A. J. Simmons, *et al.*, *Quart. J. Roy. Meteorol. Soc.* **131** (2005) 2961–3012.

Advances in Geosciences
Vol. 16: Atmospheric Science (2008)
Eds. Jai Ho Oh *et al.*

SIMULATION OF RECORD RAINFALL EVENT OVER MUMBAI ON 26 JULY, 2005

G. P. SINGH
Department of Geophysics, Banaras Hindu University,
Varanasi-221 005, India
singgpin@yahoo.co.in

JAI-HO OH
Department of Environment and Atmospheric Sciences,
Pukyong National University, Busan, South Korea

H. K. CHAUDHARY
Indian Institute of Tropical Meteorology, Pune, India
hemantkumar@tropmet.res.in

The Santa Cruz observatory at Mumbai airport recorded a very heavy precipitation of 94.4 cm (in less than 24 hours) on 26 July 2005. The country important commercial city came to a complete standstill due to severe flooding. This flooding situation caused a severe damage of life and poverty. The Financial Times and Economics Times news on 4 August 2005 have reported that the number of dead in the Maharashtra floods could well be above 1000 and around Rs. 5000 crores estimated loss in the state. We have simulated a case of extremely high precipitation using a National Centre for Atmospheric Research regional climate model (RegCM3) at 20 km horizontal resolution. Results indicate that the model captures well the well-marked cyclonic circulation (low) and the simulated precipitation is more close to observed value of precipitation when FC, KUO and AS convective cumulus parametrizations schemes are used.

1. Introduction

Mumbai (Metropolitan city) is located on the windward side of the Western Ghats of India. Most of the meteorological stations which are located on windward side of the Western Ghats generally receive high precipitation normally in the month of July and August due to the orographic effects.

It is well known that the southwesterly moist laden wind from the Arabian Sea strikes perpendicularly to the north-south oriented Western Ghats. It lifted the moist wind vertically during active monsoon season and gives high precipitation on the windward side of the Western Ghats. George (1956) has mentioned that because of this orography, strong southwesterly/westerly monsoonal flow over the Arabian Sea results in the formation of offshore trough/mesoscale vortices over the Arabian Sea off the west coast. It leads to a heavy precipitation generally along the west coast of India including Mumbai.

Simulation of Indian summer monsoon circulation features and associated rainfall by a numerical model have been the most challenging problem so far. There have been some attempts to simulate the seasonal summer monsoon features by regional climate models. Bhaskaran *et al.* (1996) simulated the Indian summer monsoon using RCM with a horizontal resolution of 50 km nested with global atmospheric model (GCM). Their study showed that RCM derived precipitation is larger by 20% than GCMs. A modified version of RegCM3 has been used in the present study. Although the RegCM3 has been widely used for various meso-scale studies (Qian and Giorgi, 1999 and Pal *et al.*, 2000), it has not been tested to study heavy precipitation events like Mumbai of 2005 over India. Dash *et al.* (2006) have simulated the seasonal ISMR with RegCM (version 2) using lateral boundary conditions from European Centre for Medium Range Weather Forecasts (ECMWF). Recently, Singh and Oh (2007) have examined the effect of warm SST surrounding the Indian peninsula on seasonal monsoon circulation and precipitation with RegCM3. It has not been tested to study the extra ordinary precipitation events like over Mumbai in 2005.

2. Model Descriptions and Experiment Design

The modified version of RgCM3 used in the present study is originally developed by Giorgi *et al.* (1993 a,b) and then augmented and described by Pal *et al.* (2000). The dynamical core of the RegCM3 is similar to the hydrostatic version of the National Centre for Atmospheric Research (NCAR)/Pennsylvaia State University mesoscale model MM5 (Grell *et al.* 1994). In this paper, we have analyzed the model results using three convective cumulus parametrization schemes. The first one is the Emanuel convective scheme. In this scheme, fundamental entities are sub-cloud scale draft rather than the cloud themselves. The transport of small scale drafts are idealized as follows. The air from sub-cloud layer is lifted to each level between cloud base and the level of neutral buoyancy for diluting the air. A fraction of condensed water is then converted into precipitation which precipitates and partially or completely evaporates in an unsaturated downdraft. The remaining cloudy air is then assumed to form a uniform spectrum of mixtures with the environment. These then ascend or descend according to their buoyancy. Detail descriptions of this scheme are given in Emanuel (1991). The second scheme used is a simplified Kuo-type cumulus paramterization described in Grell *et al.* (1994) and has been widely used in many years. This scheme uses the convective instability and moisture convergence as a measure of cumulus convection. The precipitation is initiated when the moisture convergence in vertical exceeds a certain threshold and the vertical sounding is convectively unstable. Once the convection is initiated, a fraction of the moisture convergence goes into precipitation while the remaining fraction moistening the atmospheric column following a prescribed vertical profile. The vertical moistening depends on the local relative humidity (i.e., more moisture is allowed at drier points). This scheme produces much convective rainfall but less resolved scale rainfall. The third scheme used in this study was developed by Grell (1993). This is a mass flux scheme that includes the moistening and heating effects of penetrative updrafts and corresponding downdrafts. Due to the simplicity of the mass flux scheme, any closure assumption can be adopted to complete the scheme. This scheme has two closure assumptions namely Arakawa and Schubert (hereafter referred to as AS) and Fritsch and Chappell (hereafter referred to as FC) type closures. In the AS scheme, available buoyant energy is assumed to be released by the cumulus cloud systems instantaneously at each time step. While in FC, the buoyant energy release occurs with a temporal scale of 30 minutes.

RegCM3 has been integrated at 20 km horizontal resolution. There are 180 points along latitude circle and 160 points along longitude circle. The domain covers approximately 60–90E and 1–28N. The model has been integrated from the period of 2005072312 to 2005072800 (approximately 5 days). The time step of the model is 60s.

3. Observed Synoptic Situation Over Mumbai

Due to record precipitation on 26 July 2005, the life in Mumbai was totally stopped. This heavy rainfall event was localized over a region of only 20–30 km. Santa Cruz meteorological observatory in Mumbai recorded rainfall of 94.4 cm within a span of less than 24 hours. Another weather station close to Santa Cruz has recorded rainfall of 104.5 cm (Salt Lake) within same period. According to India Meteorological Department, an observatory (Coloba) in South Mumbai merely 27 km away, received only 7.3 cm of precipitation during the same period. The distance between Coloba and Bombay is hardly 27 km. The event shows a high spatial variability in precipitation.

During the heavy precipitation episode, the summer monsoon over major parts of the country including west cost of India was in active phase. A low pressure area formed over the North Bay of Bengal on 24 July. It intensified into a well-marked low and it moved inland and brought the monsoon trough to the south of its normal position. There was a strong cross equatorial flow. As the system moved westward, the low level jet gained strength and strong westerly winds lashed over the west cost. Finally, the rainfall belt moved further north towards Gujrat, but in mean time it severally devastated the Mumbai commercial city. Though the overall strong monsoon conditions of heavy precipitation was predicted by IMD over the Mubmbai (Jenamani *et al.*, 2006), localized nature of the events and the high intensity was not anticipated at all (Bohra *et al.*, 2006; Jenamani *et al.*, 2006.

4. Model Results

Figure 1 presents the model simulated wind patterns at 850 hPa using AS, FC and Emanuel convective parametrization schemes from the periods of 24 to 26 July 2005. Figure 1 also shows that model captures well the cyclonic circulation formed over the head Bay of Bengal on 24 July 2005 (left column of Fig. 1). A strong monsoonal south westerly flow can be seen over the Arabian Sea during same period.

Figure 1 also shows a well-marked cyclonic circulation covers the north Bay of Bengal, Gangetic West Bengal and Orrisa coast. This cyclonic circulation persisted over the same area for 2 subsequent days and become slightly stronger on 25th July. It moves gradually over land on 26 (next day). By then Arabian Sea branch becomes strong. On 26 July, Bay of Bengal branch also becomes strong and covers large parts of peninsular India and east coast of India. Arabian Sea branch also becomes strong and

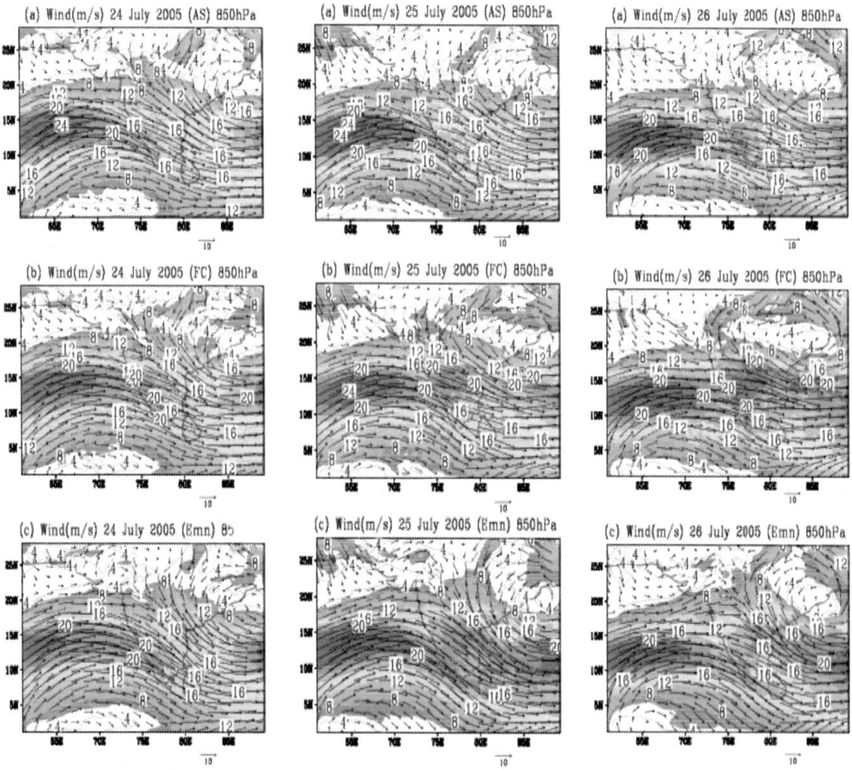

Fig. 1. Left column shows wind (m/s) patterns at 850 hPa on 24 July 2005 for (a) AS, (b) FC and (c) Emanuel convective schemes. Central and right columns are same as in left column but on 25 and 26 July 2005 respectively.

covers good parts of the south peninsular India. It can be clearly seen in model simulated wind fields. The strengthening of the Arabian Sea branch of the monsoonal southwesterly flow in association with cyclonic circulation over the Bay of Bengal may be one of the causes of heavy precipitation over Mumbai.

Figure 2 shows the model simulated precipitation using different convective cumulus parametrizations schemes. The results show that the maximum precipitation simulated by FC schemes are more than 80 cm followed by Kuo 80 cm, AS, 60 cm and Emanuel, nearly 40 cm. This results indicate that the FC and KUO convective schemes performed slightly better.

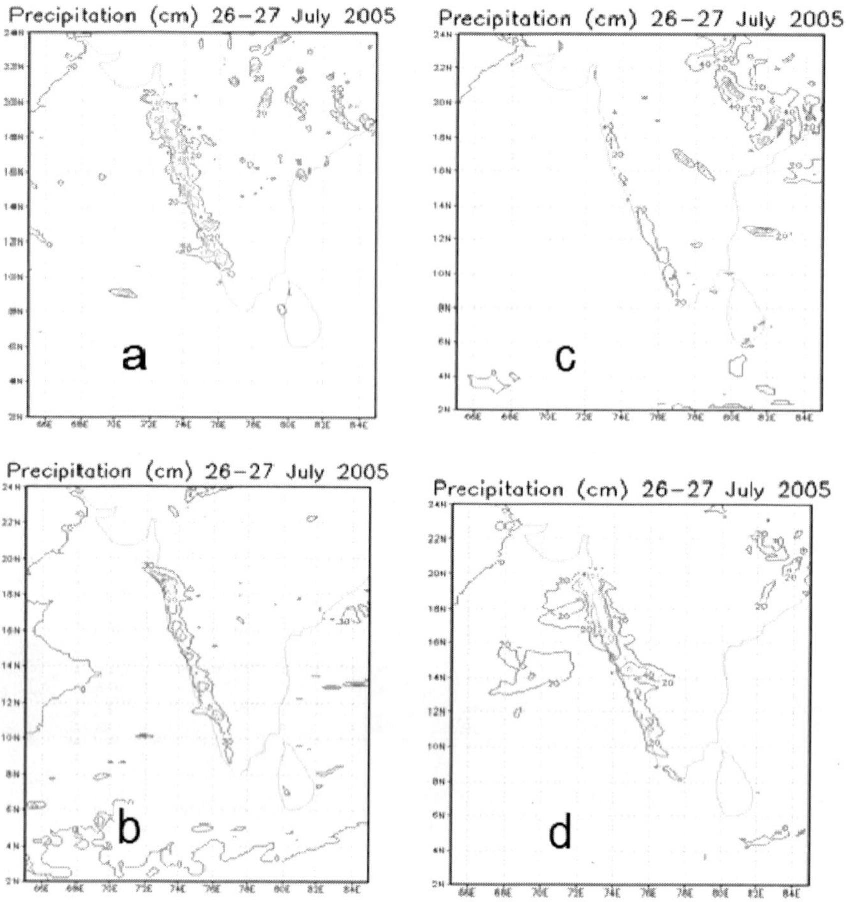

Fig. 2. Simulated precipitation using (a) AS, (b) FC, (c) Emanuel and (d) KUO convective scheme, respectively.

5. Conclusions

Results indicate that RegCM3 captures well the cyclonic circulation formed over the Bay of Bengal and strong south westerly flow over the Arabian Sea. Results also indicate that RegCM3 simulated precipitation is slightly below and more close to the observed one when using the FC and Kuo schemes. Our preliminarily results indicates that the RegCM3 can be effectively used to study the extreme events. We plan to consider more cases for supporting our conclusions.

Acknowledgment

The authors are very much grateful to Dr. F. Giorgi and Jeremy Pal of the Physics of weather and climate group, ICTP, Trieste for their encouragement and suggestion. The authors also want to thank the Department of Science and Technology, Government of India, for providing the Project (SR/S4/AS:05/2008).

References

1. B. R. Bhaskaran, G. Jones, J. M. Murphy and M. Noguer, *Clim. Dyn.* **12** (1996) 573–578.
2. A. Bohra, K. Swati Basu, E. N. Rajagopalan, G. R. Iyengar, M. Das Gupta, R. Ashrit and A. Athiyaman, *Curr. Sci.* **90** (2006) 1188–1194.
3. S. K. Dash, M. S. Shekhar and G. P. Singh, *Theor. Appl. Climatol.* **86** (2006) 161–172.
4. K. A. Emanuel, *J. Atmos. Sci.* **46** (1994) 2313–2335.
5. F. Giorgi, M. R. Marinucci and G. T. Bates, *Mon. Weather Rev.* **121** (1994a) 2794–2813.
6. F. Giorgi, M. R. Marinucci and G. T. Bates, *Mon. Weather Rev.* **121** (1994b) 2814–2832.
7. P. A. George, *Indian J. Geophys.* **7** (1956) 225–240.
8. G. A. Grell, *Mon. Weather Rev.* **121** (1993) 764–787.
9. G. A. Grell, J. Dudhia and D. R. Stauffer, Technical Notes NCAR/TN-398+STR, Vol. 21 (1994).
10. R. K. Jenamani, S. C. Bhan and S. R. Kalsi, *Curr. Sci.* **90** (2006) 1345–1362.
11. J. Pal and S. E. Eltahir, *J. Geophys. Res.* **105** (2000) 29579–29594.
12. Y. Qian and F. Giorgi, *J. Geophys. Res.* **104** 6477–6499.
13. G. P. Singh and J. H. Oh, *Inter. J. Climatol.* **27** (2007) 1455–1465.

Advances in Geosciences
Vol. 16: Atmospheric Science (2008)
Eds. Jai Ho Oh *et al.*

THE LEADING MODE OF VARIABILITY IN THE INDIAN MONSOON REGION IN THE ABSENCE OF THE ENSO VARIABILITY IN THE NCEP CFS

DEEPTHI ACHUTHAVARIER
Department of Atmospheric, Oceanic and Earth Sciences
George Mason University, Fairfax, Virginia
deepthi@cola.iges.org

V. KRISHNAMURTHY
Center for Ocean-Land-Atmosphere Studies,
Institute of Global Environment and Society,
Calverton, Maryland and Department of Atmospheric,
Oceanic and Earth Sciences, George Mason University,
Fairfax, Virginia
krishna@cola.iges.org

The leading mode of precipitation in the Indian monsoon region in the absence of the ENSO variability is examined using a regionally coupled simulation of the NCEP CFS, where Pacific SST is reduced to climatology. The results show that the leading mode has a dipole structure in the equatorial Indian Ocean and has significant correlation with the Indian Ocean dipole mode index of the SST. Regressed fields of SST and low-level winds suggest that this is a coupled mode of variability with maximum amplitude in the boreal fall, similar to the observations. This study indicates that a dipole-like variability in precipitation and SST can exist in the Indian Ocean even in the absence of the ENSO variability. However, unlike the observations, the variance of this dipole mode over the Indian subcontinent is negligible, which may be due to model errors.

1. Introduction

A part of the interannual variability in the Indian Ocean has a dipole-like pattern in the tropical sea surface temperature (SST), accompanied by similar precipitation pattern and strong equatorial zonal winds (Saji *et al.*, 1999; Webster *et al.*, 1999). This pattern, known as the Indian Ocean dipole (IOD) mode or zonal mode is characterized by cold (warm) SST anomalies (SSTA) over the eastern Indian Ocean off the coast of Sumatra and warm (cold) anomalies over the rest of the tropical Indian Ocean. The dipole mode exhibits seasonal character such that the east-west contrast in the SSTA is strongest during the northern fall (Saji *et al.*, 1999; Krishnamurthy and Kirtman, 2003). The dipole variability is commonly identified by a dipole mode index (DMI) defined as the SSTA average over the western basin (50°E–70°E, 10°S–10°N) minus the average over the eastern basin (90°E–110°E, 10°S–0°) (Saji *et al.*, 1999).

Since its discovery, whether the dipole mode is an independent variability of the Indian Ocean or a manifestation of the remote forcing due to El Niño-Southern Oscillation (ENSO) in the Pacific has been examined by several studies (e.g., Reason *et al.*, 2000; Krishnamurthy and Kirtman, 2003). Using observed SST for the period 1870–1998, Krishnamurthy and Kirtman (2003) showed that the dipole mode has a robust correlation with the ENSO when its seasonality is considered. Numerical model studies have shown evidence of a tropical "atmospheric bridge" mechanism (Klein *et al.*, 1999), which provides an explanation for the combined variability of the ENSO and the IOD (Lau and Nath, 2004; Fischer *et al.*, 2005; Huang and Shukla, 2007). There is a general consensus among these studies that the ENSO-related SST anomalies in the Pacific can induce atmospheric circulation anomalies that extend over to the Indian Ocean basin, which in turn change the air sea fluxes and therefore generate SST anomalies in that region. This way, the ENSO can be considered as a trigger for the dipole events.

Although many dipole events can be accounted for as varying together with ENSO, there are a few years in recent history (e.g., 1961, 1967 and 1994) when the dipole event did not coincide with an evolving ENSO in the Pacific. General circulation model (GCM) simulations have also shown that a dipole-like variability can exist in the Indian Ocean SST even without Pacific SST interannual variability (Fischer *et al.*, 2005; Huang and Shukla, 2007). However, the development of dipole events during the monsoon season and its daily variability has not been investigated.

A recent observational study has shown that an atmospheric dipole mode persists throughout the monsoon season and has varying interannual relation with another persistent mode related to ENSO (Krishnamurthy and Shukla 2008). The relative role of the IOD and ENSO modes is important in determining the mean monsoon. Whether such IOD and ENSO modes are properly simulated in a state-of-the-art coupled GCM was investigated by analyzing the simulations of the Climate Forecast System (CFS) of the National Centers for Environmental Predictions (NCEP) in USA (Achuthavarier, 2009; Achuthavarier and Krishnamurthy, 2009).

However, no dipole mode, separate from the ENSO mode, was found. In this study, the existence of the IOD mode in the coupled model has been further examined by performing a simulation of the CFS in the absence of the ENSO variability. The numerical experiment consists of allowing the SST variability in the Indian and Atlantic Oceans but prescribing climatological SST in the Pacific Ocean, an approach similar to some earlier studies (e.g., Fischer *et al.*, 2005; Huang and Shukla, 2007). This study shows that a dipole-like mode in precipitation and SST can exist in the Indian monsoon region over the Indian Ocean, possibly due to air-sea interaction. This mode is shown to persist throughout the monsoon season in a coherent and non-oscillatory manner and reveal interannual variability.

2. Methodology

2.1. *Model experiment*

In the CFS, the atmospheric component is a coarse resolution version of NCEP's Global Forecast System (GFS) and the ocean component is the Modular Ocean Model version 3 (MOM3) developed by the Geophysical Fluid Dynamics Laboratory (GFDL) in Princeton. The GFS has a spectral triangular truncation of 62 waves in the horizontal and a finite differencing in the vertical with 64 sigma layers. The MOM3 has a zonal resolution of 1° and a meridional resolution of 1/3° between 10°S and 10°N which gradually decreases poleward till 30°N and 30°S. The atmospheric and ocean components are coupled in the region 65°S–50°N, and observed climatological SST is used in the regions poleward of 74°S and 64°N. A detailed description of the CFS is provided by Saha *et al.* (2006).

The main numerical experiment conducted in this study is a regionally coupled simulation of the CFS in which the Pacific SST variability is reduced to climatology while retaining the interannual variability and coupled interaction in the Indian and Atlantic basins (Pacific climatology run or PCLM hereafter). The prescribed climatological SST is obtained from a fully coupled simulation of the CFS (Pegion and Kirtman, 2008), which will be referred to as the "control". The control and PLCM are both 30 year-long single realizations of the CFS. In the PCLM, climatological SST is prescribed in all ocean grid points in the region (120°E–90°W, 30°S–50°N). Additionally, a 10°-wide buffer zone is employed along the north and south boundaries of the prescribed region, where the SST is computed by linear interpolation of the prescribed and model-produced values.

2.2. *Analysis method*

The leading modes of precipitation variability are determined by performing a multi-channel singular spectrum analysis (MSSA) of the daily anomalies. The MSSA is also known as extended empirical orthogonal functions (EOF) analysis. A detailed review of the method is provided by Ghil *et al.* (2002). The mathematical procedure of the MSSA can be summarized as an eigenanalysis of the lagged covariance matrix, where the lag window length is chosen depending on the timescale of interest. The lagged covariance matrix is constructed by using M lagged copies of the L spatial points in the data at N time intervals. The eigenvectors contain M sequences of spatial maps, referred as space-time EOFs (ST-EOFs). The space-time principal components (ST-PCs), each of length $N - M + 1$, are obtained by projecting the original data matrix on the corresponding ST-EOFs. The component of the original data corresponding to each eigenvalue can be reconstructed by combining the ST-PC and its respective ST-EOF in a least square sense, and is referred as the reconstructed component (RC). Thus, the modes of variability are obtained in a data-adaptive manner.

The MSSA has been successfully used to examine the intraseasonal and interannual modes of the Indian monsoon in observations (Krishnamurthy and Shukla, 2007; 2008). The intraseasonal signals are obtained as oscillatory modes in pairs while any lower frequency variability or trend in the data are resolved as non-oscillatory or persistent modes. By the term "persistent mode", it is implied that the mode is non-oscillatory within the lag window applied in the analysis. The persistent components in the observations were found to be related to ENSO and IOD, accounting for significant interannual variability. Since the focus of the present study is to examine the interannual variability, the analysis is limited to the persistent modes in the MSSA of the model data. In this study, a lag window of length 181 days is used in the MSSA of the daily anomalies of precipitation over the region (40°E–100°E, 20°S–35°N) for the 30 years of the PCLM simulation. The daily data used here is unfiltered, except for the application of a 5-day running mean.

3. Results

By analyzing the ST-EOF maps and the ST-PC time series of the first MSSA mode, it is concluded that the first mode is seasonally persistent in nature. The power spectrum of its space-time amplitude (ST-PC1) is mostly red in nature (figure not shown), which confirms its non-oscillatory behavior within the lag window of 181 days. Since the focus of this paper is to document the spatial and temporal behavior of this mode, the rest of the analysis is on the RC of the first mode (RC1) that is computed by combining ST-EOF1 and ST-PC1. It is important to note that RC1 has the same spatial and time extents of the total anomalies.

Since the persistent mode varies in a coherent and non-oscillatory manner throughout the monsoon season, the best way to describe its space-time variability is to perform an ordinary spatial EOF analysis of the RC1. In order to capture the spatial pattern that exists during the boreal summer, the EOF analysis is performed on daily RC1 during June–September of the 30 years of the model simulation. The first EOF explains 88% of the variance in RC1, and its spatial pattern is shown in Fig. 1(a). The first EOF shows a dipole-like structure with positive (negative) anomalies over the western (eastern) equatorial Indian Ocean, which is somewhat similar to the observed dipole mode. However, it should be noted that the variance of this mode is mainly over the equatorial Indian Ocean whereas the precipitation anomalies extend to the Indian subcontinent as well in observations (see Fig. 7 in Krishnamurthy and Shukla, 2008).

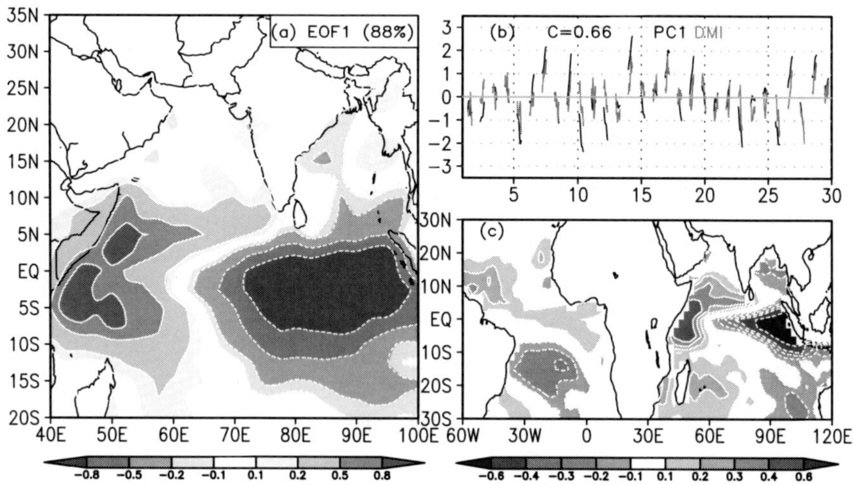

Fig. 1. (a) EOF1 of the RC1 (first MSSA mode) of daily precipitation anomalies. The variance explained is 88%. Units are in mm day^{-1}. (b) Standardized values of PC1 of RC1 (black) and daily DMI of total SST anomaly (red). (c) Point correlation between the PC1 and daily anomalies of the SST.

The interannual variability of this mode is examined in Fig. 1(b), where PC1 is plotted along with the daily DMI values of the total SST anomalies. The SST DMI is defined in Sec. 1. Since the EOF was performed on the daily data of the summer season, the corresponding PC1 is obtained for 122 days for each of the 30 years. It is clear from Fig. 1(b) that PC1 varies very little within the season while its year-to-year variability is noticeable. There is a good correspondence between PC1 and total DMI with a daily correlation of 0.6. As evident from Fig. 1(b), the interannual correlation between the seasonal means of PC1 and DMI is even stronger and has a value of 0.7. Additionally, to verify the relation between this precipitation dipole mode and the model's SST, daily point correlation between PC1 (shown in Fig. 1(b)) and daily anomalies of the SST in the tropical Indian and Atlantic Oceans is computed (Fig. 1(c)). The correlation map shows a clear dipole pattern in the Indian Ocean with negative values in the eastern basin and positive values in the western basin. The dipole is most pronounced in the tropical Indian Ocean north of 10°S where magnitude of the correlation values reach up to 0.6. This mode also has weak but statistically significant correlations in the Atlantic. Weak positive and negative correlations are present in the tropical and south Atlantic, respectively. These results suggest that a dipole-like variability can be present in the Indian Ocean in the absence of ENSO variability.

The evolution of this mode is examined by presenting lead/lag regressions of monthly anomalies of precipitation, SST and winds at 850 hPa on the monthly mean of the PC1 of RC1 in Fig. 2. Lag 0 indicates simultaneous regression for the months of June–September while lag +3 indicates that September–December precipitation fields are regressed on PC1 and so on. The regressed fields of precipitation show that a dipole mode is formed during the monsoon season, matures by September, and weakens by December. This life cycle, where the anomalies are the strongest in boreal fall, is similar to observations. The regressed fields also show that precipitation anomalies, although weak, are present over Africa and the tropical Atlantic, suggesting that the dipole or the Indian Ocean variability may have impacts over Africa.

The dynamics of this mode is further examined in Fig. 3 where similar regression maps of SST and horizontal winds at 850 hPa are shown. The precipitation and SST form a consistent dynamic picture where positive (negative) anomalies of precipitation coincide with warm (cold) SST. Easterly anomalies are present in the equatorial Indian Ocean extending from the cold SST anomalies in the eastern Indian Ocean to the western

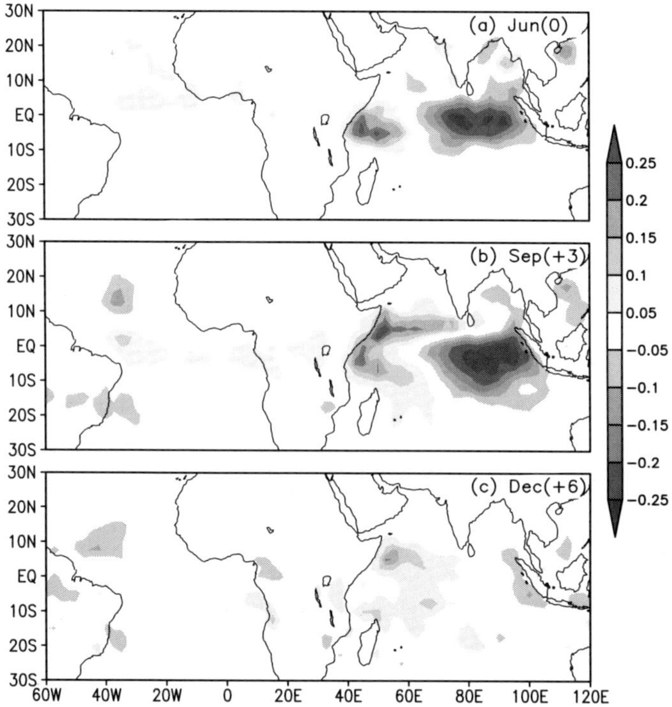

Fig. 2. Lead/lag regression of monthly total precipitation anomalies on the monthly means of PC1 of June–September RC1. Lag 0 denotes simultaneous regression and lag +3 (+6) denotes that September–December (December–March) total precipitation anomalies are regressed on the PC1 of June–September RC1. Units are mm day^{-1} per unit standard deviation of PC1.

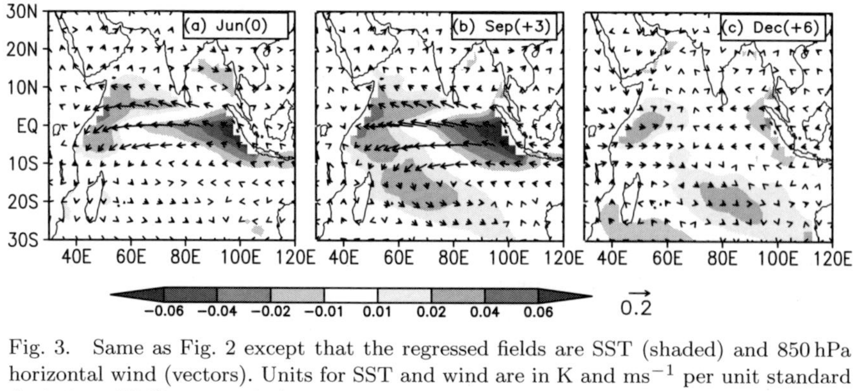

Fig. 3. Same as Fig. 2 except that the regressed fields are SST (shaded) and 850 hPa horizontal wind (vectors). Units for SST and wind are in K and ms^{-1} per unit standard deviation of PC1, respectively.

part. Similar to the precipitation pattern, the SST and the winds are strongest in lag +3. From this preliminary analysis, it is inferred that the easterly wind anomalies, once triggered in response to cooling in the eastern Indian Ocean, help to amplify the dipole in SST through wind mixing. It can be argued that the easterly wind anomalies enhance the total wind speed in the eastern Indian Ocean where summer climatological winds are easterly while reduces the total wind in the western part where climatological winds are westerly. This can lead to more mixing in the eastern parts of the Indian Ocean and less mixing in the western parts, which helps to maintain the dipole pattern.

4. Summary and Discussion

This study has examined the leading mode of precipitation in the Indian region during the monsoon season in a numerical simulation of CFS where the Pacific SST variability is reduced to climatology. It is found that this leading mode has a dipole structure in the tropical Indian Ocean. However, its negligible variance over the Indian subcontinent is a weak point of the model. The correlation analysis with SST and winds shows that this mode is closely related to the dipole mode in the SST. This study has shown that a dipole-like variability can be manifested in the Indian Ocean even in the absence of the ENSO, possibly by air-sea interaction mechanisms. However, further analysis is required to understand whether ocean dynamics is critical for its existence. An independent dipole mode was not obtained in the fully coupled control simulation (Achuthavarier, 2009; Achuthavarier and Krishnamurthy, 2009), possibly due to the dominance of the ENSO in the model, Future analysis should explore what triggers a dipole-like pattern in the Indian Ocean in the absence of the ENSO and whether the climatological SST in the Pacific has any role. Similarly, the impact of this mode over Africa and Atlantic is also worth examining.

Acknowledgments

This research was partially supported by grants from the National Science Foundation (0334910), the National Oceanic and Atmospheric Administration (NA040AR4310034), and the National Aeronautics and Space Administration (NNG04GG46G). This work formed a part of the Ph.D. thesis of DA submitted to George Mason University.

References

1. D. Achuthavarier, Role of the Indian and Pacific Oceans in the Indian summer monsoon variability, Ph.D. Thesis, George Mason University (2009), p. 193.
2. D. Achuthavarier and V. Krishnamurthy, Dominant daily modes of South Asian summer monsoon variability in the NCEP Climate Forecast System, 2009, *Clim. Dyn.* (2009).
3. A. S. Fischer, P. Terray, E. Guilyardi, S. Gualdi and P. Delecluse, Two independent triggers for the Indian Ocean Dipole/zonal mode in a coupled GCM, *J. Climate* **18** (2005) 3428–3449.
4. M. Ghil, M. R. Allen, M. D. Dettinger, K. Ide, D. Kondrashov, M. E. Mann, A. W. Robertson, A. Saunders, Y. Tian, F. Varadi and P. Yiou, Advanced spectral methods for climatic time series, *Rev. Geophys.* Vol. 40 (2002), doi:10.1029/2000RG000092
5. B. Huang and J. Shukla, Mechanisms for the interannual variability in the tropical Indian Ocean. Part II: Regional processes, *J. Climate* **20** (2007) 2937–2960.
6. S. A. Klein, B. J. Soden and N.-C. Lau, Remote sea surface temperature variations during ENSO: Evidence for a tropical atmospheric bridge, *J. Climate* **12** (1999) 917–932.
7. V. Krishnamurthy and B. P. Kirtman, Variability of the Indian Ocean: Relation to monsoon and ENSO, *Quart. J. Roy. Meteorol. Soc.* **129** (2003) 1623–1646.
8. V. Krishnamurthy and J. Shukla, Intraseasonal and seasonally persisting patterns of Indian monsoon rainfall, *J. Climate* **20** (2007) 3–20.
9. V. Krishnamurthy and J. Shukla, Seasonal persistence and propagation of intraseasonal patterns over the Indian monsoon region, *Clim. Dyn.* **30** (2008) 353–369.
10. K. Pegion and B. P. Kirtman, The impact of air-sea interactions on the simulation of tropical intraseasonal variability, *J. Climate* **21** (2008) 6616–6635.
11. C. J. C. Reason, R. J. Allan, J. A. Lindesay and T. J. Ansell: ENSO and climatic signals across the Indian Ocean basin in the global context: Part I: Interannual composite patterns, *Int. J. Climatol.* **20** (2000) 1285–1327.
12. S. Saha, *et al.*, The NCEP Climate Forecast System, *J. Climate* **15** (2005) 3483–3517.
13. N. H. Saji, B. N. Goswami, P. N. Vinayachandran and T. Yamagata, A dipole mode in the tropical Indian Ocean, *Nature* **401** (1999) 360–363.
14. P. J. Webster, A. M. Moor, J. P. Loschnigg and R. R. Leben, Coupled ocean-atmosphere dynamics in the Indian Ocean during 1997–98, *Nature* **401** (1999) 356–360.

Advances in Geosciences
Vol. 16: Atmospheric Science (2008)
Eds. Jai Ho Oh *et al.*

LONGITUDINAL OSCILLATIONS OF THE SOUTH ASIAN HIGH AND THE SUBTROPICAL WESTERN PACIFIC HIGH DURING BOREAL SUMMER

LING ZHANG and XIEFEI ZHI*
Key Laboratory of Meteorological Disaster of Ministry of Education,
Nanjing University of Information Science and Technology,
Nanjing, Jiangsu 210044, China
**zhi@nuist.edu.cn, xiefeizhi@yahoo.com.cn*

Based on the NCEP/NCAR monthly reanalyses, the interannual variation of the longitudinal locations of the South Asian high (SAH) and the subtropical western Pacific high (SWPH) in summer and its association with the diabatic heating anomalies in the atmosphere has been investigated. Results show that the longitudinal location of the SAH is significantly correlated with that of the SWPH on an interannual basis in summer, especially in July. The two anticyclonic systems move towards each other or move away from each other in terms of their longitudinal oscillations. Further study suggests that the heat source (sink) strengthens (weakens) the ascending motion of the atmosphere in the East of the Tibetan Plateau, which reinforces (reduces) the high-level divergence. Consequently, the intensity of the SAH increases (decreases) with its ridge extending eastward (withdrawing westward). In addition, the abnormal descending (ascending) motion of the atmosphere occurs in middle and low latitudes over the western Pacific as a result of the anomalous northerly in upper (lower) troposphere, which is accompanied by adiabatic heating (cooling). Therefore, the deep and warm SWPH is intensified (declined) and extends westward (withdraws eastward).

1. Introduction

The monsoon is a global circulation system caused by the differential heating of land and sea (Krishnamurti 1985; Lau *et al.*, 1988). The Asian summer monsoon is the most important one and it is closely related to large-scale circulations (Lau *et al.*, 1984; Matsumoto, 1992; Nagazawa, 1992; Tanaka, 1992, among others). Tao *et al.* (1987) indicated that the lower-level summer monsoon circulation over East Asia is controlled by the cross equatorial flow, the Mei-yu front, subtropical western Pacific high (SWPH) and the mid-latitude weather system (e.g., Siberian high). In the upper troposphere, the South Asian high (SAH, also referred to as Tibetan anticyclone), the easterly jet over South Asia, and the westerly jet north of the SAH are the primary characteristics of the Asian summer monsoon (Krishnamurti, 1971; Lau *et al.*, 1984). The East Asian summer monsoon is strongly influenced by the SWPH. The movement of the major rainband associated with the East Asian summer monsoon is related to the variation in the SWPH (Lau *et al.*, 1988; Murakami and Matsumoto, 1994; among others). And the longitudinal oscillation of the SWPH plays an important part in the transition of the moisture. Some studies suggest that the convection over the warm pool and the offshore SSTA are associated with the longitudinal position of the SWPH (Lu *et al.*, 2001; He *et al.*, 2001; Zhou *et al.*, 2008). In addition, the complementary cooling and descending motion in the western Pacific are related to anomalous longitudinal circulation associated with ENSO (Sui *et al.*, 2007).

The SAH is the most intense and persistent circulation found at 100 hPa over the Northern Hemisphere. Its influence extends from the Atlantic coast of Africa across southern Asia to the Pacific Ocean (Mason *et al.*, 1963). Zhang *et al.* (2002) found the bimodality in the longitudinal location of the SAH. The SAH is classified into the Tibetan Mode and the Iranian Mode in terms of the preferred region of the anticyclonic system. The former is closely related to the diabatic heating of the Tibetan Plateau, whereas the latter is more associated with the adiabatic heating in the free atmosphere as well as the diabatic heating near the surface.

Tao *et al.* (1964) indicated that the 100 hPa summer flow regime over southern Asia is intimately associated to the advance and retreat of the 500 hPa subtropical high over western Pacific with both anticyclonic systems moving towards each other or withdrawing from each other in zonal direction. Although statistical analysis has shown that the longitudinal location of the SAH is correlated with that of the SWPH in terms of the short period and interannual variations, the longitudinal locations of the SAH and the SWPH as well as the mechanism of the interaction between these two anticyclonic systems are not yet very clear. This study investigates the interannual variation of the SAH in terms of its longitudinal location and its relation to that of the SWPH. The mechanism of the interactions between the two anticyclonic systems at different levels is discussed by analyzing the diabatic and adiabatic heating anomalies in the atmosphere.

The description of the dataset used in this study is given in Sec. 2. In addition, the calculations of the apparent heat source and apparent moisture sink are described in Sec. 2. The climatic features of the SAH and SWPH in summer are discussed in Sec. 3. Interannual variations of the longitudinal locations of the SAH and the SWPH are analyzed in Sec. 4. The diabatic heating anomalies and their associations with the longitudinal oscillations of the SAH and the SWPH are examined in Sec. 5. The results of this study are summarized in Sec. 6.

2. Data and Methods

Considering that the data quality of the early NCEP/NCAR reanalyses (Kalnay *et al.*, 1996; Kistler *et al.*, 2001) over Asia may be low prior to 1968 (Yang *et al.*, 2002) and the NCEP/NCAR reanalysis has discontinuity in 1979 (Sturaro, 2003; Chen *et al.*, 2008) due to the introduction of satellite infrared and microwave retrievals. In this study, monthly geopotential heights of the NCEP/NCAR reanalyses since 1980 are applied to analyze the interannual variations of the SAH and the SWPH.

The apparent heat source Q_1 and apparent moisture sink Q_2 are calculated for each level by the u, v, T, ω and q of the NCEP/NCAR daily reanalyses of 12 levels from 1000 hPa to 100 hPa, and then vertical integration of the Q_1 and Q_2 over the whole layers in the troposphere $<Q_1>$, $<Q_2>$ is performed and the NCEP/NCAR daily pressure at surface reanalyses is also used. After Yanai *et al.* (1973), Q_1 and Q_2 are calculated as follows.

$$Q_1 = C_p \left[\frac{\partial T}{\partial t} + \vec{v} \cdot \nabla T + \omega \frac{\partial \theta}{\partial p} \left(\frac{p}{p_0} \right)^k \right], \tag{1}$$

$$Q_2 = -L \left[\frac{\partial q}{\partial t} + \vec{v} \cdot \nabla q + \omega \frac{\partial q}{\partial p} \right], \tag{2}$$

where Cp is the specific heat at constant pressure and $Cp = 1004\, J/K{\cdot}kg$, T is temperature, L is the condensation latent and its unit is $J/g, q$ is specific humidity, ω is vertical velocity in p-coordinate system, $p_0 = 1000\,\text{hPa}$, $k = R/Cp$, $\vec{v}$ is the horizontal wind vector.

The vertical integration of the Q_1 and Q_2 over the whole layers in the troposphere, $<Q_1>$, $<Q_2>$ is calculated as follows.

$$<Q_1> = \frac{1}{g}\int_{p_s}^{p_t} Q_1 dp, \tag{3}$$

$$<Q_2> = \frac{1}{g}\int_{p_s}^{p_t} Q_2 dp, \tag{4}$$

where p_s is the surface pressure and p_t refers to 100 hPa for $<Q_1>$ and 300 hPa for $<Q_2>$.

3. Climatic Features of the SAH and SWPH in Summer

The SWPH mainly occurs in the middle and lower troposphere and its center is located over the Pacific. The intensity of the SWPH becomes the largest during boreal summer. The large scale anticyclonic system over the Tibetan Plateau and its adjacent area, the South Asian high (SAH) is located in the upper troposphere. In this paper, the location of the SAH is expressed by the longitude of the east ridge-point of 1672 dagpm, 1680 dagpm and 1680 dagpm contour line at 100 hPa in June, July and August, respectively. Accordingly, the longitudinal locations of the east ridge-point of the SAH are calculated for June, July and August during 1976–2006. The longitudinal location of the SWPH is characterized by its west ridge-point index, namely, the longitude of the west ridge-point of the 588 dagpm contour line at 500 hPa. Figure 1 shows the features of monthly mean geopotential heights at 100 hPa and 500 hPa in summer. The west ridge-point of the SWPH is situated at 127.5°E, 125.0°E and 132.5°E, and its average ridge line is located at 20.0°N, 25.0°N and 30.0°N in June, July and August, respectively. The monthly mean ridge line of the SAH is located further northward than that of the SWPH in summer. The east ridge-point of the SAH is situated at 120.0°E, 112.5°E and 107.5°E in June, July and August, respectively. And its mean ridge line is located at 25.0°N, 30.0°N and 32.5°N in June, July and August, correspondingly. In addition, the mean ridge line of the SAH extends from Northwest to Southeast in June and approximately from West to East in July and August, while that of the SWPH extends from Northeast to Southwest in June, approximately from East to West in July, and from Southeast to Northwest in August. The distance between the ridge points of the SAH and the SWPH is smallest in June, and reaches the largest in August in zonal direction. In July, the

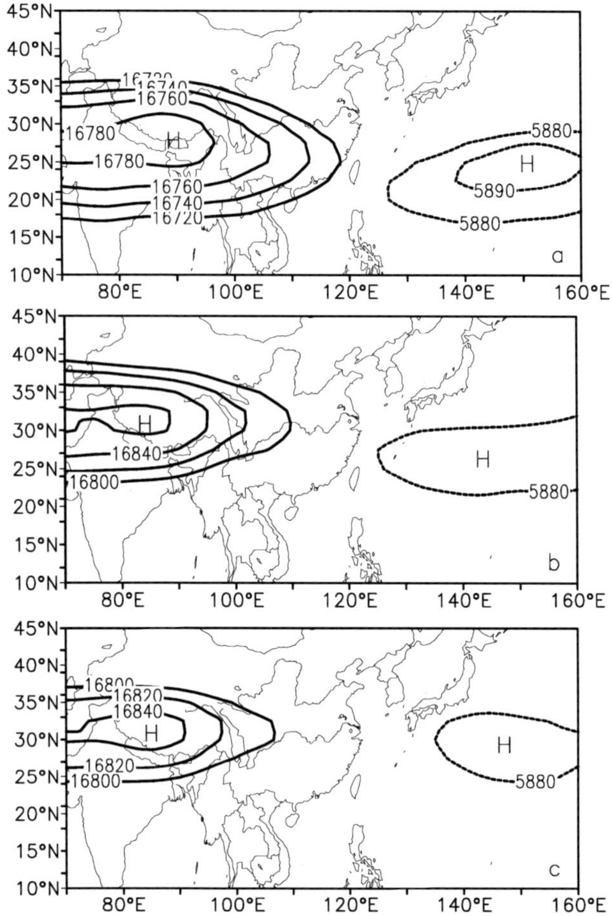

Fig. 1. Monthly mean geopotential heights at 100 hPa (solid line) and 500 hPa (dashed line) in June (a), July (b) and August (c) during 1976–2006 (units: gpm).

ridge lines of the SAH and the SWPH stretch approximately from East to West. Therefore, the average values of the ridge line location of the SAH and the SWPH are more reasonable in July than in June and August in terms of the latitude locations of the two anticyclonic systems. In Sec. 5, the longitudinal oscillations of the SAH and the SWPH are discussed mainly in July.

As shown in Fig. 1, the SWPH always maintains high intensity during summer, and the 588 dagpm contour line persistently exists, while the intensity of the SAH varies from month to month. From June to August, the SAH undergoes the weak-to-strong transition from June to July and the strong-to-weak transition from July to August. The maximum value of the monthly mean geopotential height at the SAH center gradually increases from 1672 dagpm to 1684 dagpm in June and July, and the intensity of the SAH reaches the highest in July, while the SAH becomes weaker in August.

4. Interannual Variations of the SAH and SWPH in terms of their Longitudinal Locations

The East Asian summer monsoon and associated circulations have noticeable interdecadal variations (Zeng *et al.*, 2007; Zhang *et al.*, 2000). In this study, we focused on the interannual variation features of the SAH and the SWPH in terms of the longitudinal locations. In order to fetch the interannual variation of the two anticyclonic systems, the decadal and the interdecadal variations are removed from the original series which include interannual, decadal and interdecadal variations. The 9-point running mean of the data is applied to represent the decadal and the interdecadal variations of the time series. The interannual variation features of the SAH and the SWPH in terms of their longitudinal locations are discussed for the period of 1980–2002 in this study. Figure 2 shows that the interannual variations of the longitudinal locations of the SAH and the SWPH are evident, and the east ridge-point index of the SAH and the west ridge-point index of the SWPH are negatively correlated with each other in summer. The correlation coefficients between the two indices in June, July and August are −0.414, −0.453 and −0.35, respectively. The former two correlation coefficients are statistically significant at 95% confidence level, while the latter one is statistically significant at 90% confidence level. Consequently, in JJA, especially in July, the two anticyclonic systems move towards each other or move away from each other in terms of their longitudinal oscillations.

5. The Longitudinal Oscillation of the SAH and the SWPH and its Relation to Diabatic Heating Anomalies in July

5.1. *Definition of the east and west patterns of the SAH and SWPH*

As discussed in previous section, the SAH and the SWPH are strong and stable, and the ridge lines of the two anticyclonic systems extend

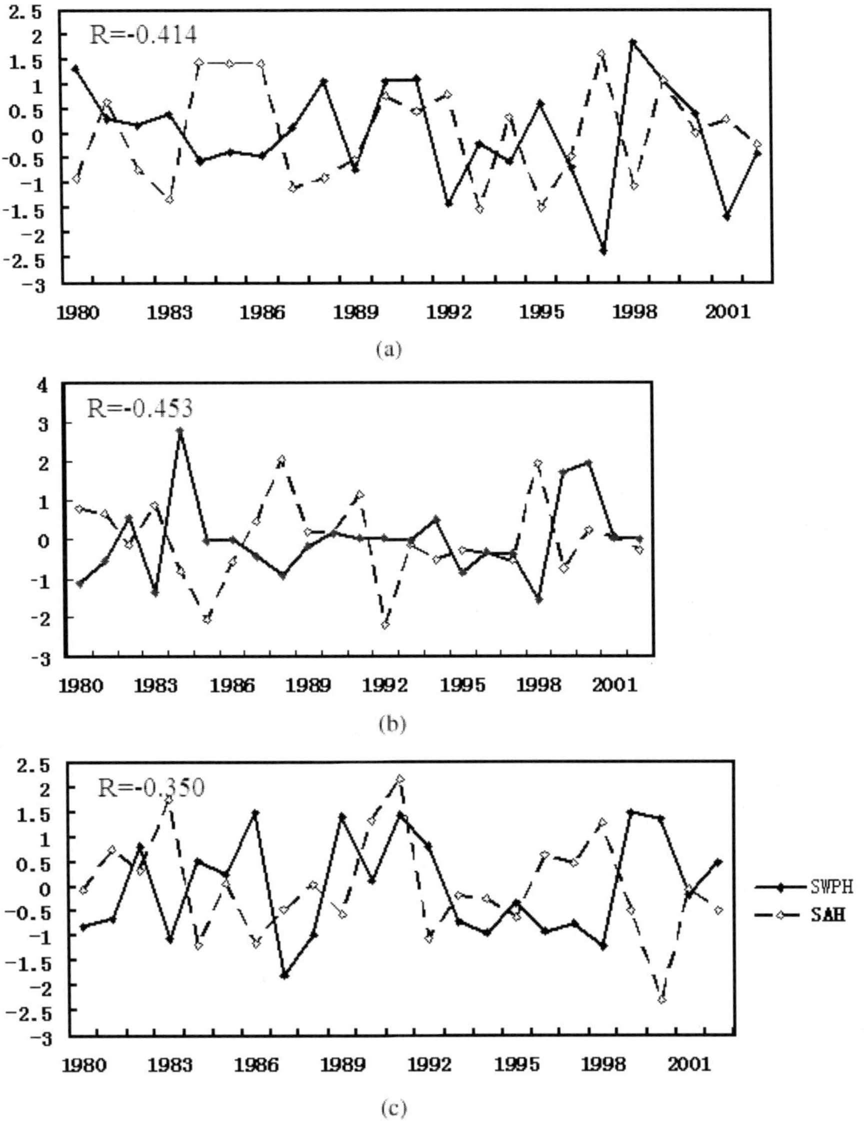

Fig. 2. The interannual variation of the standardized west ridge-point index of the SWPH (solid line) and standardized eastward extending ridge-point index of the SAH (dashed line) in June (a), July (b) and August (c). R represents the correlation coefficients between the two indices.

approximately in the east-west direction in July. In order to investigate the longitudinal oscillation features of the SAH and the SWPH in terms of their longitudinal location anomalies, composite analysis is performed for anomalous years of the two anticyclonic systems further eastward or westward than normal. In this study, we define the anomalous year of the SAH as that with the location of the SAH further eastward (westward) than normal, or east (west) pattern of the SAH, if the anomaly of the eastward extending ridge-point index of the SAH is greater (less) than 0.5σ (-0.5σ) in a particular year. With the criteria mentioned above, 1980, 1981, 1983, 1988, 1991 and 1998 are identified as anomalous years with the east pattern of the SAH, whereas 1984, 1985, 1992 and 1999 as the west pattern of the SAH.

Similarly, the anomalous year of the SWPH is defined as that with the location of the SWPH further eastward (westward) than normal, or east (west) pattern of the SWPH, if the anomaly of the west ridge-point index of the SWPH is greater (less) than $0.5\,\sigma$ ($-0.5\,\sigma$) in a particular year. With these criteria, 1980, 1983, 1988, 1995 and 1998 are identified as anomalous years with the west pattern of the SWPH, whereas 1982, 1984, 1999 and 2000 as the east pattern of the SWPH.

5.2. *Features of geopotential heights of the SAH and the SWPH for anomalous years*

As shown in Fig. 3, the composite map of the SAH at 100 hPa for anomalous years indicate that the east ridge-point of the SAH at 100 hPa is situated at 122.5°E, which extends about 10.0° longitudes eastward than normal (the normal location of the SAH is located at 112.5°E in July). At the same time, the west ridge-point of the SWPH is located at 117.5°E over the South China Sea, which extends about 7.5° longitudes westward than normal (the normal location of the SWPH is located at 125.0°E in July). Thereby, the SAH in upper troposphere and the SWPH in middle troposphere overlap over southern China. The overlap area of the two anticyclonic systems is dominated by drought in southern China (Guo *et al.*, 2008). When the west pattern of the SAH occurs, the east ridge-point of the 1680 dagpm contour line at 100 hPa is situated at 97.5°E, which withdraws about 15° longitudes westward than normal. At the same time, the main body of the SWPH withdraws eastward to the western Pacific. These two anticyclonic systems are approximately 23° longitudes apart.

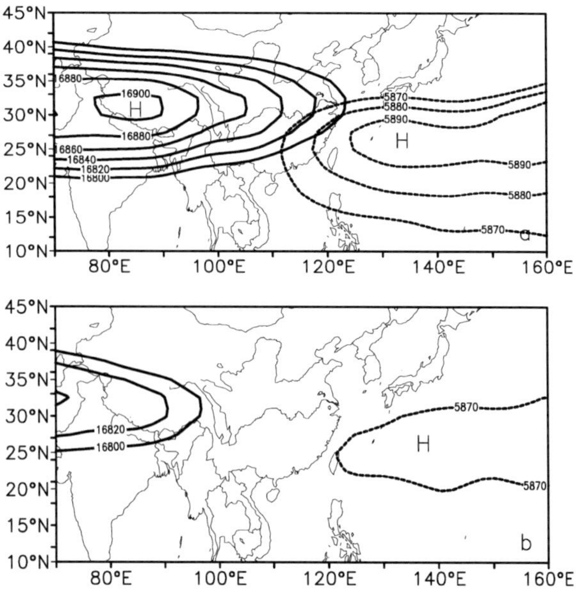

Fig. 3. Composite map of the geopotential heights at 100 hPa (solid line) and 500 hPa (dashed line) in July for the east pattern (a) and west pattern (b) of the SAH (units: gpm).

5.3. *The relationship between the locations of the SAH and the SWPH and diabatic heating anomalies*

The formation of the SAH is not only related with the latent heat release of the monsoon precipitation, but also with the sensible heat over the Tibetan Plateau, and both played a major role in the maintenance of the SAH (Wu *et al.*, 2002). The strong land surface sensible heating on the west and condensation heating on the east over each continent may generate cyclones in the lower layers and anticyclones in the upper layers, whereas radiative cooling over oceans generates the lower-layer anticyclone and upper-layer cyclone circulations (Liu *et al.*, 2004). This may partly interpret the mechanism of the formation of the SAH and the SWPH. In this study, the singular value decomposition (SVD) is used to analyze the relationship between the apparent heat source and geopotential heights at 100 hPa as well as 500 hPa. Figure 4 shows the second SVD mode of the total apparent heat source (the sum of the $<Q_1>$ and $<Q_2>$) and the geopotential heights at 100 hPa and 500 hPa. The second mode of the SVD explains 23.63% and 14.91% of the total covariance of the geopotential heights at 100 hPa and 500 hPa, respectively. Positive correlations are found

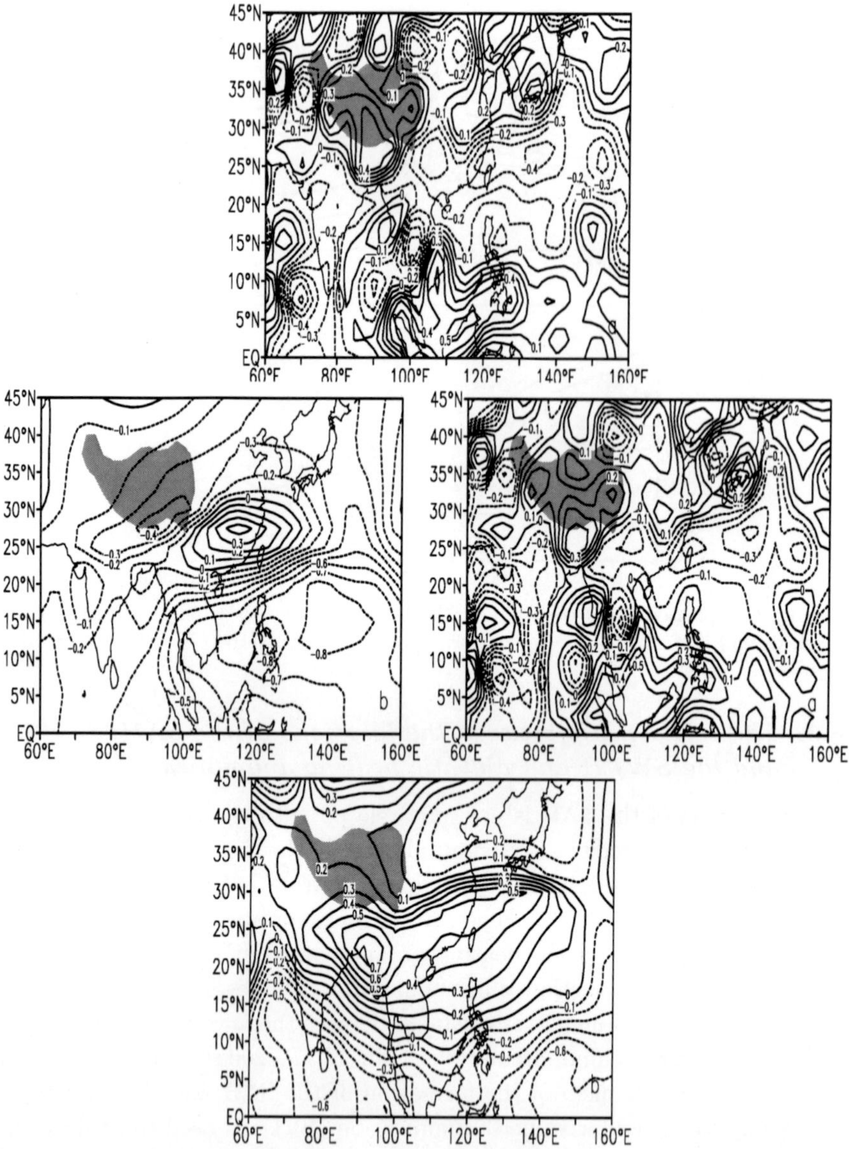

Fig. 4. The second SVD mode of the total apparent heat source (left panel) and the geopotential heights (right panel) at 100 hPa (top panel) and 500 hPa (bottom panel).

between the diabatic heating over the Tibetan Plateau together with the east of the Plateau (Fig. 4(a) and Fig. 4(c)) and the geopotential heights at 100 hPa over the southeast of the Tibetan Plateau (Fig. 4(b)) as well as those at 500 hPa over a vast area from the south of the Yangtze River to the South China Sea, from the Indochina Peninsula to the western Pacific (Fig. 4(d)). To further explain the relationship between the apparent heat source and abnormal locations of the SAH and the SWPH, 1980, 1983, 1988 and 1998 are chosen as typical years of the east pattern of the SAH, while 1984 and 1999 as typical years of the west pattern of the SAH for the composite analysis.

As shown in Fig. 5, large positive anomalies of the $<Q1>$ and $<Q2>$ extend from the east of the Tibetan Plateau to the eastern coast of China for the east pattern of the SAH, while large negative anomalies of the

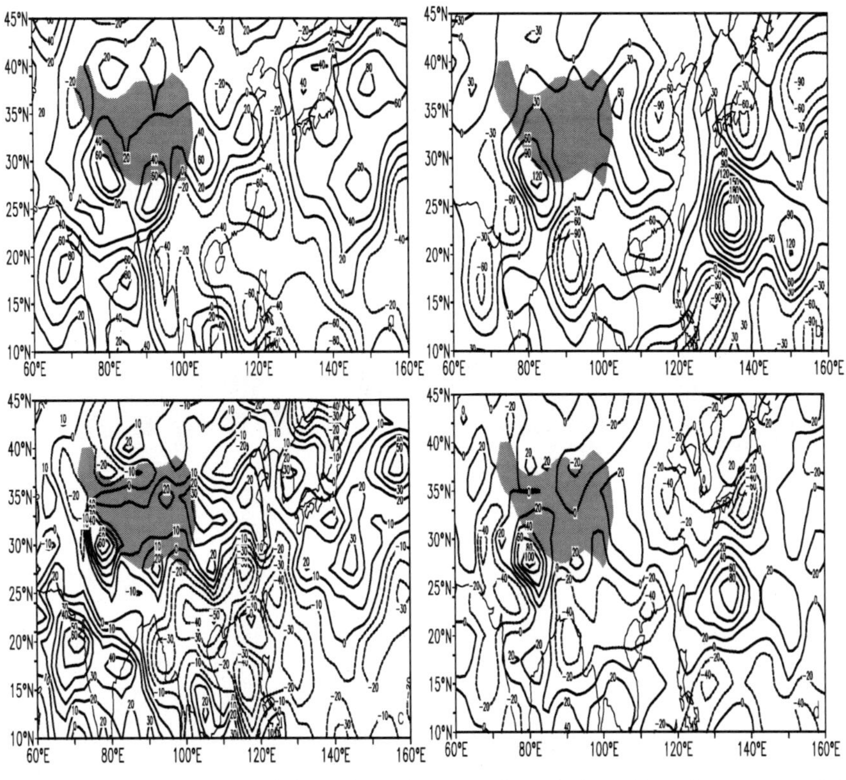

Fig. 5. Composite map of the $<Q_1>$ (top panel) and the $<Q_2>$ (bottom panel) anomalies (units: W/m^2) in July of the years with the east (left panel) and west (right panel) pattern of the SAH.

$<Q1>$ and $<Q2>$ are found over East China for the west pattern of the SAH, which is opposite to the situation over the east of the Plateau, where there remain the large positive anomalies of the $<Q1>$ and $<Q2>$. The maximum value of the $<Q1>$ anomaly is about $-90\,\mathrm{w/m^2}$, much larger than that of the $<Q2>$ anomaly (about $-40\,\mathrm{w/m^2}$).

The climatic average of the $<Q_1>$ in July illustrates that the Tibetan Plateau and the east of the Plateau are dominated by the sensible heat source and the ascending motion (figure omitted). When the east pattern of the SAH occurs, the heat source anomaly over the Yellow River and Huaihe River Valley strengthens the lifting motion of the atmosphere over the east of the Tibetan Plateau (Fig. 6(a)), which reinforces the high level divergence. Consequently, the intensity of the SAH increases and the ridge of the SAH extends further eastward. Fig. 6(c) shows the ω anomaly along the ridge line of the SWPH (at 25°N). From this figure, it is found that an anomalous descending motion dominates over the southeastern coast of China and an anomalous ascending motion around 145°E, while the anomalous sinking motion occurs over the east of 155°E. The configuration

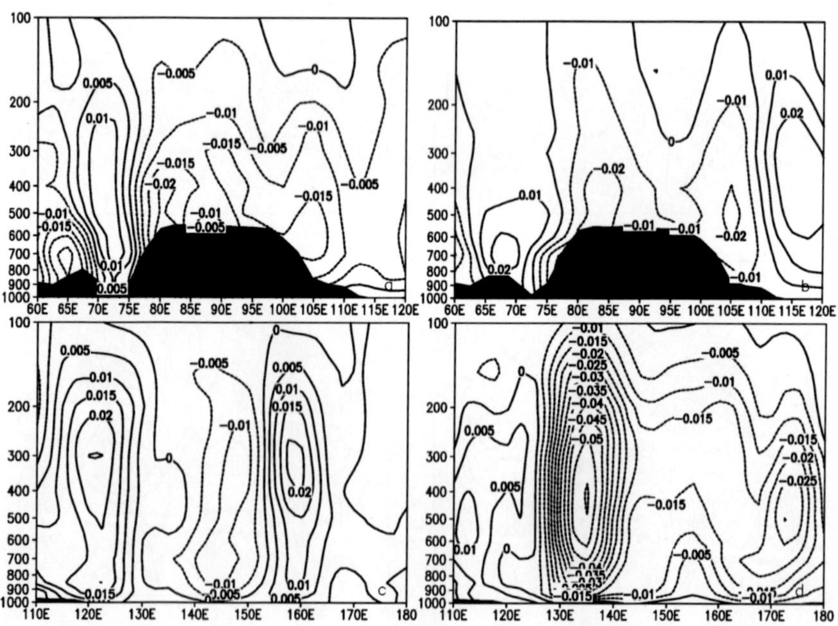

Fig. 6. Composite map of ω anomalies (units: Pa/s) along 32.5°N (top panel) and 25.0°N (bottom panel) in July of the years with the east (left panel) and west (right panel) pattern of the SAH. The shaded area denotes terrain.

of the vertical movements is favorable to the occurrence of the east (west) pattern of the SAH (SWPH). The anomalous descending motion over the sea reinforces the intensity of the SWPH. Although the anomalous ascending motion occurs around 145°E and the center of the SWPH tends to break into two parts (figure not shown), the abnormal sinking motion dominates over the southeastern coast of China, which leads to the intensification of the SWPH and its extension further into the Southeast China. When the west pattern of the SAH occurs, the anomalous heat sink resides from the east of the Tibetan Plateau to the eastern coast of China, which reduces the ascending motion in this area (Fig. 6(b)), and the intensity of the SAH decreases and withdraws westward. The anomalous ascending motion occurs in the east of 125°E along the ridge line of the SWPH (Fig. 6(d)), which reduces the intensity of the SWPH, and consequently it retreats eastward.

6. Conclusions

During boreal summer, especially in July, the interannual variations of the SAH and SWPH in terms of their longitudinal locations are evident. The two anticyclonic systems in middle and upper troposphere tend to move towards each other or move away from each other on an interannual basis.

There exist positive correlations between the diabatic heating (the sum of the apparent heat source and apparent moisture sink) over the Tibetan Plateau together with the east of the Plateau and the geopotential heights at 100 hPa over the southeast of the Tibetan Plateau as well as those at 500 hPa over a vast area from the south of the Yangtze River to the South China Sea, and from the Indochina Peninsula to the western Pacific.

The distribution of the diabatic heating and the associated vertical movements in the atmosphere over the east of the Tibetan Plateau may explain the formation and the maintenance of the longitudinal oscillation of the SAH and the SWPH. The anomalous heat source (sink) strengthens (weakens) the ascending motion of the atmosphere over the east of the Tibetan Plateau for the east (west) pattern of the SAH, which reinforces (reduces) the high-level divergence. Consequently, the intensity of the SAH increases (decreases) and the ridge of the SAH extends further eastward (withdraws westward). In addition, the abnormal sinking (rising) motion of the atmosphere occurs in middle and low latitudes over the western Pacific owing to the anomalous northerly in upper (lower) troposphere (figure omitted), which is accompanied by adiabatic heating (cooling). As a result, the deep and warm SWPH is strengthened (weakened) and extends further westward (withdraws eastward).

Numerical experiments using a climate model with anomalous heating will be the focus of future studies in an attempt to verify the diagnostic analysis mentioned above.

Acknowledgments

This work is jointly funded by Key Laboratory of Meteorological Disaster of Ministry of Education, NUIST (KLME050210) and the Scientific Research Foundation for the Returned Overseas Chinese Scholars, Ministry of Education.

References

1. W. Chen and X. F. Zhi, Comparisons of the West Pacific subtropical high and of the South Asian high between NCEP/NCAR and ECMWF reanalysis datasets, *J. Trop. Meteorol.* **14** (2008) 121–124.
2. R. Guo and X. F. Zhi, Synoptic analysis of severe droughts during the summer 2003 in southern China, *J. Nanjing Institute Meteorol.* **31** (2008) 234–241, in Chinese.
3. J. H. He, B. Zhou, M. Wen and F. Li, Vertical circulation structure, interannual variation features and variation mechanism of western pacific subtropical high, *Adv. Atmos. Sci.* **18** (2001) 497–509.
4. E. Kalnay, *et al.*, The NCEP/NCAR 40-year Reanalysis Project, *Bull. Am. Meteorol. Soc.* **77** (1996) 437–471.
5. R. Kistler and Coauthors, The NCEP-NCAR 50-year reanalysis: Monthly means CD-ROM and documentation, *Bull. Am. Meteorol. Soc.* **82** (2001) 247–268.
6. T. N. Krishnamurti, Tropical east-west circulations during northern summer, *J. Atmos. Sci.* **28** (1971) 1342–1347.
7. T. N. Krishnamurti, Summer Monsoon Experiment — A review, *Mon. Weather Rev.* **113** (1985) 1590–1626.
8. K. M. Lau and M. T. Li, The monsoon of East Asia and its global associations — A survey, *Bull. Am. Meteorol. Soc.* **65** (1984) 114–125.
9. K. M. Lau, G. T. Yang and S. H. Shen, Seasonal and intraseasonal climatology of summer monsoon rainfall over East Asia, *Mon. Weather Rev.* **116** (1988) 18–37.
10. Y. M. Liu, G. X. Wu and R. C. Ren, Relationship between the subtropical anticyclone and diabatic heating, *J. Climate* **17** (2004) 682–698.
11. R. Y. Lu and B. W. Dong, Westward extension of North Pacific subtropical high in summer, *J. Meteorol. Soc. Jap.* **79** (2001) 1229–1241.

12. R. B. Mason and C. E. Anderson, The Development and decay of the 100 mb summertime anticyclone over Southern Asia, *Mon. Weather Rev.* **91** (1963) 3–12.
13. J. Matsumoto, The seasonal changes in Asian and Australian monsoon regions, *J. Meteorol. Soc. Jap.* **70** (1992) 257–273.
14. T. Murakami, and J. Matsumoto, Summer monsoon over the Asian continent and western North Pacific, *J. Meteorol. Soc. Jap.* **72** (1994) 719–745.
15. T. Nagazawa, Seasonal phase lock of intraseasonal variation during the Asian summer monsoon, *J. Meteorol. Soc. Jap.* **70** (1992) 597–611.
16. G. Sturaro, A closer look at the climatological discontinuities present in the NCEP/NCAR reanalysis temperature due to the introduction of satellite data, *Clim. Dyn.* **21** (2003) 309–316.
17. C. H. Sui, P. H. Chung and T. Li, Interannual and interdecadal variability of the summertime western north pacific subtropical high, *Geophys. Res. Lett.* **34** (2007) L11701.1–L11701.6.
18. M. Tanaka, Intraseasonal oscillation and the onset and retreat dates of the summer monsoon over East, Southeast Asia and the western Pacific region using GMS high cloud amount data, *J. Meteorol. Soc. Jap.* **70** (1992) 613–629.
19. S. Y. Tao and F. K. Zhu, The 100 mb flow patterns in Southern Asia in summer and its relation to the advance and retreat of the west-Pacific subtropical anticyclone over the Far East (in Chinese), *Acta Meteorologica Sinica* **34** (1964) 385–395.
20. S. Y. Tao and L. X. Chen, A review of recent research of the East Asian summer monsoon in China, in *Monsoon Meteorology*, eds. C.-P. Chang and T. N. Krishnamurti (Oxford University Press, 1987), pp. 60–92.
21. G. X. Wu, J. F. Chou, Y. M. Liu and J. H. He, *The Dynamical Questions about the Formation and Variation of the Subtropical High* (Science Publisher, Beijing, 2002), pp. 108–244.
22. M. Yanai, S. Esbensen and J. H. Chu, Determination of bulk properties of tropical cloud clusters from large-scale heat and moisture budgets, *J. Atmos. Sci.* **30** (1973) 611–627.
23. S. Yang, K. M. Lau and K. M. Kim, Variations of the East Asian and Asian-Pacific-American winter climate anomalies, *J. Climate* **15** (2002) 306–325.
24. G. Zeng, Z. B. Sun, W. C. Wang and J. Z. Min, Interdecadal variability of the East Asian Summer Monsoon and associated atmospheric circulations, *Adv. Atmos. Sci.* **24** (2007) 915–926.
25. Q. Zhang and Y. F. Qian, Interannual and interdecadal variations of the South Asia high, *Chinese J. Atmos. Sci.* **24** (2000) 67–78, in Chinese.
26. Q. Zhang, G. X. Wu and Y. F. Qian, The bimodality of the 100 hPa South Asia High and its relationship to the climate anomaly over East Asia in summer, *J. Meteorol. Soc. Jap.* **80** (2002) 733–744.
27. T. J. Zhou, *et al.*, Why the western Pacific subtropical high has extended westward since the late 1970s, *J. Climate* (2008), doi:10.1175/2008JCLI2527.1.

Advances in Geosciences
Vol. 16: Atmospheric Science (2008)
Eds. Jai Ho Oh *et al.*

SINGULAR VECTOR AND ENSO PREDICTABILITY IN A HYBRID COUPLED MODEL*

XIAOBING ZHOU[†,‡] and YOUMIN TANG[†,§]
[†] *Environmental Science and Engineering,*
University of Northern British Columbia,
3333 University Way, Prince George, BC, V2N 4Z9, Canada
[§] *ytang@unbc.ca*
[‡] *Centre for Australian Weather and*
Climate Research (CAWCR), Bureau of Meteorology,
700 Collins St, Melbourne, VIC 3001, Australia

In this study, singular vector (SV) and retrospective ENSO (El Niño and Southern Oscillation) predictions were performed respectively for the period from 1876 to 2000 using a hybrid coupled model. Emphasis was placed on exploring the relationship between SV and ENSO predictability. It is found that a defined Niño3 index from the first singular vector of sea surface temperature anomaly (SSTA) is highly correlated with the predicted Niño3 SSTA index of 6-month leads and that the first singular value (FSV) is positively correlated with the predictive skill. These results and findings improve our knowledge and understanding to the relationship between SV and predictability. It was thought that the fastest growth rate of errors to be inversely related to the prediction skill. The reasons why there is such a relationship between SV and realistic predictability include: (1) the strong signals of ENSO variability that favour the growth of initial uncertainties also have significant contributions to the predictability; (2) the averaged climate state of the tropical Pacific Ocean simultaneously effects both SV and predictability.

1. Introduction

The earliest report on the singular vector can be found in Lorenz's paper (1965) which introduced singular vector (SV) analysis into meteorology to study atmospheric predictability [1]. However, SV analysis had not been used to investigate ENSO predictability until the 1990s. Since then a lot of work has been approached on ENSO predictability studies using SV (e.g., Blumenthal 1991; Xue *et al.*, 1997a,b; Chen *et al.*, 1997; Thompson

*This work was supported by Canadian Foundation for Climate and Atmospheric Sciences grant to Y. Tang.

1998; Moore *et al.* 1996, 1997a,b, 2003; Fan *et al.*, 2000; Tang *et al.*, 2006; Zhou *et al.*, 2007) [2–13]. Despite different coupled models and norms used, these researches derived similar findings and conclusions such as: (1) the optimal initial and final patterns have large-scale features in the tropical Pacific Ocean. The initial pattern is insensitive to initial conditions while the final pattern depends on initial conditions; (2) the perturbation growth in coupled models is usually controlled by one dominant growing mode and its final pattern resembles ENSO-like pattern; (3) the optimal growth rates vary with the seasonal cycle and the phase of ENSO.

A wide perception in ENSO predictability and SV is an inverse relationship existing between them, namely, when the leading singular value is large, the predictability is low and vice versa (e.g., Moore *et al.*, 1996) [7]. The physical interpretation behind the perception is that the largest singular value describes the fastest growth rate of errors based on the definition of SV. Such an understanding and recognition to SV and ENSO predictability has been widely accepted. However, this theoretical perception has not been validated effectively using real ENSO prediction skill and SV because a long-term retrospective ENSO prediction and SV analysis was not available due to the limited length of observations, leading to difficulties to drive stable and robust conclusions and findings. Almost all SV analyses have to date, focused on a period around 20–30 years (e.g., Xue *et al.*, 1997b; Fan *et al.*, 2000) [4,11]. Recently we performed a long-term retrospective ENSO events and SV analysis for over 100 years from 1876–2000 (Zhou *et al.*, 2009) [14], which allows us to examine the relationship between SV and real predictability.

The motivation of this work is related to some recent studies on ENSO predictability. It has been found that ENSO predictability is highly dependent on the strength of ENSO variability. As the anomalies (signals) present in initial conditions are strong, the predictions are likely to reliable (e.g., Chen *et al.*, 2004; Tang *et al.*, 2005; Tang *et al.*, 2007) [18, 13]. On the other hand, large singular values are also often associated with large anomalies. Thus the relationship between SV and predictability is most likely complicated, and probably unable to be simply characterized by the aforementioned theoretical perception. A detailed examination from long-term real prediction and SV analysis should be required.

In this paper, we will explore the relationship between predictability and SV through real prediction and SV for the past 125 years (1876–2000). Emphasis will be placed on the decadal variations in the predictability and in the SV. This paper is structured as follows: the model is introduced in

Sec. 2; the results on SV and ENSO predictability as well as their decadal variations will be presented in Sec. 3; finally a discussion and conclusion can be found in Sec. 4.

2. The Hybrid Coupled Model and the Computation of Singular Vector

The Hybrid Coupled Model (HCM) used in this study is composed of a nonlinear dynamical ocean model and a nonlinear empirical atmospheric model, identical to Tang (2002) and Zhou *et al.* (2007) [18, 13] The ocean model is one of intermediate complexity, derived from Anderson and McCreary (1985) and Balmaseda *et al.* (1994, 1995) and extended to six activate layers [19, 20-21]. It covers the tropical Pacific Ocean 30°N–30°S in latitude and from 123°E–69°W in longitude with a horizontal resolution of $1.5° \times 1.5°$, and consist of the depth averaged primitive equations in six layers (with reference thickness of 100 m, 175 m, 250 m, 320 m, 400 m and 500 m from top to bottom). The atmospheric model is built by the nonlinear regression method, a neural network (NN).

The SVs were calculated using the tangent linear model (TLM) and Adjoint Model (AM) of the HCM ARPACK (ARnoldi PACKage) software package (Lehoucq *et al.*, 1998) based on the Lanczős algorithm was used to solve the singular value problem [22]. A L_2-norm is adopted for the computation of SVs in this study. Details of the SVs computation can be found in Zhou *et al.* (2007) [13].

3. Result Analysis

The ocean model was forced by the reconstructed wind stress and integrated for 125 years from 1876 to 2000 (Zhou *et al.*, 2009) [14]. The observed SSTA –ERSST data (Extended Reconstructed SST; Smith and Reynolds, 2003, 2004 [23–24]) was assimilated into the oceanic model, leading to good ENSO simulation and predictions up to lead times of 12 months from 1876–2000 (Zhou *et al.*, 2009) [14]. The correlation between analyzed Niño3 SSTA index and the observed counterpart is up to 0.98 at the past 125-year period. The SV and each prediction were performed from same initial conditions, at a three-month interval starting from April 1, 1876 to January 1, 2001. This coupled model was able to successfully predict the major ENSO signals at a 6-month lead in the past 125 years. Hence we will focus on the ENSO prediction at a 6-month lead and SV at a 6-month optimal period in following analysis.

3.1. *Singular vector and predictability*

The first singular vector (SV1) has the largest singular value that is much larger than others. Its typical spatial distribution in this coupled model consists of an east-west dipole spanning the entire tropical Pacific basin (not shown), which is similar to those of previous researches (e.g., Chen *et al.*, 1997; Xue *et al.*, 1997a; Zhou *et al.*, 2007) [5, 3, 14]. As can be seen in Fig. 1, the positive maximum center in SV1 along the equator moves eastward during El Niño and westward during La Niña. Its moving track is similar to the continuous redistribution of the warm surface waters. These waters are driven westward by intense westward trade wind during La Niña, and eastward by relaxed winds during El Niño (Fedorov and Philander, 2000) [25].

To study the relationship between SV1 and predictability, we defined a Niño3 SV1 index similar to the traditional Niño3 SSTA index, i.e. the averaged SV1 over Niño3 region. This region has the strongest interannual variability, and the largest forecast error growth in ENSO models. Thus, the SV1 index reflects well the variation in strength of SV1 pattern with time. Figure 2 shows the variation of Niño3 SV1 index and the predicted Niño3 SSTA index at a 6-month lead from 1876–2000. As can be seen, the Niño3 SV1 index is highly correlated with the predicted Niño3 SSTA at a 6-month lead with the correlation coefficient of 0.75 during the past 125 years. Furthermore, the variations of the amplitude of Niño3 SV1 index are closely associated with variations in the strength of ENSO events. For example, a strong ENSO event often has a large amplitude of Niño3 SV1 index and vice versa. Figure 2 sheds light on some important issues on SV: (1) the SV1 spatial pattern actually varies with initial conditions, as suggested by temporary variation of SV1 index, although its large scale structure is always characterized by the east-west dipole spanning the whole tropical Pacific for all initial condition like other models (e.g., Chen *et al.*, 1997; Xue *et al.*, 1997a,b; Zhou *et al.*, 2007) [5, 3-4, 13]; (2) the Niño3 region is a center of the dipole pattern of SV1, therefore Niño3 SV1 index measures the uncertainties most favourable for the error growth of prediction. Based on the concept of SV, a large value of Niño3 SV1 index should correspond with a poor prediction. However, it is not true here. As shown in Fig. 2, a large value of Niño3 SV1 index often corresponds with a good prediction. This is because a large Niño3 SV1 index often occurs at a strong ENSO event.

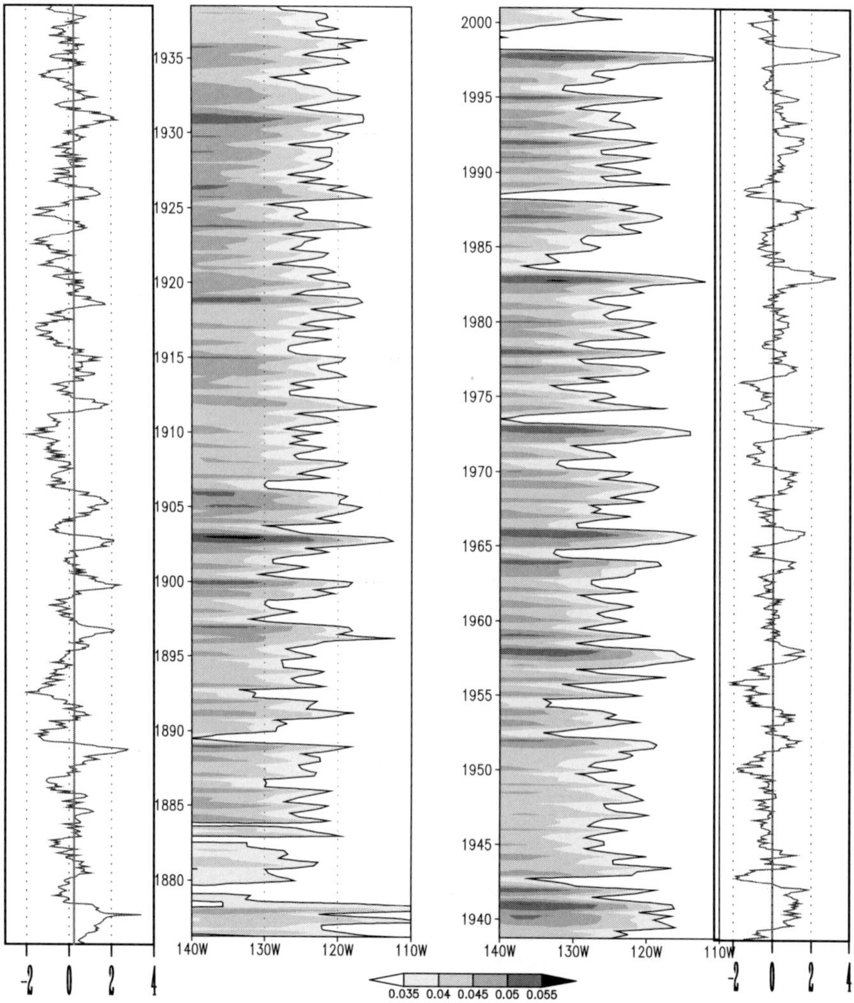

Fig. 1. Time-longitude diagram of the first singular vector (SV1) along the equator (shaded graphs) and the observed Niño3 SSTA index (line figures), from 1976–2000. Contour line in shaded graphs is 0.035°C and the values less than 0.035°C are blank in the SV pattern.

It has been found that the strength of ENSO signals significantly impact the prediction skill (Chen *et al.*, 2004; Tang *et al.*, 2005, 2007) [5, 16–17]. Therefore one should take caution to use SV to interpret predictability; (3) the Niño3 SV1 index can be used as a precursor for the ENSO event since it is quite consistent with the observed SSTA index of 6 months later.

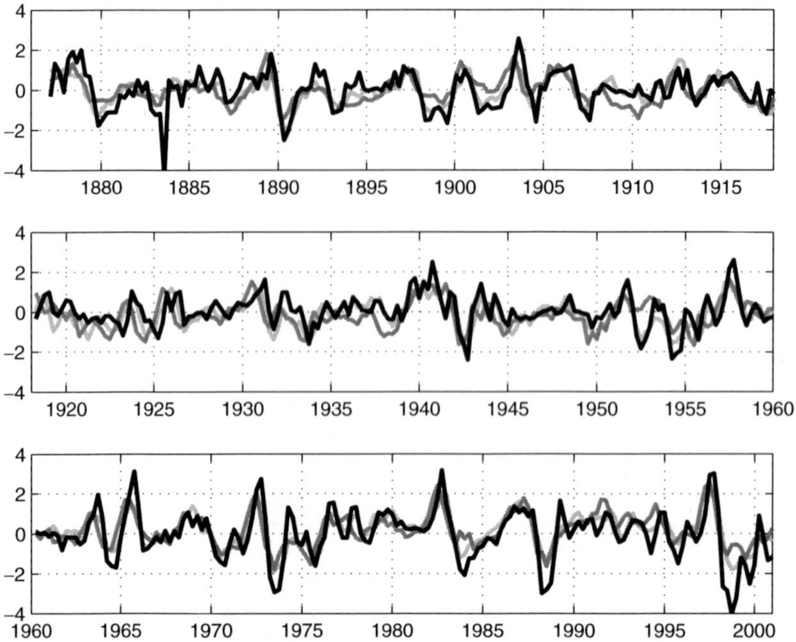

Fig. 2. Variations of the predicted Niño3 SSTA index of a 6-month lead (cyan), the observed Niño3 SSTA index (magenta) and normalized Niño3 SV1 index (black).

3.2. *Singular value and predictability*

3.2.1. *Interdecadal variations of singular values*

We have discussed SV1 in preceding section. In this section, we will turn to analyze the singular values, which reflect the optimal perturbation growth rates of initial uncertainties in a specified period. SVs are computed by adding small perturbations to the background state. The singular values are theoretically independent with perturbations but dependent on the background state with seasonal cycle, ENSO phase, and the decadal variability of ENSO. The previous studies only explored the impact of the seasonal cycle and ENSO phase on SVs in a relatively short period (e.g., Chen *et al.*, 1997; Xue *et al.*, 1997b; Moore and Kleeman, 1996; Tang *et al.*, 2006) [5, 4, 7, 12], which prohibited investigating the relationship between singular values and interdecadal variations of background mean state.

The 20-yr running mean of the first singular value (referred to as MSV1), model's spatial-average SST between 15°S–15°N (referred to as MSST) and the variance of Niño3 SSTA index (referred to as VNINO).

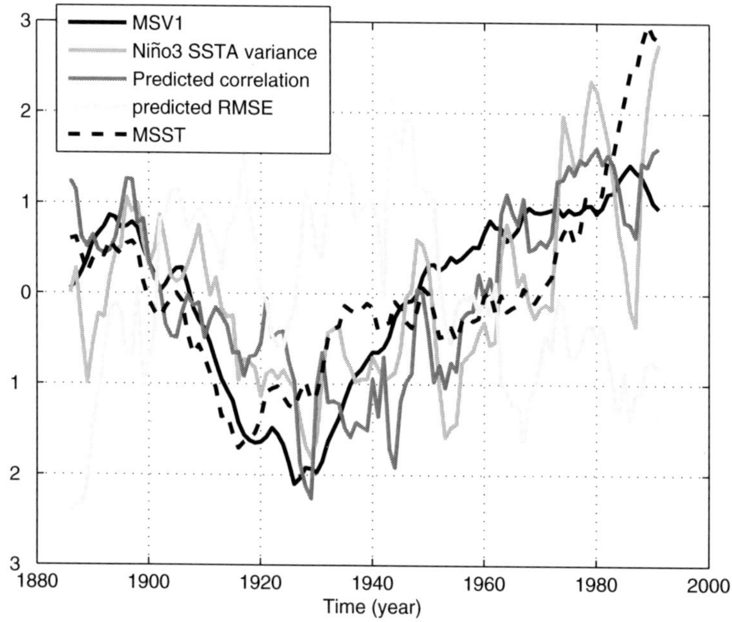

Fig. 3. Variation of the mean SST over 15°–15°S and 123°E–69°W, the variance (cyan) of Niño3 SSTA, MSV1 (solid black), predictive correlation (magenta) and RMSE (yellow) skill from 1876 to 2000. All were normalized prior to plotting.

This was obtained using 20-yr running windows,[a] as is shown in Fig. 3. As can be seen, MSST reached a minimum in the 1920s, and then increased gradually till 1940. During the period from the 1940s to 1960s, it seemed stable. After the 1960s, MSST continuously increased, being consistent with the global warming. The correlation between MSST and VNINO is up to 0.68, indicating a good relationship between both. In other words, the signal strength (amplitude) and the mean state are highly related each other. The variation of MSV1 is in a very good agreement with that of MSST and VNINO. The good relationship between MSV1 and MSST (VNINO) is probably due to the fact that both were driven by the wind stress which showed significant interdecadal variations (not shown). Moore and Kleeman

[a]We calculate the skill in the first 20-yr window, saying from 1876–1895, denoted at 1885. Then we move the window forward one year, saying from 1877–1896, to calculate the skill again, denoted at 1886. This process will repeat until the skill is calculated in the last window, 1981–2000, denoted at 1990. Consequently, we can obtain a time series of 105 samples, which characterizes the decadal and interdecadal variations in prediction skills.

(1996) found that the first singular varied approximately in proportion to $|W|$ [7], a representative value of the mean wind speed, which acted as a coupling coefficient between the ocean and atmosphere. This is also true for the model used here. We found the singular values in this model are highly dependent on the coupling coefficient which determines the wind anomalies.

3.2.2. *Interdecadal variations of singular value and predictability*

In this subsection we will examine the relationship between decadal variation of the first singular value (FSV) and decadal variation of real prediction skill of ENSO, which is measured by correlation and root mean square error (RMSE) of predicted Niño3 SSTA index against the observed counterpart.

To study the interdecadal changes of prediction skills, the prediction skills are computed at each running window of 20-yr from 1876–2000. Figure 3 shows variations of the normalized correlation coefficient and RMSE at a 6-month lead from 1876–2000. As can be seen, the scores measured by RMSE are not always consistent with those by correlation (RMSE and correlation is quite different so they are always inconsistent). It is interesting to see that the trend of MSV1 is consistent with that of the correlation skill which has a correlation coefficient of 0.77. An inverse relationship also exists between MSV1 and RMSE and their correlation is −0.53. This indicates that a large MSV1 has a high prediction skill and vice versa. Theoretically SV1 indicates the fastest growth rate of the initial uncertainties, leading to a lower prediction skill. However, the real situation in this model is totally different as the evidence above suggested. The reason is probably because strong signal anomalies (e.g., El Niño or La Niña) that control prediction skills also favour the growth of initial uncertainties. Some evidence can be found for this argument. For example, previous studies suggested that the final pattern of SV1, which described the final uncertainty growth, resembled the mature ENSO pattern. It should be noted that ENSO is the strongest interannual signal in the tropical Pacific Ocean. In Sec. 3.1, the Niño3 SV1 index is large before strong ENSO events and small before weak ENSO events. In Sec. 3.2, the variation of MSV1 follows well the variation of the strength of ENSO events. Therefore the contribution of stronger signal anomalies to predictability can be two sided. It can enhance predictability due to bringing more information or lowering predictability due to faster error growth. When the former exceeds the latter, a positive relationship between SV1 and predictability could be shown, as in Fig. 3. One should not understand their relationship as a counterexample of the concept of SV. Instead, it should be interpreted as a net contribution of signal anomalies of ENSO variability to its predictability.

Figure 3 also shows that the change of the correlation skill is associated with the variation of MSST and VNINO. The correlation skill is lower during the relative "cold" period than that of "warm" period. The lowest correlation skill occurred around the 1930s, corresponding to the coldest period. Apparently the trend of the correlation skill is in good agreement with that of the MSST, and the fluctuation of the skill is consistent with that of VNINO. In general, a large VNINO will have relative high prediction skill, but this is not valid in some periods. For example, the "warm" period of the 1890s had relatively small VNINO but attained a high skill. This indicates that the correlation skill is determined by both MSST and VNINO. It can be taken as additional evidence to explain why MSV1 has a high positive relationship to correlation skill; because MSV1 is closely related to both MSST and VNINO.

4. Summary and Discussion

In this study, the ENSO predictability was explored based on singular values and prediction skills in the last 125 years from 1876 to 2000. The results show that the first singular vector actually varies with different initial conditions, although its large-scale feature along the equator is not sensitive to them. The positive maximum center of SV1 in the equatorial eastern Pacific moves westward during La Niña and flows back eastward during El Niño. The correlation coefficient between the Niño3 SV1 anomaly index and predicted Niño3 SSTA at a 6-month lead is up to 0.75 during a whole period of 1876–2000. Thus the Niño3 SV1 anomaly index can be used as a precursor of ENSO prediction.

The 20-yr running mean of the first singular value, measuring the mean growth rate of optimal initial perturbations in a 20-yr period, was calculated from 1876 to 2000. This running mean largely depends on the variation of 20-yr running means of model SST (MSST) and 20-yr-running variance of Niño3 SSTA index (VNINO). A large MSST will be possibly accompanied by a large MSV1 and vice versa.

The prediction skills are largely determined by both MSST and VNINO. The correlation skill is much lower during the "cold" period than during the "warm" period. The minimum correlation skill occurred in the coldest period of the 1930s. The fluctuations of correlation skill seem to coincide with those of VNINO. Since MSV1 depends on MSST, a large MSV1 does not show a low correlation skill. On the contrary, a large MSV1 generally has a high correlation skill. The correlation coefficient between predictive correlation skill and MSV1 is up to 0.77 during the past 125 years.

Generally, model prediction skills are model dependent. A comparison of prediction skill between our model and Lamont's model (Zebiak and Cane, 1987; Chen, *et al.*, 2004) was examined [26, 15]. The result shows that the interdecadal variations of prediction skills in the two models are very similar (not shown), and both are consistent with the variations of MSST. This suggests these results found in our model might be applied to the other models.

The prediction skill in the last three decades is obviously higher than that in other periods. One might argue that this is because the coupled model used the data of these periods for training. We used a cross-validation scheme for any training, therefore artificial skill should be greatly alleviated although we cannot completely exclude it. Here we argue another factor, sea surface mean temperature in the tropical Pacific, plays an important role in attaining high prediction skill after the 1970s. The mean SST of the tropical Pacific increased significantly after the 1970s, corresponding with the climate regime shift in 1976. As evidenced in this study, the ENSO events are more predictable in the "warm" ocean than in the "cold" ocean, since the former is often associated with strong signals of *SST* variability leading to more information provided by predictions (Tang *et al.*, 2005) [16]. This also explains why both our model and Lamont's model had a good prediction skill during 1876 to 1900, when the mean SST was high.

Acknowledgments

We would like to thank Dr. Ziwang Deng for providing us with the reconstructed wind stress and helpful comments.

References

1. E. N. Lorenz, A study of the predictability of a 28-variable atmospheric model, *Tellus* **17** (1965) 321–333.
2. M. B. Blumenthal, Predictability of a coupled ocean–atmosphere model, *J. Climate* **4** (1991) 766–784.
3. Y. Xue, M. A. Cane and S. E. Zebiak, Predictability of a coupled model of ENSO using singular vector analysis. Part I: Optimal growth in seasonal background and ENSO cycles, *Mon. Weather Rev.* **125** (1997a) 2043–2056.

4. Y. Xue, M. A. Cane and S. E. Zebiak, Predictability of a coupled model of ENSO using singular vector analysis. Part II: Optimal growth and forecast skill, *Mon. Weather Rev.* **125** (1997b) 2057–2073.
5. Y.-Q. Chen, D. S. Battisti, T. N. Palmer, J. Barsugli and E. S. Sarachik, A study of the predictability of tropical Pacific SST in a coupled atmosphere–ocean model using singular vector analysis: The role of the annual cycle and the ENSO cycle, *Mon. Weather Rev.* **125** (1997) 831–845.
6. C. J. Thompson, Initial conditions for optimal growth in a coupled ocean–atmosphere model of ENSO, *J. Atmos. Sci.* **55** (1998) 537–557.
7. A. M. Moore and R. Kleeman, The dynamics of error growth and predictability in a coupled model of ENSO, *Quart. J. Roy. Meteorol. Soc.* **122** (1996) 1405–1446.
8. A. M., Moore and R. Kleeman, The singular vectors of a coupled ocean–atmosphere model of ENSO, II: Sensitivity studies and dynamical interpretation, *Quart. J. Roy. Meteorol. Soc.* **123** (1997a) 983–1006.
9. A. M. Moore and R. Kleeman, The singular vectors of a coupled ocean-atmosphere model of ENSO. Part 1: Thermodynamics, energetics and error growth, *Quart. J. Roy. Meteorol. Soc.* **123** (1997b) 953–981.
10. A. M. Moore, J. Vialard, A. T. Weaver, D. L. T. Anderson, R. Kleeman and J. R Johnson, The role of air-sea interaction in controlling the optimal perturbations of low-frequency tropical coupled ocean-atmosphere modes, *J. Climate* **16** (2003) 951–968.
11. Y. Fan, M. R. Allen, D. L. T. Anderson and M. A. Balmaseda, How predictability depends on the nature of uncertainty in initial conditions in a coupled model of ENSO, *J. Climate* **13** (2000) 3298–3313.
12. Y. Tang, R. Kleeman and S. Miller, ENSO predictability of a fully coupled GCM model using singular vector analysis, *J. Climate* **19** (2006) 3361–3377.
13. X. Zhou, Y. Tang and Z. Deng, The impact of atmospheric nonlinearities on the fastest growth of ENSO prediction error, *Clim. Dyn.* (2007), doi:10.1007/s00382-007-0302-5.
14. X. Zhou, Y. Tang and Z. Deng, Assimilation of historical SST data for long-term ENSO retrospective forecasts, *Ocean Modelling* (2009).
15. D. Chen, M. A. Cane, A. Kaplan, S. E. Zebiak and D. Huang, Predictability of El Niño in the past 148 years, *Nature* **428** (2004) 733–736.
16. Y. Tang, R. Kleeman and A. Moore, On the reliability of ENSO dynamical predictions, *J. Atmos. Sci.* **62** (2005) 1770–1791.
17. Y. Tang, Z. Deng, X. Zhou, Y. Cheng and D. Chen, Interdecadal variation of ENSO predictability in multiple models, *J. Climate* **21** (2008) 4811–4833.
18. Y. Tang, Hybrid coupled models of the tropical Pacific? — Interannual variability, *Clim. Dyn.* **19** (2002) 331–342.
19. D. L. T. Anderson and J. P. McCreary, Slowly propagating disturbances in a coupled ocean–atmosphere model, *J. Atmos. Sci.* **42** (1985) 615–629.
20. M. Balmaseda, D. L. T. Anderson and M. K. Davey, ENSO prediction using a dynamical ocean model coupled to statistical atmospheres, *Tellus* **46A** (1994) 497–511.

21. M. Balmaseda, M. K. Davey and D. L. T. Anderson, Decadal and seasonal dependence of ENSO prediction skill, *J. Climate* **8** (1995) 2705–2715.
22. R. B. Lehoucq, D. C. Sorensen and C. Yang, *ARPACK Users' Guide* (SIAM, Philadelphia, USA, 1998).
23. T. M. Smith and R. W. Reynolds, Extended reconstruction of global sea surface temperatures based on COADS data (1854–1997), *J. Climate* **16** (2003) 1495–1510.
24. T. M. Smith and R. W. Reynolds, Improved extended reconstruction of SST (1854–1997), *J. Climate* **17** (2004) 2466–2477.
25. A. V. Fedorov and S. G. H. Philander, Is El Niño changing? *Science* **288** (2000) 1997–2002.
26. S. E. Zebiak and M. A. Cane, A model El Niño–Southern Oscillation, *Mon. Weather Rev.* **115** (1987) 2262–2278.

Advances in Geosciences
Vol. 16: Atmospheric Science (2008)
Eds. Jai Ho Oh *et al.*

STUDY OF LIGHTNING ACTIVITY OVER INDIAN SUBCONTINENT

H. S. CHAUDHARI
Indian Institute of Tropical Meteorology, Pashan, Pune-411008, India
hemantkumar@tropmet.res.in

M. R. RANALKAR
India Meteorological Department, Pune-411008, India

Y. V. KUMKAR, JAI HO OH
Pukyong National University, Busan, South Korea

G. P. SINGH
Banaras Hindu University, Varanasi, India
singgpin@yahoo.co.in

The seasonal distribution of lightning flash activity over the Indian subcontinent (Equator-35°N and 60°E–100°E) is studied using the quality checked monthly lightning flash data obtained from Lightning Imaging Sensor (LIS) on board the Tropical Rainfall Measuring Mission (TRMM) satellite. This paper presents results of diurnal and seasonal variation of the lightning activity over Indian subcontinent. The diurnal variation of maximum flash rate is observed to peak at 11 UTC and minimum at 04 UTC. Thus it clearly shows that maximum thunderstorms activity occur during late afternoon to evening hours. Seasonal total flash counts during monsoon season are more compared to pre-monsoon season but normalized seasonal total flash counts are observed to be more during pre-monsoon season. Thus thunderstorms during pre-monsoon season are more intense.

1. Introduction

The availability of lightning data and its utilization in the operational weather forecasting has seen phenomenal increase during last 2 decades (Aiya and Sonde, 1963; Toracinta *et al.*, 2002; Khandalgaonkar *et al.*, 2005; Manohar *et al.*, 1999). The recent developments in space technologies enabled detection and spatial location of the lightning from the space using optical sensors. NASA Marshall Space Flight Center in Huntsville, Alabama had developed an optical lightning detection sensor. A prototype (OTD,

Optical Transient Detector) was launched in 1995 (Boccippio *et al.*, 2000) and its improved version LIS (Lightning Imaging Sensor) was launched on board the Tropical Rainfall Measuring Mission (TRMM) in 1997 (Christain *et al.*, 1999). The LIS instrument is revolving at an altitude of 350 km with an orbit inclination of 35°. Therefore it has short sampling time and which in turn results in snapshots of few minutes of the lightning activity within a given area of interest. Data obtained from this system is freely available for scientific studies and is useful for lightning analysis especially in the tropics.

Lightning is generally associated with the meso-scale convective systems and it possesses very high temporal and spatial variability. Therefore, it would be appropriate to study such activity with higher spatial resolution. In this study, LIS flash data is utilized for the study of spatial and temporal distribution of lightning activity over the Indian sub continent with the higher spatial grid resolution of $0.2° \times 0.2°$.

2. Data and Methodology

The quality checked monthly LIS data obtained from Marshall Space Flight Centre for the period of 1998–2005 is used in this study. LIS observes lightning activity over the tropical region bounded by 35°N–35°S. This instrument detects total lightning which include cloud-to-ground, intra-cloud and cloud-to-cloud discharges. The LIS is useful for identifying spatial location of lightning, time of lightning events, and radiant energy from lightning activity. Figure 1 shows the extent of the Indian subcontinent selected for this study.

For this study, we have gridded the monthly total lightning flash counts with spatial resolution of $0.2° \times 0.2°$. We used 0.2 degree resolution data to indicate two separate simultaneous events of thunderstorms, which would give accurate representative values of localized lightning flash densities. It is also believed that, there is no need to mask ocean in coastal regions as high grid resolution will lead to much less error in the prepared datasets.

3. Diurnal Variation of Lightning Activity

The diurnal variation of the mean total number of lightning flashes over Indian Subcontinent (0° to 35°N and 60°E–100°E) is presented in Fig. 2. It is seen that maximum lightning flash counts occur at 11 UTC (1630 IST). The maximum flashes occur during late afternoon and evening hours. This result is in agreement with the diurnal variation thunderstorms over Indian

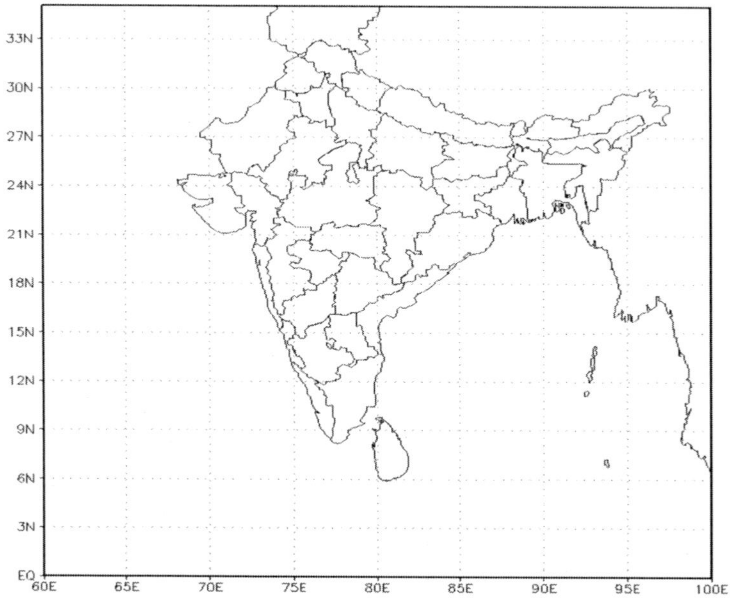

Fig. 1. Map showing region of Indian sub-continent considered in this study.

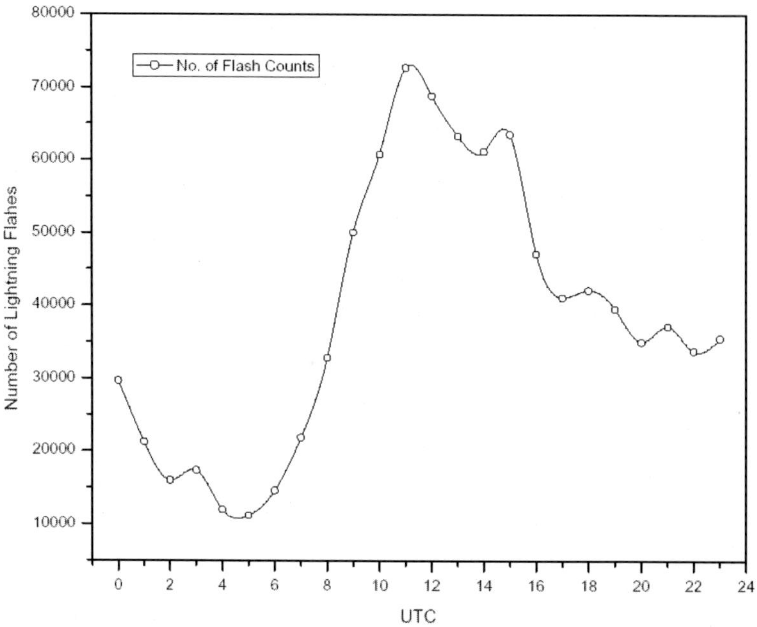

Fig. 2. Diurnal Variation of mean total number of lightning flashes over Indian Subcontinent (0°–35° N and 60° E–100° E) for the period of 1998–2005.

region. The lightning flash counts during morning hours are observed to be low with minima at 04 UTC (1030 IST). The maximum flash counts during late afternoon and evening hours are the result of solar forcing. The Sun heats up the surface during the day. Solar radiations are absorbed within few millimeters of the surface during the day. At night this radiation is easily radiated to space from the land surfaces as they cool. Thus the diurnal temperature range is much more over the land surfaces which affect the diurnal variation of thunderstorms and hence lightning.

4. Latitudinal Variation of Lightning Activity

Indian sub-continent region shows east-west contrast in thunderstorm activity (Mahohar and Kesarkar, 2003). In this viewpoint, we have divided entire Indian subcontinent across the longitude of 80°E. The west region is represented by 60°E–80°E and the east region is represented by 80°E–100°E. Previous studies (Kandalgaonkar *et al.*, 2003, 2005) indicate that latitudinal distribution of thunderstorm activity over the Indian region is different from the rest of the tropics. Therefore it is interesting to study seasonal latitudinal variation of lightning activity. In this context, we have computed the zonal averages for these two regions for different season viz. winter (Jan–Feb), pre-monsoon (Mar–May), Southwest monsoon (Jun–Sep) and post-monsoon (Oct–Dec) (as shown in Figs. 3(a)–3(d), and 4(a)–4(d)).

4.1. *Winter season*

The lightning activity is observed to increase with latitude. Figure 3(a) shows the latitudinal variation of lightning flash density over Indian subcontinent (0° to 35°N and 60°E–80°E) for the period of 1998–2005. It can be seen from Fig. 3(a) that the lightning flash density during winter is more over the northern parts of India with a maximum of $0.85\,\text{km}^{-2}\,\text{month}^{-1}$ over a latitudinal belt 31.8°N–32.0°N. This may be attributed to lower tropospheric low pressure system called western disturbances that pass through the northern latitudes of India during winter. Western disturbance systems acquire fresh moisture supply from Arabian Sea and Bay of Bengal. This convection along with supply of moisture is favourable for the genesis of thunderstorms and may be the reason for increased lightning activity over the northern India.

Figure 4(a) shows that the latitudinal variation of lightning activity over the longitudinal belt 80°E–100°E is high in the equatorial region

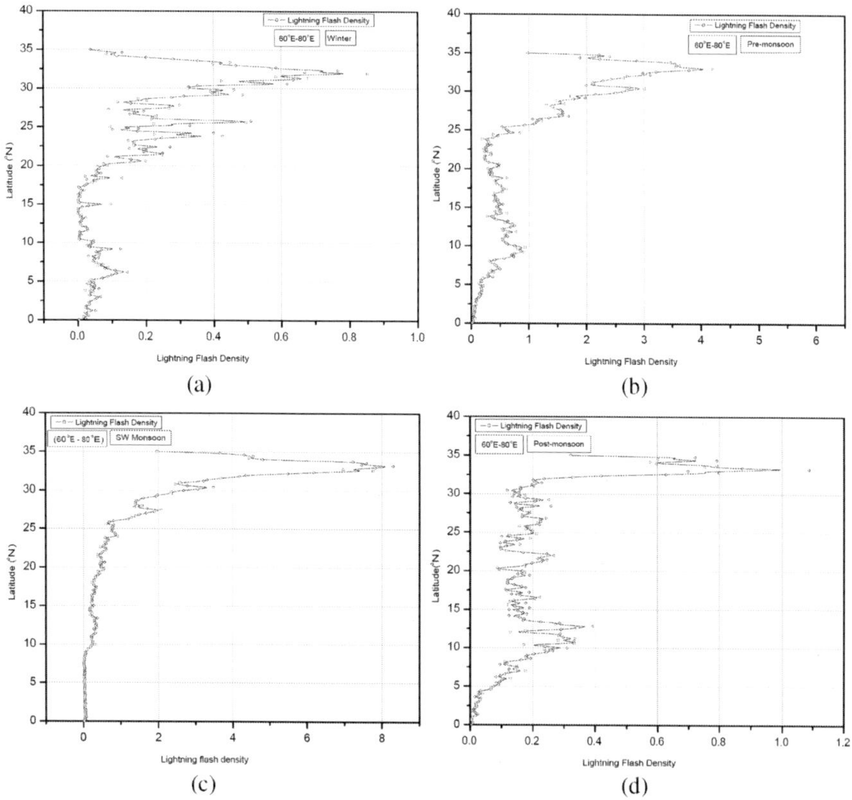

Fig. 3. Latitudinal variation of lightning flash density (Units: km^{-2} month^{-1}) over Indian Subcontinent (0°–35°N and 60°E–80°E) for the period of 1998–2005 during (a) Winter season (Jan–Feb) (b) Pre-monsoon season (Mar–May) (c) Southwest monsoon season (Jun–Sep) and (d) Post-monsoon season (Oct–Dec).

of Equator-5°N with a peak value of 0.58 km^{-2} month^{-1} at 4.0°N–4.2°N. Thereafter, it gradually decreases northward. At north of 20°N, it again increases and attains maxima (at 25°N) of 0.5 km^{-2} month^{-1}.

4.2. *Pre-monsoon season*

The lightning flash density in the region 60°E–80°E depicts secondary maxima of 0.93 km^{-2} month^{-1} at 9.8°N (Fig. 3(b)). The flash density has a decreasing trend up to 24°N, thereafter it increases and principle maxima of 4.2 km^{-2} month^{-1} occurs at 33°N. Figure 4(b) shows the latitudinal variation of lightning flash density over Indian subcontinent (0° to 35°N and 80°E–100°E). It is seen that lightning activity increases steadily at north

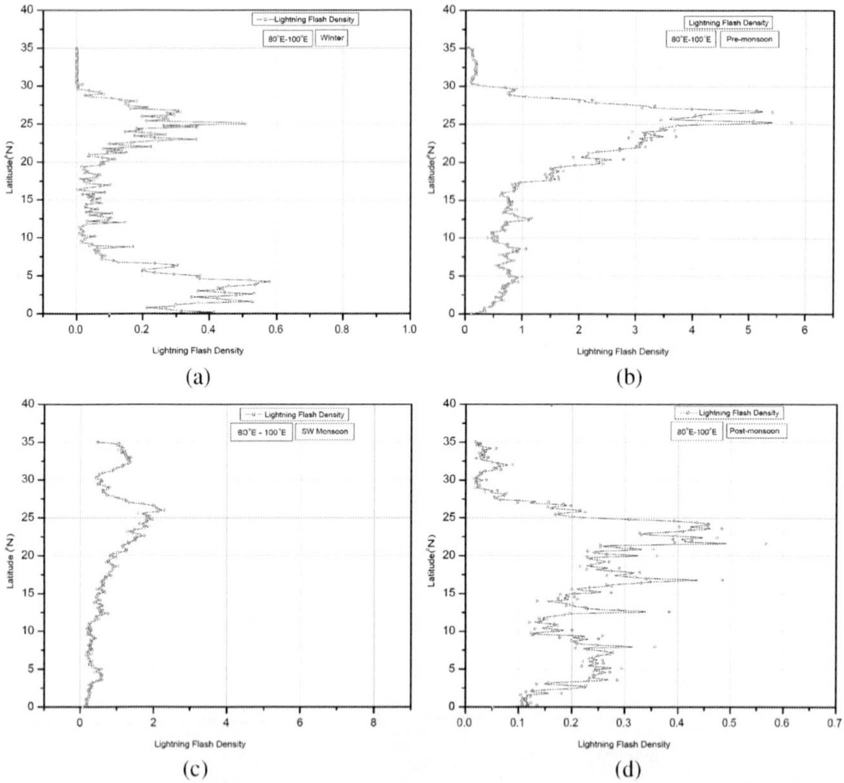

Fig. 4. Latitudinal variation of lightning flash density (Units: km^{-2} month^{-1}) over Indian subcontinent (0°–35°N and 80°E–100°E) for the period of 1998–2005 during (a) Winter season (Jan–Feb) (b) Pre-monsoon season (Mar–May) (c) Southwest monsoon season (Jun–Sep) and (d) Post-monsoon season (Oct–Dec).

of 15°N and peaks in the latitudinal belt of 25.2°N–26.6°N with mean flash density of 4.56 km^{-2} month^{-1}. At north of 28°N, the flash density decreases steeply.

4.3. *Southwest monsoon season*

During the Southwest monsoon season, lightning flash density over the northern latitudinal belts is significant than the peninsular latitudinal belts (Fig. 3(c) and 4(c)). Figure 3(c) shows peak values of 8.3 km^{-2} month^{-1} for the latitudinal belt of 33.0°N–33.2°N (for the region 60°E–80°E). Figure 4(c) denotes peak value of 2.3 km^{-2} month^{-1} for the latitudinal belt of 25.8°N–26.0°N (for the region 80°E–100°E).

In monsoon season, a trough of low pressure at surface runs from head of Bay of Bengal in the east to Ganganagar in the northwest India. This is called the monsoon trough and is a semi-permanent feature of the monsoon circulation. It is the region of convection, which favours the formation of the thunderstorm and lightning in the north.

4.4. *Post-monsoon season*

From Fig. 3(d), it is seen that the lightning flash density is maximum in the northern latitudinal belts with a peak of 1.1 $\text{km}^{-2}\,\text{month}^{-1}$ at 33.2°N. It may be attributed to passing western disturbances over the northern India. The flash density then decreases southward and again increases over the latitudinal belt of 20°N–22°N i.e. central parts of India with peak value of 0.3 $\text{km}^{-2}\,\text{month}^{-1}$ at about 22°N. Figure 4(d) also shows maximum flash density in this region. Kandalgaonkar *et al.* (2005) attributed this maximum flash density to the retreating of Inter-tropical Conversion Zone (ITCZ) and convection associated with large scale convection.

5. Seasonal Variation of Lightning Flash

The seasonal variation of lightning flash counts is presented in Fig. 5. It is observed that, maxima in the seasonal total lightning flash (STLF) counts occur in the monsoon season (47.2%). This result is in agreement with previous studies made by Manohar and Kesarkar (2005). They studied the climatology of thunderstorm activity over Indian region and observed that the maximum numbers of annual total thunderstorms occur in the monsoon season. Figure 5 also depicts the seasonal total lightning flash counts normalized using number of days of the season. It is seen from this figure that, maxima of normalized seasonal total lightning flash (NSTLF) counts occur in pre-monsoon season with 45.9% of total flashes as compared to 39.3% of total flashes in the monsoon season. It may be inferred that the thunderstorms in pre-monsoon season are more intense than in monsoon season.

6. Annual Variation of Lightning Activity

A spatial plot for annual distribution of lightning density over the Indian sub-continent is shown in Fig. 6. The flash density distribution is in qualitative agreement with the climatology of thunderstorm days.

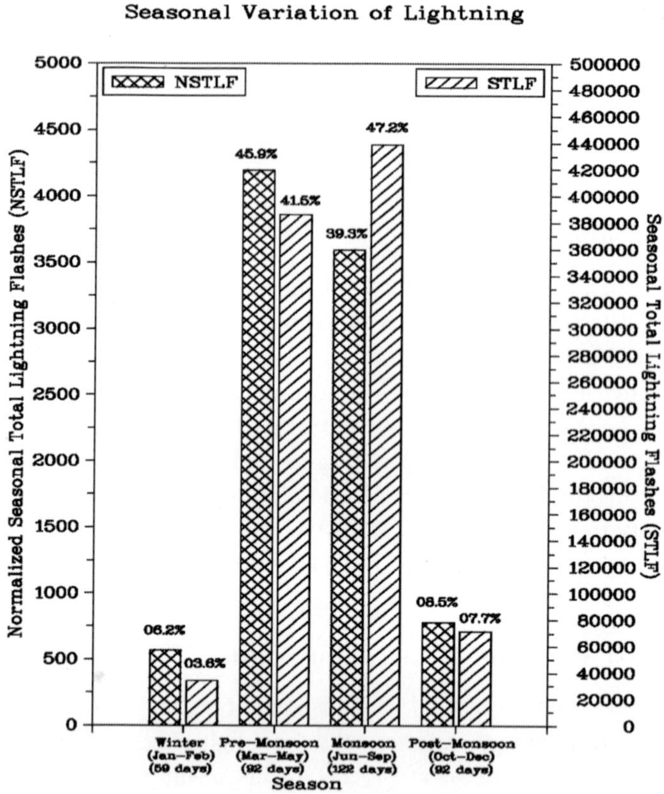

Fig. 5. Seasonal variation of lightning flash counts and normalized lightning flash counts.

The peak lightning flash density of $28.2\,\text{km}^{-2}\,\text{year}^{-1}$ is observed at 33.2°N/74.6°E. Previous studies have reported annual flash density maxima of $33.3\,\text{km}^{-2}\,\text{year}^{-1}$ at 33.75°N/73.25°E over the Indian sub-continent. The high flash density over northern Pakistan is attributed to the Himalayan orographic barrier. It is clear from Fig. 6 that there is a sharp decrease in the lightning flash density to the north of the Himalayan ranges.

Studies over dynamics of convective systems over ocean are limited. The meteorological features that delineates convective activity over ocean and land are: There are no orographic lifting mechanism over oceanic areas, the Sea Surface temperature (SST) is fairly homogeneous hence pressure gradient is uniform over large oceanic areas in the absence of any synoptic situation, Convective Available Potential Energy (CAPE) over the ocean is relatively weak for severe thunderstorm to occur, convection over ocean

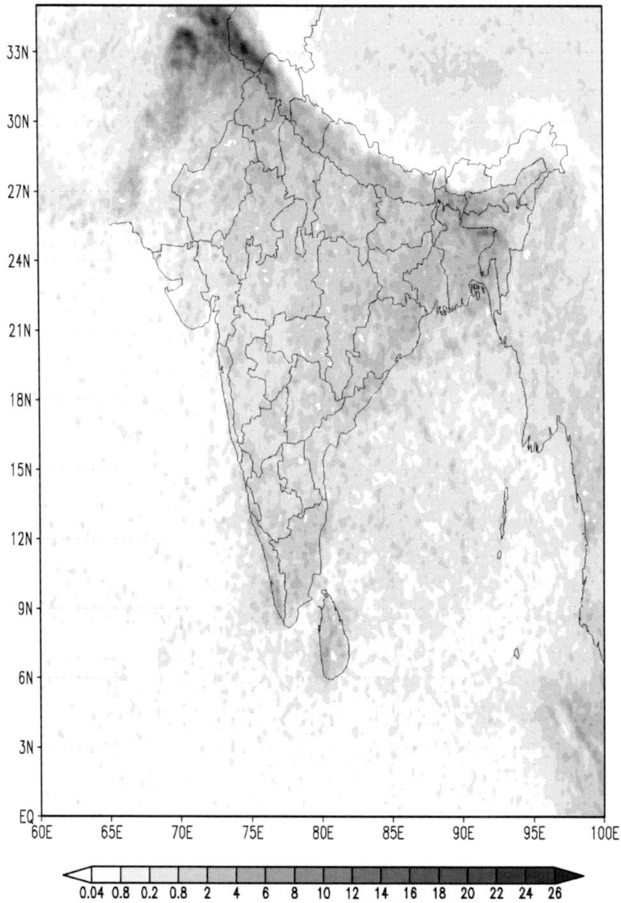

Fig. 6. Annual distribution of lightning activity (Units: km^{-2} year^{-1}) over Indian sub-continent (0° E–35° N and 60° E–100° E) for the period of 1998–2005.

is very weak due to low external trigger energy to overcome Convective Inhibition Energy (CINE), specific heat capacity of the ocean is very high, insolation of short wave energy is mostly reflected back from calm seas, from rough seas the incident energy is reflected at various angles, some of the incoming energy is absorbed up to the mixed layer depth. These features lead to significantly low lightning flash density over Arabian Sea and Bay of Bengal as compared to adjoining land areas (Fig. 6). Most of the lightning activity over the ocean areas is associated with cyclonic circulations.

7. Conclusions

The diurnal variation of maximum flash rate is observed to peak at 11 UTC and minimum at 04 UTC. Thus it clearly shows that maximum thunderstorms activity occur during late afternoon to evening hours. However, it was observed that over the oceanic areas maximum flash rate generally occur during late night and early morning hours.

Seasonal latitudinal variation of lightning flash density during winter season has maxima in the northernmost latitudes of Indian region. During pre-monsoon season the lightning flash density in the region 60°E–80°E increases northward from the equator and peaks up at 9.8°N. It then decreases northward up to 24°N. North of 24°N it increases steeply and peaks at 33°N. In the region 80°E–100°E the flash density peaks in the belt 25.2°N–26.6°N which is the region of intense convective activity over northeast India. In the monsoon season the flash density in the southern parts of India is much low but to the north of 25°N the flash density increases gradually and peaks at 33°N. During monsoon season in the region 80°E–100°E the flash density gradually increase from 20°N to 25°N. Thus along the monsoon trough flash density is more. During the post-monsoon season in the region 60°E–80°E the lightning flash density in the northern latitudes and peaks at 33.2°N. Then it decreases southward. The flash density is more again in the belt 20°N–22°N. The flash density over the peninsular latitudinal belt from 10°N–13°N is also high during post-monsoon season. It has been seen that, the seasonal total flash counts are maximum during the monsoon season and is in agreement with the propagation of Inter-tropical Conversion Zone (ITCZ) over this region.

Seasonal total flash counts during monsoon season are more compared to pre-monsoon season but normalized seasonal total flash counts are observed to be more during pre-monsoon season. Thus thunderstorms during pre-monsoon season are more intense.

Acknowledgments

This work was funded by the Korea Meteorological Administration Research and Development Program under Grant CATER 2006–1101. The authors wish to thank Prof. B.N. Goswami, Director, Indian Institute of Tropical Meteorology for providing facilities at the institute. Authors also thank Dr. R. Krishnan, Head, Climate and Global Modeling Division and Dr. A. SuryaChadra Rao, for providing the encouragement to carry out this work.

References

1. S. V. C. Aiya and B. C. Sonde, *Proc. Inst. Electr. Electron. Eng.* **51** (1963) 1493–1501.
2. R. E. Toracinta, D. J. Cecil, E. J. Zipser and S. W. Nesbitt, *Mon. Weather Rev.* **130** (2002) 802–824.
3. S. S. Kandalgaonkar, M. I. R. Tinmaker, J. R. Kulkarni, A Nath, M. K. Kulkarni and H. K. Trimbake, *J. Geophys. Res.* (2005), doi: 10.1029/2004JD005631.
4. G. K. Manohar, S. S. Kandalgaonkar and M. I. R. Tinmaker, *J. Geophys. Res.* **104** (1999) 4169–4188.
5. D. J. Boccippio, K. Driscoll, W. Koshak, R. Blakeslee, W. Boeck, D. Mach, D. Buechler, H. J. Christian and S. J.Goodman, *J. Atmos: Oceanic Technol.* **17** (2000) 441–458.
6. H. J. Christain, *11th Int. Conf. Atmospheric Electricity*, Natl. Aeronaut. and Space Admin., Guntersville, Al (1999).
7. G. K. Manohar and A. P. Kesarkar, *Mausam* **54** (2003) 819–828.
8. S. S. Kandalgaonkar, M. I. R. Tinmaker, J. R. Kulkarni and A. Nath, *Geophys. Res. Lett.* (2003), doi:10.1029/2003GL018005.
9. G. K. Manohar and A. P. Kesarkar, *Mausam.* **56** (2005) 581–592.

Advances in Geosciences
Vol. 16: Atmospheric Science (2008)
Eds. Jai Ho Oh *et al.*

IMPROVING NEURAL NETWORK MODELS FOR FORECASTING SEASONAL PRECIPITATION IN SOUTHWESTERN IRAN: THE EVALUATION OF OCEANIC–ATMOSPHERIC INDICES

MOHAMMAD J. NAZEMOSADAT
School of Agriculture, Shiraz University, Shiraz, Iran
jafar@shirazu.ac.ir
mjnazemosadat@yahoo.com

PEYMAN SETOODEH and ALI A. SAFAVI
School of Electrical and Computer Engineering,
Shiraz University, Shiraz, Iran

Most parts of southern Iran have frequently experienced extreme climate conditions including drought and floods. Seasonal prediction of dry and wet episodes is essential for competent management of limited water resources during these extreme events. The capability of artificial neural network (ANN) models for forecasting seasonal precipitation was examined for two key stations (Shiraz and Bushehr) in southwestern Iran. Besides precipitation time series, historical records of three climate indicators including the Persian Gulf Sea Surface Temperature (PGSST), North Atlantic Oscillation (NAO), and Southern Oscillation Index (SOI) were used as the predictors. The Auto-Regression with eXtra inputs (ARX) model was firstly used as a linear approach to predict seasonal precipitation one season ahead. The neural network-based extension of the ARX model (NNARX) for nonlinear systems was trained and optimized as the next step. Results confirmed the ability of the employed ARX family models in general and the optimized NNARX in particular for successful prediction of seasonal precipitation.

1. Introduction

Drought and floods have long been particular climate hazards for most parts of Iran, especially in the southern regions (Fig. 1). For the periods that temperature is above normal, demand for water increases for both domestic and irrigation purposes. Severe drought is therefore expected when prolonged shortage of rainfall coincides with hot weather. Considering the ever-increasing population of Iran, which is putting an increasing strain on supplies of fresh water, competent management of water resources

Fig. 1. Iran map.

is needed to mitigate the hazards of drought and flooding. In order for such management to occur, accurate precipitation forecasts are becoming increasingly necessary. Such forecasts are also important for resource management in agriculture, tourism, housing, and industry.

Understanding of the behavior of the atmosphere at seasonal and inter-annual time-scales has improved considerably during the past decade. It has been shown that rainfall variability in Iran has been accentuated by the occurrence of the El Niño-Southern Oscillation (ENSO), the NAO, and the PGSST.[1–7] Compared to other parts of Iran, the impact of the PGSST on surface climate is more meaningful for southwestern portion of the country.[4] These studies have shown that the variations in the PGSST account for about 17% of the total variance of winter rainfall in southwestern Iran. For this region, the influence of ENSO and NAO phenomena on rainfall was, however, found to be less than the corresponding values related to PGSST.

Two sites including Shiraz and Bushehr cities (the capital cities of Fars and Bushehr provinces, respectively) that have the longest record length

Table 1. Seasonal variations of total rainfall (millimeters) and average temperature (degree centigrade) for Shiraz and Bushehr.

	Shiraz		Bushehr	
Season	Rainfall	Temperature	Rainfall	Temperature
Spring	60	22.5	20.2	28.6
Summer	2	28.1	0.3	32.8
Autumn	88	13.6	108.6	21.8
Winter	221	8.7	149	16.4
Annual	371	18.2	278.1	24.9

of precipitation data in the southwestern region were selected for further analysis (Fig. 1). For these cities, precipitation is restricted to cold months generally from November to April. Some climatic attributes of the selected sites are shown in Table 1. Due to the rapid population growth in this region, shortage of drinking as well as irrigation water during dry spells is becoming acute in both provinces. Precipitation forecasting is therefore essential for comprehensive management of available water resources and for reducing the risk of climatic hazards. Although the influence of the ENSO and PGSST on rainfall variability in southern regions has already been investigated, a model has not yet been developed to understand the impact of combined indices on the rainfall.

The complex interactions between climatological phenomena make it very difficult and often intractable to build models based on first principles. Therefore, building black-box models based on the recorded data is a reasonable approach for climate modeling. Among different black-box models, Neural Networks (NNs) have been shown to be powerful computational tools for learning the complex behavior of climatological time series and for modeling complicated systems in general.[8,9] In order to develop an NN model, both the relevant time series and a judicious selection of the network structure are needed. There is a substantial body of literature available on various NN-based models and their applications.[9,10] There is also a rich literature on applying these methods to the rainfall forecasting issues.[11–13]

In this paper, we investigate the possibility of precipitation forecasting in Shiraz and Bushehr by a season ahead using linear and nonlinear black-box models. Three predictors including the SOI, NAO, and PGSST were used as the inputs of a multi-input model. A model-independent order identification method was employed to choose the proper order for the models. The order of the model determines the memory of the model with respect to the input and output time series. Both linear and nonlinear models from the ARX family were then examined and their results were compared.

2. Data Preparation

The time series of precipitation, PGSST, SOI, and NAO data all in monthly time scales were used for the period 1951–1993. While precipitation time series were essentially considered as the predictant, the three other variables were used as the predictors. It should also be mentioned that the historical precipitation data (trained records) were also used as a predictor during the test period. Monthly time series of all variables were then transformed to their seasonal series to construct a vector of 172 data points for each variable. Winter, spring, summer, and autumn were considered as January–March, April–June, July–September, and October–December, respectively.

The precipitation data sets were derived from the website of the Iranian Meteorological Organization (http://www.irimo.ir). The PGSST data were extracted from the Comprehensive Oceanic and Atmospheric Data Set (COADS). The SOI data, which were used as the ENSO indicator, were supplied from the website of the Australian Bureau of Meteorology (http://www.bom.au). Lastly the NAO data were extracted from the website of the National Oceanic and Atmospheric Administration (NOAA), USA. Seasonal time series of all variables were constructed by considering their values related to winter, spring, summer, and autumn sequentially from 1951 to 1993. Models were trained based on the data from 1951–1990 and tested based on the data from 1991–1993.

3. Methodology

3.1. *Building a forecast model*

In order to predict the value of $y(t)$ from time series $y(1), \ldots, y(t-1)$, the previous values of the series as well as values of other time series (or external variables), $u(t)$, that have a causal relationship with the series to be forecasted, are usually chosen as the inputs to the model. In other words, it is required to define a regression vector:

$$x(t) = [y(t-1) \cdots y(t-n_a) u(t-n_k) \cdots u(t-n_b-n_k)]^T. \tag{1}$$

Two key parameters, n_a and n_b, determine the dimension of the regression vector. The vector shows that the value of the time series at time t depends on the n_a previous values of the predictant time series as well as the current and n_b previous values of the predictor time series. Depending on the system, the values of the predictors at time t may not be available to be used as the inputs to the model. In this case the most recent values of the predictors, which are available, are used. The delay term, n_k, in (1) shows that the most recent available values of the predictors belong to time instant $t - n_k$. The parameters n_a and n_b are estimated by order identification, which is explained in the following subsection.

Since the regression vector contains previous values of the output, we are dealing with a dynamic system. There are two ways to model dynamic systems using static artificial neurons, which differ in the way that we build the regression vector. In the first approach, we use feedback in which the output values of the model are sent to the input through multiple delay lines. Introducing feedback through the choice of regressor, makes the NN recurrent. Although feedback helps to build more powerful models, it may cause instability problems. In the second approach, values of the target time series at previous time instants are used as the inputs of the model. Therefore, the need for feedback is avoided by using memory elements. As will be clear in the following, the regression vector of NNARX model is built without feedback and therefore, this model does not face instability problems.

3.2. *Order identification*

As mentioned previously, in a dynamic system the current outputs depend not only on the present inputs but also on the inputs and outputs at the previous time instants. In general, the input-output model of the dynamic system can be represented by a nonlinear mapping:

$$y = f(x), \tag{2}$$

where x is the regression vector (1). In black-box modeling, an approximation of the nonlinear function $f(x)$ should be obtained using the available data. A reasonable estimate of the order of the system under consideration, i.e., the n_a and n_b in (1), should be obtained. In other words, we should reasonably estimate the degree to which previous inputs and outputs have a measurable effect on the current output. If the chosen order is not large enough (i.e., missing data in the input or short length of data), the employed approximation method will not be able to achieve satisfactory learning results. On the other hand, redundancy in the input variables will also increase the estimation error. This means that over-estimating the order or choosing an order which is higher than what is necessary, increases the error. Therefore, it is very important to find the optimal dimension of the lag space or the optimal number of input variables in (1).[14] Different methods such as Akaike Information theoretic Criterion (AIC), Final Prediction Error (FPE) criterion, and Minimum Description Length (MDL) have all been proposed for order estimation.[15] By using these approaches, one needs to train many models with different orders and then compare their respective performances to find out what model order is most appropriate. However, these methods can be employed only after the model has been constructed and validated, which means that in the case of NN-based models, several NNs must be trained and compared to find the best model order.[16]

An adapted order estimation method based on Lipschitz quotients, however, depends only on measured input and output data.[14] In contrast to AIC, FPE, or MDL approaches, these quotients are independent of implementation of any chosen model and can be employed before model building. Since instead of constructing and validating models, Lipschitz quotients just use the input and output data, it is very computationally efficient compared to the mentioned methods. A Lipschitz quotient is defined as the ratio of the distance between two output points and the distance between two input points:[14]

$$L_{ij}^{n_a n_b} = \frac{|y(i) - y(j)|}{|x(i) - x(j)|} \quad (i \neq j). \tag{3}$$

Since the dimension of external variables, $u(t)$, is three (PGSST, SOI, and NAO), for $i \neq j$ and a chosen delay, n_k, the Lipschitz quotient can be rewritten as:[16]

$$L_{ij}^{n_a n_b} = \sqrt{\frac{\|y(i) - y(j)\|^2}{\frac{1}{n_a}\sum_{p=1}^{n_a}\|y(i-p) - y(j-p)\|^2 + \frac{1}{3(n_b+1)}\sum_{q=1}^{n_b}\|u(i-q-n_k) - u(j-q-n_k)\|^2}}, \tag{4}$$

where $\|.\|$ denotes the Euclidean norm. The following index is used to find the optimal order:

$$L^{n_a n_b} = \left[\prod_{k=1}^{p}\sqrt{n_a + 3(n_b+1)}L^{n_a n_b}(k)\right]^{1/p}, \tag{5}$$

where $L^{n_a n_b}(k)$ is the k-th largest Lipschitz quotient among all $L_{ij}^{n_a n_b}$ ($i \neq j$; $i, j = 1, \ldots, N$) with $\sqrt{n_a + 3(n_b + 1)}$ input variables and $p \in [0.01N, 0.02N]$. In order to reduce the sensitivity to measurement noise, the geometric mean of the sequence $L^{n_a n_b}(1), \ldots, L^{n_a n_b}(p)$ with a factor $\sqrt{n_a + 3(n_b + 1)}$ was used to calculate the Lipschitz index in (5).[14]

For order estimation, first, scaled versions of time series in the training set with zero mean and unit variance were obtained. In order to determine a proper lag space structure, a matrix of Lipschitz indices was calculated from (4) and (5), using the scaled time series corresponding to output and external variables.[17] An insufficient lag space structure (i.e., small n_a and n_b) leads to a large Lipschitz index. By increasing n_a and n_b, the Lipschitz index enters a saturated range, in which further expansion of the lag space will not change the Lipschitz index significantly suggesting data redundancy. Therefore, the knee-point of the plot of $L^{n_a n_b}$ vs n_a and n_b, determines the optimal lag space structure (i.e., the optimal order).[14,17]

3.3. *Model evaluation*

After obtaining a reasonable estimate of the order of the system, several black-box modeling techniques were employed and their forecasting capabilities were compared. Two different criteria including the Normalized Root Mean Square Error (NRMSE) and the correlation coefficient (R) between model outputs and targets were used to evaluate the performance of the models. Smaller values of NRMSE and greater values of R indicate better performance of the model. If R is equal to 1, then there is perfect correlation between targets and outputs.

3.4. *Model fitting*

3.4.1. *ARX linear model*

The golden rule in model fitting is to try simple models first. If a linear model is able to capture the underlying dynamics, then there is no point investing time and effort on complicated nonlinear models such as NN-based models. In the simplest case, the relationship between the input, $u(t)$, and the output, $y(t)$, of a system is governed by a linear difference equation:

$$\begin{aligned} y(t) + a_1 y(t-1) + \cdots + a_{n_a} y(t - n_a) \\ = b_0 u(t - n_k) + \cdots + b_{n_b} u(t - n_b - n_k) + e(t). \end{aligned} \tag{6}$$

This relation can be written in the following form to view it as a way of determining the next output value from previous observations:

$$y(t) = -a_1 y(t-1) - \cdots - a_{n_a} y(t-n_a) + b_0 u(t-n_k) + \cdots + b_{n_b} u(t-n_b-n_k) + e(t). \tag{7}$$

For more compact notation we introduce the weight vector:

$$\theta = [-a_1 \cdots - a_{n_a} b_0 \cdots b_{n_b}]^T \tag{8}$$

and the regression vector as in (1):

$$x(t) = [y(t-1) \cdots y(t-n_a) u(t-n_k) \cdots u(t-n_b-n_k)]^T. \tag{9}$$

Then the predictor (7) can be rewritten as:

$$y(t) = \theta^T x(t) = x^T(t)\,\theta. \tag{10}$$

To emphasize that the calculation of $y(t)$ from past data (7) depends on the parameters in θ, the output of the model is written as:

$$\hat{y}(t|\theta) = \theta^T x(t) = x^T(t)\,\theta. \tag{11}$$

The above model is called ARX (Auto Regression with eXtra inputs, also called eXogeneous variables in econometrics).[15]

The ARX model was used to develop a linear relationship between three external variables (PGSST, NAO and SOI) and precipitation. As shown in (10) the previous values from external variables and output time series are also used as inputs. Due to its ability to consider the combined impact of the climate indices on the precipitation, such a linear model could improve predictions compared to models that use only one climate index for rainfall forecasting.

3.4.2. *Neural network extension of ARX*

Among various nonlinear black-box modeling techniques, neural networks because of their learning ability with or without a teacher, deserve special attention. Implementation of a neural network-based extension of the ARX model (NNARX) for nonlinear systems has been proposed in the literature.[17] NNARX provides a nonlinear ARX model of a dynamic system by training a one-hidden-layer NN with the Levenberg-Marquardt algorithm.[17] For this model the predictor will be:

$$\hat{y}(t|\theta) = g(x(t), \theta), \tag{12}$$

where g is a neural network realized function. Instead of feeding back the delayed outputs to the input, the regressor, $x(t)$, is built by using previous elements of the target time series. Therefore, NNARX model is a predictor without feedback and will not face instability problems.

3.5. *Model optimization*

Initially, fully connected network architectures were selected. The trained network was then examined in order to find and remove any superfluous connections. This was done in order to improve the model's generalization ability and also to avoid overfitting the network. The optimal network architecture was obtained by pruning the network using the Optimal Brain Surgeon (OBS) strategy.[17] Although this stratgy is quite slow, it is the safest approach for network pruning. Weights were eliminated one after another, and network retraining was applied after each elimination. The network with the smallest test error and best generalization ability was finally chosen as the optimal network. When the optimal network architecture was found, it was retrained without weight decay.[17]

4. Results

4.1. *Order identification*

The Lipschitz indices were calculated for up to 40 lags for both input and output. Figure 2 shows the results for the case that Shiraz rainfall and PGSST are considered as the output and input, respectively. Since the

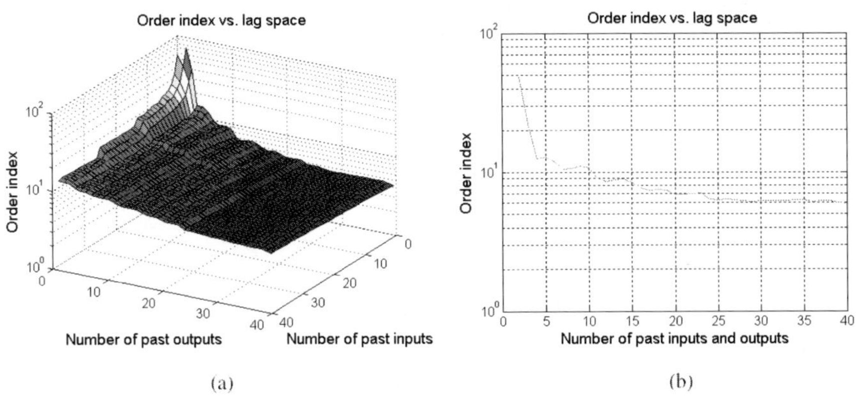

(a) (b)

Fig. 2. (a) Order identification considering PGSST as input and Shiraz rainfall as output. (b) a cross-section of (a), where input and output delays are equal, $n_a = n_b$.

slope of the curve is mostly flat for orders equal to or greater than 7, a seventh-order model was selected for further analysis. This means that, in addition to the values of predictors related to the current season, the values of the predictors and the predictant related to the 7 preceding seasons were considered. Also, Fig. 2(b) suggests that it might be a good idea to examine a fourth-order model for Shiraz station.

4.2. *Forecast models*

Both ARX and NNARX models were used to develop a seventh-order model with three external variables (PGSST, NAO and SOI) and one output (precipitation). For the NNARX a fully connected network architecture with 15 hyperbolic tangent hidden nodes, one linear output node, and 496 parameters (weights and biases) was initially chosen. An ensemble of 30 NNs, using random initial parameters, was then trained for 100 iterations with a small weight decay of 0.001. The NN that attained the lowest NRMSE and the highest R was finally chosen as the best solution. The results are shown in Table 2. The ability of the ARX model to consider the combined effects of the three climate indices on rainfall has considerably improved the reliability of predictions compared to regression models that consider only one index. Also, the statistics in Table 2 show a noticeable improvement in the performance of the NNARX compared to that of the ARX.

The diagram of Lipschitz indices shown in Fig. 2(b), suggests that both 4 and 7 can be considered as the knee points of the curve. Since a lower order model leads to a less complex network architecture, a fourth-order model was also examined for Shiraz rainfall forecasting. As indicated in Table 2, the fourth-order model performs better in terms of both correlation and NRMSE, when compared to the seventh-order model. This may be the

Table 2. Rainfall forecasting results.

			Train		Test	
City	Model structure	Model order	NRMSE	R	NRMSE	R
Shiraz	ARX	7	0.4432	0.815	0.5774	0.674
	NNARX	4	3.0249e-6	1	0.4249	0.87
	NNARX	7	3.0097e-4	1	0.5147	0.789
	Pruned-NNARX	7	1.9945e-5	1	0.3939	0.907
Bushehr	ARX	7	0.5239	0.775	0.5411	0.721
	NNARX	7	2.3971e-5	1	0.4224	0.86

result of two factors: the limited number of data points in the available dataset and data overfitting during the training process. This latter problem can be remedied by pruning the network, which is discussed in the following subsection.

4.3. *Network optimization*

The pruning approach to remove the superfluous weights from a trained network is an efficient way for reducing complexity of NN-based models and addressing the overfitting issue.[17] The trained seventh-order NNARX model was optimized by eliminating the weights one by one and retraining the network after each elimination for maximum 50 iterations. Figure 3(a) shows the training error, test error, and FPE. FPE reflects the prediction-error variance when the model is applied as a predictor to datasets other than the training set (in this case the test dataset).[15] Therefore, FPE is a measure of the generalization ability of the NN, for all intermediate networks.[17] When data overfitting happens, the network performance is very good for the training dataset, but poor for the test dataset. Therefore, it makes sense if we pick the network with minimum test error as the optimal network.

Figure 3(b), which shows a zoomed portion of Fig. 3(a), reveals that the minimum test error occurs when there are only 478 weights and biases left in the network. As indicated in Table 2, the NRMSE is substantially reduced

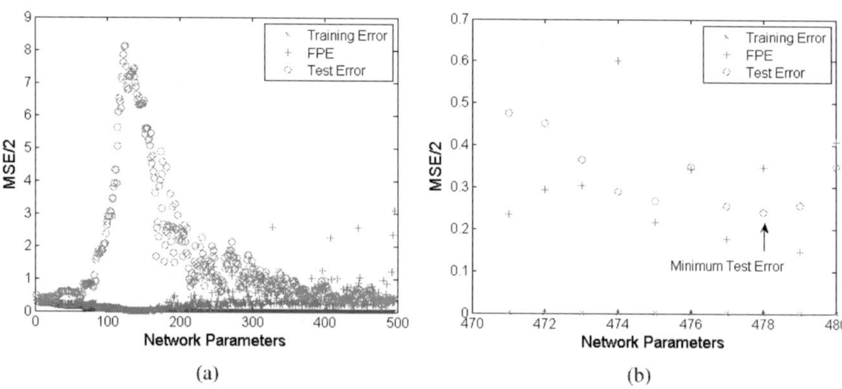

Fig. 3. Network pruning result: (a) training error, test error and final prediction error (FPE) estimate of generalization for each of the intermediate networks, where half of mean square error (MSE/2) was chosen as the performance index. (b) a close-up of (a), which shows the minimum test error.

for this optimized network. Thus, pruning has considerably improved the performance of the network.

It should be noted that this method is not so effective in all situations. By pruning the obtained neural network model for forecasting Bushehr rainfall, we did not get satisfactory results. However, results depend on the local minimum of the error surface in the network parameter space that was achieved during the training process and from which the pruning process starts out.[17] Since Bushehr is adjacent to the sea, local synoptic conditions associated with short-term convective activities play an influential role on the amount and distribution of precipitation there. Forecasting of seasonal rainfall for this coastal site is therefore generally more complicated than that for the interior parts of the country.

5. Concluding Remarks

The forecasting abilities of the different applied models are compared in Table 2. The optimized NNARX model has been very successful in rainfall forecasting with an uncertainty band of 50 millimeters for nine out of twelve simulated predictions. The three exceptional points exceeding the 50 mm band are associated with fall 1992 as well as with the winter and fall seasons of 1993. While the predicted precipitation data were less than the measured values in fall 1992 and winter 1993, the predicted precipitation was higher than the observed value for fall 1993. It is noteworthy that during both the fall of 1992 and 1993, strong and weak El Niño events were, respectively, recorded over the equatorial Pacific Ocean (SOI was −10.0 and −3.8, respectively). Above and below normal precipitation is mostly expected when the SOI is highly negative (El Niño) or positive (La Niña), respectively.

The Shiraz precipitation was above average in fall 1992 (246 mm) but it was astonishingly low during fall 1993 (9.8 mm). The NAO index was positive and almost similar during these two periods, indicating a reduced influence of this index on precipitation variability for these years. The developed model does not explain the reason for such unexpected wet and dry events during these two autumnal warm ENSO events. Besides autumn, the above average precipitation during winter 1993 (El Niño) was also not predicted by our model. It is worth mentioning that the Shiraz rainfall tends to be slightly below average during warm ENSO periods.[6] The increase in winter precipitation during 1993, was, therefore, not foreseeable by our model. Overall, the given results indicate that besides the considered indicators, rainfall variability in southwestern Iran is also modulated by other phenomena. Further research is therefore needed to identify these unknown phenomena and their impact on rainfall variability.

Acknowledgments

The authors would like to thank Karl Wiklund for many helpful suggestions and feedback on this paper.

References

1. M. J. Nazemosadat and I. Cordery, *Int. J. Climatol.* **20** (2000) 47–61.
2. M. J. Nazemosadat and I. Cordery, *26th Nati. and 3rd Int. Hydrology and Water Resources Symp.* (2000), pp. 538–543.
3. M. J. Nazemosadat, *Iranian J. Sci. Technol.* **25** (2001) 611–624.
4. M. J. Nazemosadat, *Drought News Network* **10** (1998) 10–12.
5. A. Payedar, M.Sc. Thesis, Shiraz University, Shiraz, Iran (2001).
6. M. J. Nazemosadat and A. R. Ghasemi, *J. Climate* **17** (2004) 4005–4018.
7. M. J. Nazemosadat, N. Samani, D. A. Barry, and M. Molaii Niko, *Iranian J. Sci. Technol.* **30** (2006) 555–565.
8. L. Sjöberg, Q. Zhang, L. Ljung, A. Benveniste, B. Deylon, P. Y. Glorennec, H. Hjalmarsson and A. Juditsky, *Automatica* **31** (1995) 1691–1724.
9. S. Haykin, *Neural Networks — A Comprehensive Foundation*, 2nd ed. (Prentice-Hall, 1999).
10. X. Ding, S. Canu, and T. Denoeux, in *Neural Networks and their Applications*, ed. J. G. Taylor (John Wiley and Sons, 1996), pp. 153–167.
11. N. Ochiai, H. Suzuki, K. Shinozawa, M. Fujii and N. Sonehara, *IEEE Int. Conf. Neural Networks* (1995), pp. 1182–1187.
12. J. N. K. Liu and R. S. T. Lee, *IEEE Int. Conf. Systems, Man, and Cybernetics* **3** (1999) 429–434.
13. P. Setoodeh, A. A. Safavi, and M. J. Nazemosadat, *Iranian J. Sci. Technol.* **28** (2004) 165–174.
14. X. He and H. Asada, *American Control Conf.* (1993), pp. 2520–2523.
15. L. Ljung, *System Identification — Theory for the User*, 2nd ed. (Prentice-Hall, 1999).
16. L. Sragner and G. Horwath, *IEEE Int. Workshop on Intelligent Data Acquisition and Advanced Computing Systems: Technology and Application* (2003), pp. 266–271.
17. N. Nørgaard, Technical Report 00-E-891, Technical University of Denmark (2000), http://www.iau.dtu.dk/research/control/nnsysid.html.

Advances in Geosciences
Vol. 16: Atmospheric Science (2008)
Eds. Jai Ho Oh *et al.*

LAND SURFACE ENERGY BUDGET OVER THE TIBETAN PLATEAU BASED ON SATELLITE REMOTE SENSING DATA

YUICHIRO OKU* and HIROHIKO ISHIKAWA
Disaster Prevention Research Institute, Kyoto University,
Gokasho, Uji, Kyoto 611-0011, Japan
** oku@storm.dpri.kyoto-u.ac.jp*

This study is an attempt to apply the surface energy flux computation algorithm to geostationary satellite data for the Tibetan Plateau to understand the surface energy budget distribution of sub-plateau scale. It is generally considered that the sensible heat flux accordingly decreases and the latent heat flux exceeds the sensible heat flux after the onset of the monsoons since surface moisture increased gradually. It is apparent from our result that this dramatic change cannot be seen all over the plateau, but only in the southeast part of the plateau. In the northwest part of the plateau, sensible heat flux is always greater than latent heat flux all through the year.

1. Introduction

The Tibetan Plateau is important in the formation of the global climate and atmospheric circulation in terms of both orographic and thermal forcing mechanisms.[1,2,3,4,5] The plateau surface absorbs the larger amount of incoming solar energy than its surrounding area, so that it directly heats the middle of troposphere above it. It is essential to study energy balance between land surface and atmosphere over the plateau. The quantitative estimation of thermal effect is required to understand its influence on regional and global climate. Land surface energy fluxes are important parameters to discuss thermal effect. Numerous researches[6,7] have reported land surface-atmosphere interaction of the plateau surface based on the meteorological measurement data obtained form the Global Energy and Water Cycle Experiment (GEWEX) Asian Monsoon Experiment (GAME) project and Coordinated Enhanced Observing Period (CEOP) Asia-Australia Monsoon Project (CAMP) on the Tibetan Plateau. However these studies contribute to point or patch scale understanding of the energy

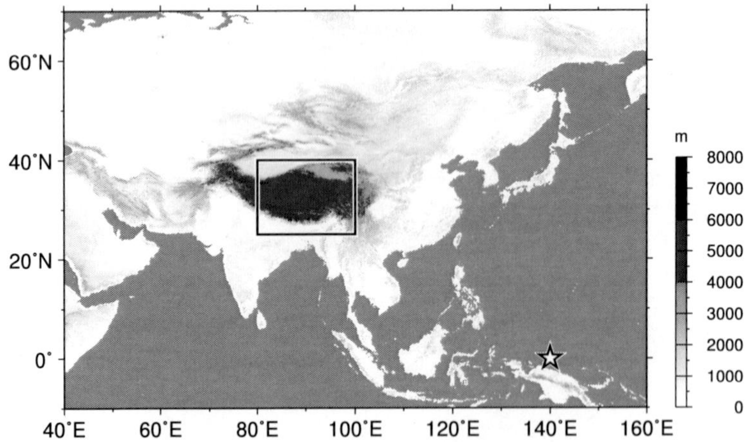

Fig. 1. Location of the study area shown as the dark shaded (above 4,000 m) inside the thick square (25–40° E, 80–100° E). The star indicates a sub-satellite point (equator at 140° E) of GMS-5.

budget in the limited area over the plateau. Patch scale knowledge from observation needs to be integrated with a regional scale understanding of the Tibetan Plateau. Satellite remote sensing offers the possibility of determining regional distributions of surface meteorological properties. Furthermore, a strong diurnal variation of those fluxes exists over the Tibetan Plateau. Therefore, to measure the diurnal cycle, the continuous data stream of a geostationary satellite is required. We have developed retrieval algorithm for land surface energy fluxes (net radiation R_n, soil heat flux G_0, sensible heat flux H and latent heat flux λE) using Geostationary Meteorological Satellite (GMS)-5 data.[8] In this study, the horizontal distribution of these fluxes across the Tibetan Plateau were calculated by using this method every an hour in 1998, and the seasonal variation of the land surface energy budget was presented. The area above 4,000 m inside the thick square depicted in Fig. 1 is defined as the Tibetan Plateau.

2. Data and Retrieval Method

To estimate land surface energy fluxes, Surface Energy Balance System (SEBS) developed for NOAA[9] is applied to GMS data. R_n is estimated as a budget of shortwave radiation, incoming longwave radiation from the air at the surface and outgoing longwave radiation from the ground surface. R_n is estimated as

$$R_n = (1 - \alpha)R_{\mathrm{S}}^{\downarrow} + R_{\mathrm{L}}^{\downarrow} - \varepsilon_{\mathrm{sfc}}\sigma T_{\mathrm{sfc}}^4, \tag{1}$$

where $R_S^{\downarrow}$ is the downward shortwave radiation as a function of the solar constant, the atmospheric transmittance and local time, α is the surface albedo, $R_S^{\downarrow}$ is the downward longwave radiation as a function of the atmospheric temperature, σ is the Stefan–Boltzmann constant, and T_{sfc} is the land surface temperature. G_0 is estimated using R_n as follows:

$$G_0 = R_n[\Gamma_c + (1 - f_c)(\Gamma_s - \Gamma_c)], \tag{2}$$

where Γ_s is the ratio between G_0 and R_n for bare soil and Γ_c is the ratio between G_0 and R_n for surface fully covered with vegetation. The fractional vegetation cover f_c is determined using the normalized difference vegetation index (NDVI). To derive the sensible heat flux H the similarity theory is used. In a complex landscape there is a height, called the blending height, where the impact of the underlying surface diminishes. Although there are various atmospheric profiles over a single pixel of satellite measurement, the blending-height concept allows one to assume a representative profile over the complex landscape in the Tibetan Plateau. At the blending height z, wind speed u and air temperature T_{air} satisfy the general conditions given by Monin–Obukhov similarity theory, which is described as

$$u = \frac{u^*}{k}\left[\ln\left(\frac{z - d_0}{z_{0m}}\right) - \Psi_m\left(\frac{z - d_0}{L}\right) - \Psi_m\left(\frac{z_{0m}}{L}\right)\right], \tag{3}$$

$$T_{\text{sfc}} - T_{\text{air}} = \frac{H}{ku^*\rho C_p}\left[\ln\left(\frac{z - d_0}{z_{0h}}\right) - \Psi_h\left(\frac{z - d_0}{L}\right) - \Psi_h\left(\frac{z_{0h}}{L}\right)\right], \tag{4}$$

where k is von Kármán's constant, ρ is the air density, C_p is the specific heat constant, u^* is the friction velocity, d_0 is the zero-plane displacement height, z_{0m} is the roughness height for momentum transfer, z_{0h} is the roughness height for heat transfer, Ψ_m is the stability correction function for momentum transfer and Ψ_h is the stability correction function for sensible heat transfer. The Monin–Obukhov stability length L is defined as

$$L = -\frac{\rho C_p T_{\text{air}} u^{*3}}{kgH}, \tag{5}$$

where g is the gravitational acceleration. Derivation of the sensible heat flux H using Eqs. (3)–(5) requires only u and T_{air} at z, as well as T_{sfc}. The latent heat flux λE is the residual resulting from an application of the energy budget theorem to the land surface:

$$\lambda E = R_n - G_0 - H. \tag{6}$$

In these fluxes' retrieval process, T_{sfc} estimation and cloud removal have an important part. To obtain the T_{sfc} distribution over the Tibetan Plateau, the split-window technique is applied to the radiances of GMS-5 infrared channels.[10] This technique utilizes the difference in atmospheric absorption at two different wavelengths (11 and 12 μm) in a radiative transfer equation. For cloud-free atmosphere under local thermodynamic equilibrium, the radiative transfer equation is available to derive T_{sfc}. T_{sfc} is determined as follows:

$$T_{sfc} = T_{11} + A(T_{11} - T_{12}) - B - C(1 - \varepsilon) - D\Delta\varepsilon, \tag{7}$$

where T_{11} and T_{12} are GMS-5 brightness temperatures at 11 and 12 μm. Here, $\varepsilon = (\varepsilon_{11} + \varepsilon_{12})/2$ is the average emissivity over both channels, and $\Delta\varepsilon = \varepsilon_{11} - \varepsilon_{12}$ is the spectral variation in emissivity. The coefficients A through D consist functions of atmospheric transmittances and temperatures at both channels. Cloud removal has an important part in these fluxes' retrieval process. In this study, 11 μm brightness temperature with a variable threshold technique is used to remove all kinds of clouds, where as many researchers use a fixed threshold to mainly identify convective clouds.

Figure 2 shows the process flowchart of the method used in this study. Surface parameters (T_{sfc}, α, ε and NDVI) are calculated from satellite

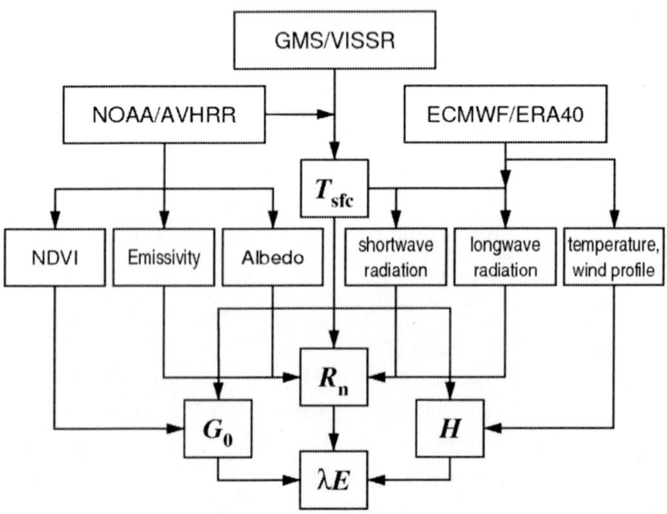

Fig. 2. Flowchart for estimating the surface energy fluxes from the GMS data.

measurements. α, ε and NDVI do not change inter-diurnal scale severely, but change intra-seasonal scale, so in this study, these are estimated from NOAA data as 10 days mean. However, atmospheric parameters (u, $T_{\rm air}$ and z) can be obtained by neither GMS nor NOAA satellite measurements. These are obtained from the 40 year of reanalysis data from the European Centre for Medium-Range Weather Forecasts (ERA-40). Both parameters in grid format with a longitude and latitude resolution 0.1° were interpolated from the original data. The detailed model evaluation has been done and presented in our previous studies.[8,10]

3. Result and Remarks

The seasonal variation of land surface energy fluxes on the Tibetan Plateau was analyzed using the hourly dataset obtained by GMS-5 observations. Spatial distributions of monthly mean the Bowen ratio (the ratio of H to λE) are shown in Fig. 3. Except for monsoon period, monthly mean H is always lager than λE (Bowen ratio is larger than 1) all over the plateau. During monsoon period, λE becomes greater than or equal to H in the southeastern part of the plateau, while H still dominates in the northwestern part of the plateau.

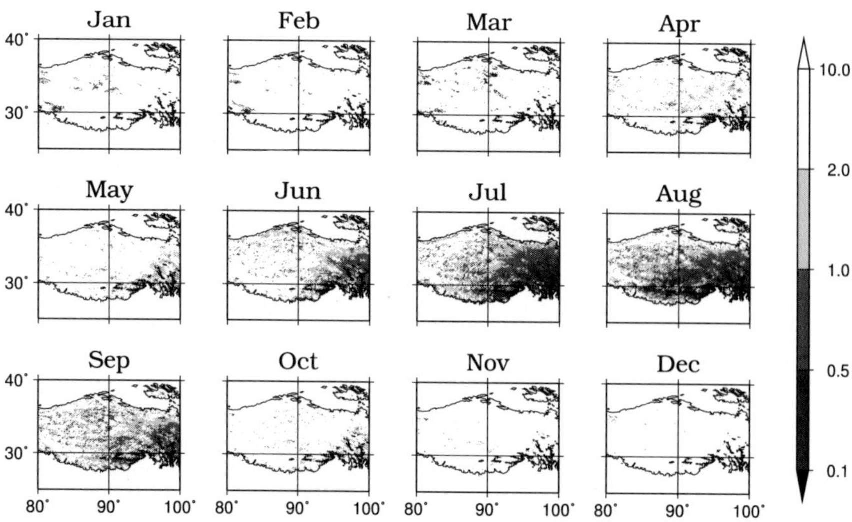

Fig. 3. Spatial distributions of the monthly mean Bowen ratio in 1998 over the Tibetan Plateau.

According to the previous studies based on surface measurement,[6,7] the daytime H is lager than λE during dry (non-monsoon) period. After the onset of the monsoon, H accordingly decreases and then λE exceeds H since surface moisture increased gradually because of increased precipitation. Finally, λE dominates during monsoon period. This dramatic change can be seen only in the surface measurement stations located across the eastern part of the plateau. The western part of the plateau belongs to arid and semi-arid areas. Since there are fewer mounts of precipitation than in the eastern part of the plateau, little water is available for evaporation, even during monsoon period. λE increases temporally when precipitation occurs, but λE might not dominate.[11] In the northern part of the plateau, few stations are available to monitor the surface energy budget, so its behavior is unknown clearly.

The present result is not only consistent with the previous works, but also significant in terms of being obtained from satellite data which are superior to ground-based measurements in spatial representativeness. The monthly mean values of Bowen ratio are thought to be distributed inhomogeneously over the complex landscape over the plateau. For example, the value of Bowen ratio becomes less than 1 over most of the southeastern part of the plateau after the onset of monsoon, but it is apparent from Fig. 3 that the area Bowen ratio greater than 1 locally exists. It might be difficult to detect this heterogeneous distribution by only using surface measurement data.

Utilizing the surface energy fluxes data used in this study, it will be expected that a better quantitative understanding of the interactions between the land surface and the atmosphere over the Tibetan Plateau can be obtained. For example, the heat source Q_1 and the moisture sink Q_2^2 can be obtained from the reanalysis data as an index of the heat budget over the troposphere and across the plateau. In comparing Q_1 and Q_2 with the land surface energy fluxes, the atmospheric heat budget from the ground surface to the atmosphere can be quantified for a diurnal-seasonal time period as the next step of our study. The resent study[12] suggested that the heated air near the surface is transported upward by moist convection with its mixing layer and it rises upper tropospheric temperature over the plateau. H is one of the most important key parameters to develop the mixing layer. Combining our products with results from various numerical simulations, it might be able to reveal the relationship between them.

References

1. T. C. Yeh and Y. X. Gao, *Meteorology of the Qinghai-Xizang (Tibet) Plateau* (Science Press, 1979).
2. M. Yanai, C. Li and Z. Song, *J. Meteorol. Soc. Jap.* **70** (1992) 319.
3. M. Yanai and C. Li, *Mon. Weather Rev.* **122** (1994) 305.
4. A. M. Duan and G. X. Wu, *Clim. Dyn.* **24** (2005) 793.
5. A. M. Duan and G. X. Wu, *J. Climate* **21** (2008) 3149.
6. K. Tanaka, H. Ishikawa, T. Hayashi, I. Tamagawa and Y. Ma, *J. Meteorol. Soc. Jap.* **79** (2001) 505.
7. M. Li, Y. Ma, W. Ma, Z. Hu, H. Ishikawa, Z. Su and F. Sun, *Adv. Atmos. Sci.* **23** (2006) 579.
8. Y. Oku, H. Ishikawa and Z. Su, *J. Appl. Meteorol. Clim.* **46** (2007) 183.
9. Z. Su, *Hydrol. Earth Syst. Sci.* **6** (2002) 85.
10. Y. Oku and H. Ishikawa, *J. Appl. Meteorol.* **43** (2004) 548.
11. J. Xu, S. Haginoya, K. Masuda and R. Suzuki, *J. Meteorol. Soc. Jap.* **83** (2005) 577.
12. K. Taniguchi and T. Koike, *J. Meteorol. Soc. Jap.* **85A** (2007) 271.

Advances in Geosciences
Vol. 16: Atmospheric Science (2008)
Eds. Jai Ho Oh *et al.*

LONG TERM VARIATIONS IN TEMPERATURE IN ASSOCIATION WITH CONVECTIVE AVAILABLE POTENTIAL ENERGY IN THE UPPER TROPOSPHERE USING RADIOSONDE DATA OVER DELHI (28.3°N, 77.1°E) AND KOLKATA (22.3°N, 88.2°E), INDIA

S. K. DHAKA*, R. SAPRA, V. PANWAR, M. KAUR and A. GOEL
Rajdhani College, University of Delhi,
New Delhi 110034, India
**skdhaka@yahoo.com*

T. K. MANDAL and A. R. JAIN
National Physical Laboratory, New Delhi 110012, India

We have shown long term variations in temperature at 100 mb pressure level using daily radiosonde at 1200 Hrs GMT from 1980 to 2006 over Delhi (28.3°N, 77.1°E) and Kolkata (22.3°N, 88.2°E), India. Association of convective available potential energy (CAPE), which is a measure of convective activity, with temperature is investigated. In general, tendency of increase (decrease) in CAPE over a given period (annual to decadal) is seen associated with decrease (increase) in temperature at 100 mb pressure level. Seasonal variation (May–June–July–August) based on daily values of CAPE and temperature in upper troposphere also confirms increase (decrease) in CAPE and decrease (increase) in temperature suggesting temperature control by CAPE. The minimum annual temperature at 100 mb level is observed almost simultaneously with enhanced annual CAPE during northern summer season showing a close association of these two parameters. Rising linear trends emerged in CAPE at Delhi and Kolkata from 1980 to 2006. However, at Kolkata, located at coast of Bay of Bengal, a close inspection shows a clear tendency of decrease in temperature (with increase in CAPE) from 1990 to 2006. CAPE shows variation, which is in accord with summer monsoon northward march appearing first at Kolkata and later at Delhi. Analysis suggests that an increasing convective activity trend in the troposphere can at least partly lead to a cooling trend in the tropopause region due to adiabatic expansion of air mass.

1. Introduction

A long-term change in convective available potential energy (CAPE) is associated with changes in convective activity in a particular region.

Therefore CAPE can act as a potential indicator of climate change. Essentially, it provides an indication of the potential energy available for moist convection. Relationships between CAPE and convective triggering frequency and its association with temperature in the upper troposphere are not clear as discussed by Emanuel *et al.* [1] and Gettleman *et al.* [2]. Gettleman *et al.* [2] have studied the trends in CAPE mainly in the western pacific region. They emphasized that the rising trend in CAPE depends upon the availability of precipitable water in a given region. However, there were a few locations too noted in their study where a decreasing trend in CAPE was observed with less precipitation water. But no effort has been made to evaluate the relationship between CAPE variability and response in temperature field in the upper troposphere. An attempt is made in this paper to examine this relationship. There are numerous studies established a strong relationship between convection and vertically propagating atmospheric wave motions in the upper troposphere and lower stratosphere on a wide temporal and spatial spectrum, especially using VHF radars over Indian and Indonesian equatorial/tropical region [3–7]. Large CAPE (i.e., large convective activity) favors the strong vertical coupling.

In a preliminary study, Dhaka *et al.* [8] have shown seasonal and annual variability in CAPE over Chennai (13.5°N, 80.2°E), Kolkata (22.3°N, 88.2°E), and Delhi (28.3°N, 77.1°E). CAPE seasonal variability showed a dependency upon the monsoon northward movement. In this paper, CAPE association with temperature field in upper troposphere is investigated over Delhi and Kolkata over a range of seasons to annual cycle and then year-to-year variability to decadal scale and longer.

Delhi is an inland station located away from tropics, while Kolkata is a coastal station facing Bay of Bengal — a highly convective region. Some differences in the relationship of CAPE and temperature could be obvious because of their specific locations. The paper is organized by mentioning data in Sec. 2, and results and discussion in Sec. 3. The summary and concluding remarks are given in Sec. 4.

2. Data

We have made use of 27 years of daily radiosonde data at Delhi and Kolkata stations from 1980 to 2006 at 1200 Hrs GMT to show variation of CAPE and temperature at 100 mb pressure level. This level is a representative of tropopause height region and a standard pressure level of measurement for atmosphere parameters. Routine radiosonde temperature data was acquired by India Meteorological Department (IMD). The error in the basic temperature measurements, due to the temperature sensor as given by IMD, is expected to be ~1 K. Radiosonde made use of thermistors for temperature measurements. These thermistors have semiconducting properties and are composed of inorganic oxides and ceramic materials and are coated with a white pigment which has a reflectivity ~0.89 and consequent absorbity of radiation ~0.11. Details of sensors and their comparison with US sensors are discussed by Jain *et al.* [9] and Schmidlin [10]. The monthly mean time series were constructed using daily values. There were very few gaps in data that were filled by averaging adjacent months. The short scale variations (seasonal) were also studied by using the daily values of CAPE and temperature at 100 mb pressure level in the summer months (May–August).

3. Results and Discussion

Composites of temporal variation of monthly mean in temperature at 100 mb pressure level and CAPE are shown in Fig. 1 over a period of 1980 to

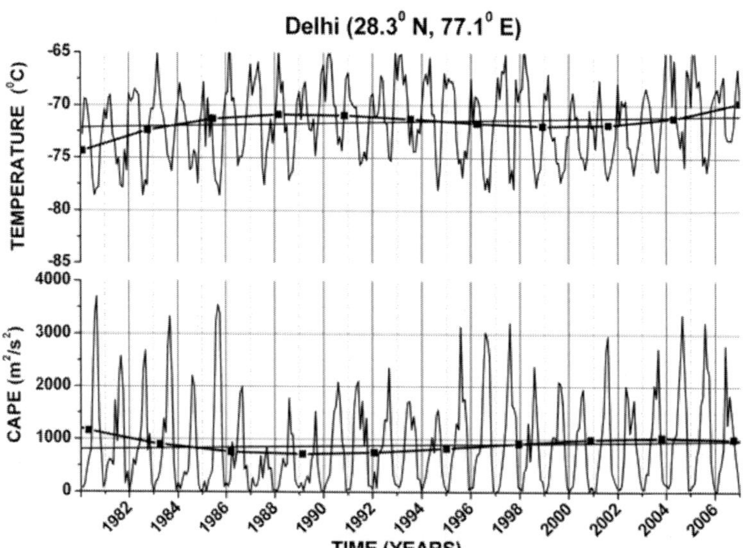

Fig. 1. Monthly mean temperature (top panel) at 100 mb level (~16 km height) and monthly mean CAPE (bottom panel) is plotted over a period 1980–2006 at Delhi. Linear trend (solid straight line) and polynomial fit (solid line with dot) of 3rd order are shown in both parameters.

2006 at Delhi. The top panel shows temperature annual variation between $-65°C$ and $-78°C$ with an average variability of $\sim$12°C. Lowest annual temperature recorded during summer seasons. Year to year variability in temperature is clearly seen. Amount of change in year-to-year annual temperatures is of the order of 3–5°C on lower as well as higher side over a period of 27 years. Change in temperature is quite gradual and seems forming a part of large scale variability. A large scale variation with period greater than 11 year (solar forcing) is noticed. Period of this slow oscillation is about 20–25 years. One can notice that data spanned over 27 years that cover about two and half solar cycles. Solar forcing may also include a component of temperature variation at 100 mb level. However, in this analysis we focus mainly with influence of CAPE on temperature in the tropopause region. Therefore, only analysis of CAPE is shown to determine the control of temperature over a wider temporal scale.

In order to discern trends and large scale variability accurately, linear trends at 95% confidence level and polynomial fit of 3rd order is employed. Polynomial fit reveals a large scale variability of more than 2 decades, which is almost twice the solar cycle period with amplitude of $\sim$2–3°C. On the other hand linear trend in the data seems biased depending upon the time selection of linear fitting. However, taking entire data into consideration, there is a marginal increase in temperature trend $\sim$1°C at 100 mb level. Nevertheless from 1990 to 2006 there is a decreasing tendency in temperature trend.

As mentioned above, one of the objectives of this analysis is to examine control of convective activity (by measuring CAPE) on the temperature. We have shown CAPE variation in the bottom panel of Fig. 1 over a parallel period. CAPE is computed based on assumptions mentioned in Bridge and Frank [11] and William and Renno [12]. Basically, CAPE is equal to the kinetic energy gained by an unmixed air parcel that ascends pseudoadiabatically:

$$CAPE(p') = -R_d \int_{p=LFC(p')}^{p=LZB(p')} (T_{V\text{parcel}} - T_{V\text{env}}) d(\ln p)$$

Here p' denotes the starting level of the parcel. The path of integration is such that the parcel rises dry adiabatically to its lifting condensation level (LCL) and moist pseudoadiabatically above that level. The LFC(p') is the level of free convection for the parcel beginning at p'; the LZB(p') is the level of zero buoyancy; R_d is the (dry) gas constant, and $T_{V\text{parcel}}$,

$T_{V\mathrm{env}}$ are, respectively, the virtual temperatures of the air parcel and the environmental sounding.

Similar to top panel of Fig. 1, polynomial fit and linear trend are plotted in CAPE data shown in bottom panel. Annual variability is quite clear in CAPE, large values occur during summer (July) that correspond to active monsoon season in north India. Year-to-year variation and large scale variability (greater than solar cycle) in CAPE is also quite significant. From 1980 to 1986, CAPE consistently showed annual larger values in comparison to duration from 1987 to 1995. It is quite noteworthy that annual peak in CAPE occurs simultaneously with peak in annual low temperature supporting a strong relationship. Trend in CAPE is also shown with 95% confidence level taking into account the entire data set. A slight increase is visible. However, a decrease in CAPE from 1980 to 1995 is much more prominent. One should note that during this interval there is a visible rise in temperature. In general, prominent trends in time segments of 27 years are such that rise in temperature and decrease in CAPE is common. That indicates a strong coupling of these parameters pointing that large convective activity (CAPE) supports a decrease in temperature at 100 mb level.

The mechanism of decrease in temperature in the upper troposphere could be understood on the basis of that vertical motions get enhanced during summer in comparison to other seasons. There are evidence based on the observations using Indian MST radar during May–June and Equatorial Atmosphere Radar (EAR) that vertical winds get enhanced associated with convective activity [3–7]. As a result adiabatic expansion of air mass at higher levels can lead to cooling. NCEP OLR data confirms that northern India is under highly convective clouds during summer monsoon season (not shown here), and after September low OLR spread is replaced by high OLR values.

Polynomial fit of CAPE is also in consonance with large scale variability (~20–25 years) observed in temperature. The most interesting part is that phase of large scale variability is opposite in both parameters, suggesting that starting from seasonal to annual and larger than decadal variability, CAPE increase (decrease) favors decrease (increase) in temperature in upper troposphere region.

Relationship is also examined between temperature and CAPE at a short scale using daily values covering a summer season. Figure 2 shows plot of temperature and CAPE as an example during May–July 1998. Linear trends are shown with 95% confidence level. Increasing trend in

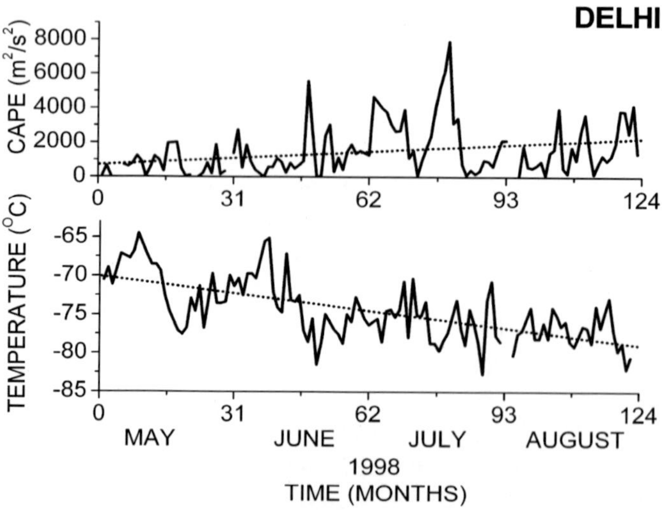

Fig. 2. Daily temperature (bottom panel) data at 100 mb level (∼16 km height) and computed daily CAPE (top panel) is plotted over a period of four months (May–July) 1998 at Delhi. Linear trend (dotted line) is shown in both parameters.

CAPE is noticed clearly with larger values during July in comparison to other months. Enhancement in CAPE during July is in accord with monsoon arrival time over Delhi region. Decreasing trend in temperature is quite significant. On average temperature has shown a decrease of ∼5–6°C from May to August. One can also notice a oscillatory nature with quasi-periodic behavior with a time period of a few days to a few weeks in both parameters. During August CAPE showed almost a stable value as convection gets weaken, temperature also showed similar response; sharp decrease in temperature noticed from May to July.

This has further strengthened the view that CAPE, which is a measure of convection, controls the temperature fluctuations and trends on a wider temporal scale starting from seasons to annual cycle and continued to higher time scale in the upper troposphere.

In this section we show data over Kolkata station. Figure 3 shows CAPE and temperature relationship. Presentation of data is similar to as shown in Fig. 1. In this case too, monthly mean data is used to observe annual variation, year-to-year variability and large scale (more than solar cycle) variation. In addition, relationship between CAPE and temperature is examined at short scale too to reveal control of temperature in the upper troposphere ranging over a wide temporal scale. Annual peak in CAPE

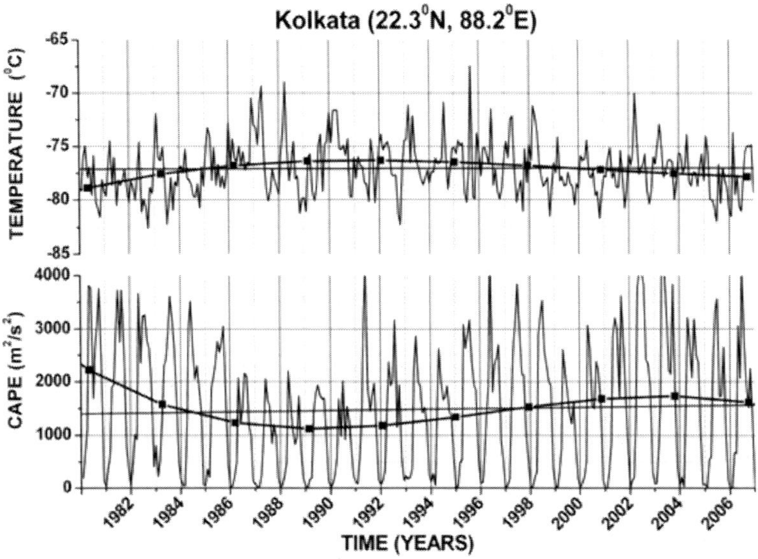

Fig. 3. Monthly mean temperature (top panel) at 100 mb level (~16 km height) and monthly mean CAPE (bottom panel) is plotted over a period 1980–2006 at Kolkata. Linear trend (solid straight line) and polynomial fit (solid line with dots) of 3rd order are shown in both parameters.

is bimodal in nature in some years; however, strong peak occurs during monsoon. Peak in CAPE also appears little early in time than at Delhi as monsoon arrives earlier at this location. Some of the features of CAPE related with seasonal dependency are discussed by Dhaka *et al.* [7]. In Fig. 3 large values of CAPE are also seen during 1980–1986, similar to at Delhi station confirming broadly high reliability of data presentation. Also comparatively lesser CAPE is observed from 1986 to 1995 with an exception around 1992 (strong ENSO year). In general, CAPE is larger at Kolkata than at Delhi.

Linear trend and polynomial fit is employed to discern information about the trend and slow variability in data. Annual peaks in temperature are not as clear as observed over Delhi, this partly seems a bimodal peak in CAPE and convection stays longer at Kolkata than at Delhi. Also, during winter returning monsoon causes rain in this area. Latitudinal variability is noticed in temperature data. Temperature at 100 mb is lower than at Delhi station by about ~5°C. This is due to transitional change in temperature moving towards equator. Taking into account entire data, trends in CAPE show a rise, and the rise is much more rapid after 1990 to 2006. It is well followed by temperature data. Decreasing trend in temperature over Kolkata after 1990 is quite significant.

Polynomial fit indicates a variability of about 20–25 years in temperature as well as in CAPE. Such variability with similar time scale is observed at both stations. This is an indicative that another forcing term exists which is different than solar forcing scale and of longer duration, which needs to be further investigated. As far as the phase of large scale variability in temperature and CAPE is concerned, it confirms similar results as seen over Delhi. Decreasing CAPE favors increase in temperature and vice-versa even at longer time scale (20–25 years).

Relationship of temperature and CAPE at short scale (in a given season) is also checked using daily values at Kolkata and shown in Fig. 4. Both temperature and CAPE is plotted from May to August during 1996 as an example. Good parallel data set in same year is not available at two stations, hence Fig. 4 shows results for year 1996, and whereas on Delhi it is shown in 1998. Trend in CAPE is clearly decreasing since maximum values of CAPE measured during May and convective activity decreases towards August. On the other hand, trend in temperature is rising during above mentioned period.

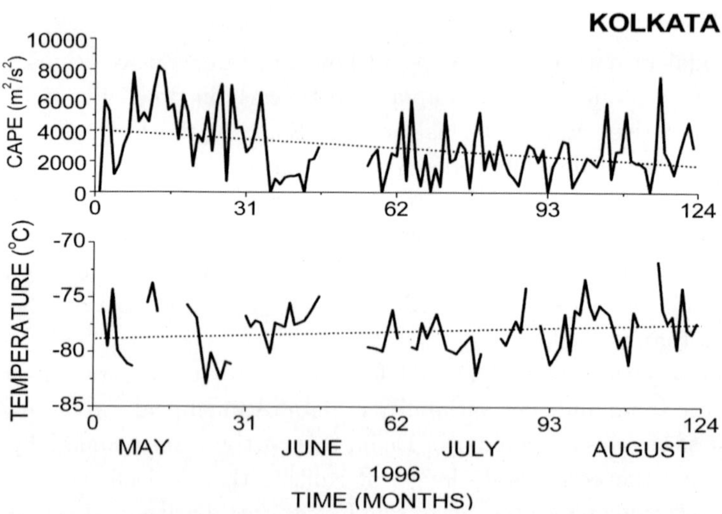

Fig. 4. Daily temperature (bottom panel) data at 100 mb level (~16 km height) and computed daily CAPE (top panel) is plotted over a period of four months (May–July) 1996 at Kolkata. Linear trend (dotted line) is shown in both parameters. Discontinuity in graphs is due to gaps in data.

Though there were a few gaps in temperature near 100 mb level, however, trend is clear. From Figs. 2 and 4 it is quite convincing that increasing convective activity (CAPE) favors a decrease in temperature at all time scales.

4. Summary and Concluding Remarks

In this paper, relationship between CAPE and temperature in the upper troposphere at 100 mb level is analyzed on a wider temporal scale ranging from a short scale (seasonal) to a longer time scale (more than solar cycle) using 27 years data at Delhi and Kolkata. Delhi is an inland station and located at higher latitude than Kolkata. On the other hand Kolkata region is highly dominated by convective activity due to its close vicinity to Bay of Bengal. It is revealed from analysis that a clear annual peak in CAPE and temperature (coldest annual peak) exists simultaneously at Delhi, where as a broader or a bimodal peak is noticeable at Kolkata. Monsoon activity and its northward march and retrieval during summer are mainly responsible components for such features. Increasing (decreasing) trend in CAPE is closely coupled with decreasing (increasing) trend in temperature. At a short time scale during summer such features are quite significant. Similar behavior is confirmed on a longer time scale using several years of data. Linear trends while taking into account entire data set, showed a small rising pattern in CAPE over Kolkata and Delhi, while temperature did not show any conclusive indication. However, at both stations decreasing trend in temperature is noticed during 1990–2006. Most significantly, polynomial fit provides a slow variability, which is longer than solar forcing scale (11 years) and that further confirms variations in CAPE controls the temperature at 100 mb level. In future work, we shall look into separately solar forcing component of temperature variability at 100 mb level. Then we can estimate individual contributions of temperature control from CAPE and solar forcing, respectively.

Acknowledgments

This work is supported by RESPOND-ISRO under CAWSES-India program. We thank Dr. Larry Oolman for providing radiosonde data of India Meteorological Department over Delhi and Kolkata.

References

1. K. A. Emanuel, J. Neelin and C. Bretherton, *Q. J. R. Meteorol. Soc.* **120** (1994) 1111–1143.
2. A. D. Gettleman, D. J. Seidel, M. C. Wheeler and R. J. Ross, *J. Geophys.* **107** (2002) 4606.
3. S. K. Dhaka, B. V. Krishna Murthy, O. P. Nagapal, R. Raghavrao, M. N. Sasi and S. Sundersen, *J. Atmos. Solar Terr. Phys.* **59** (1995) 1189–1202.
4. S. K. Dhaka, M. Takahashi, Y. Kawatani, Y. Shibagaki and S. Fukao, *J. Meteorol. Soc. Jap.* **81** (2003) 1185–1199.
5. S. K. Dhaka, M. K. Yamamoto, Y. Shibagaki, H. Hashiguchi, S. Fukao and H.-Y. Chun, *Geophys. Res. Lett.* **33** (2006) L19805, doi: 0.1029/2006 GL027026.
6. S. K. Dhaka, R. Bhatnagar, Y. Shibagaki, S. Fukao, T. Kozu, V. Malik, S. Malik and A. Dutta, *Adv. Geosci.*, Vol. 9 (WSPC, Singapore, 2006), pp. 167–173.
7. M. K. Masayuki, N. Nishi, T. Horinouchi, M. Niwano, and S. Fukao, *Radio Sci.* **42** (2007) RS3005, doi:10.1029/2006RS003538.
8. S. K. Dhaka, V. Panwar, R. Sapra and R. Bhatnagar, in *Proc. 11th Int. Workshop on Technical and Scientific Aspects of MST Radar* (2007), pp. 720–723.
9. A. R. Jain, S. S. Das, T. K. Mandal and A. P. Mitra, *J. Geophys. Res.* **111** (2006) D07106, doi:10.1029/2005JD005850.
10. F. J. Schmidlin, WMO instruments and observing methods, in WMO International Radiosonde Intercomparison Phase II, 1985, Report 29, Wallapos Island, Virginia, USA, 4 February–15 March 1985, pp. 81–109.
11. J. L. McBridge and W. M. Frank, *J. Atmos. Sci.* **56** (1999) 24–56.
12. E. William and N. Renoo, *Mon. Weather Rev.* **121** (1993) 21–35.

Advances in Geosciences
Vol. 16: Atmospheric Science (2008)
Eds. Jai Ho Oh *et al.*

IMPACT OF CLIMATE CHANGE ON THE EAST ASIA DROUGHTS*

DO-WOO KIM, KI-SEON CHOI, JI-SUN LEE and HI-RYONG BYUN†
Department of Environmental Atmospheric Sciences,
Pukyong National University, 599-1 Daeyeon 3-dong,
Nam-gu, Busan 608–737, South Korea
†*hrbyun@pknu.ac.kr*

We analyzed the changes in precipitation and drought climatology over East Asia by global warming using the daily precipitation data from 14 coupled atmosphere-ocean general circulation model simulations under the SRES A1B scenario at the end of the twenty-first century. The models were consistent in predicting an increase in the mean precipitation over East Asia. However, the increase was less significant in Southeast Asia, and was accompanied by even larger increase in precipitation variability. This predicted precipitation climatology was translated into a change in drought climatology using the effective drought index (EDI). According to the increased precipitation, East Asia tends to be wetter with a decreased frequency and duration of drought. However, because of the enhanced precipitation variability, extreme droughts are predicted to be more frequent, especially over Southeast Asia.

1. Introduction

A report from the Intergovernmental Panel on Climate Change (IPCC, 2001) indicated that East Asian water resources are threatened by an enhanced variability in the precipitation under global warming. However, few studies have tried to estimate quantitatively the hydrological disasters that we should expect. This study has an interest on drought which is caused by precipitation deficits over a prolonged period.

Several modeling studies have shown that over Asian monsoon regions, increases in greenhouse gas concentrations lead not only to an increase in mean precipitation but also to a significant enhancement in precipitation

*This work is supported the Korea Meteorological Administration Research and Development Program under Grant CATER 2006–2306.

variability on sub-seasonal to inter-annual timescales (e.g., Kitoh *et al.*, 1997; Lal *et al.*, 2000; Hu *et al.*, 2000; Kimoto *et al.*, 2005). The significance of these findings was verified by recent studies using the multi-model ensemble (MME) method (Kimoto, 2005; Kripalani *et al.*, 2006; Kripalani *et al.*, 2007). The results showed that the frequency of non-precipitation increases in a way that is similar to the frequency of heavy rainfall (Kimoto *et al.*, 2005). Monsoon excesses and deficiencies are also projected to intensify (Kripalani *et al.*, 2007). However, it remains to be understood how drought patterns are affected by the enhanced variability of precipitation.

Some of the studies on future drought were interested in the global scale dryness revealed by changes in soil moisture conditions (e.g., Wetherald and Manabe, 1999; Wetherald and Manabe, 2002; Manabe *et al.*, 2004) or Palmer Drought Severity Index (Burk *et al.*, 2006). However, the predictions for the magnitude and extent of dryness in East Asia are considerably different in each study. For example, Manabe *et al.* (2004) predicted dryer conditions while Burk *et al.* (2006) predicted wetter conditions. One reason for these different results could be the use of different global circulation models (GCMs). Due to the strong model dependence of the hydrological response to a green house gas increase, different models may predict changes with different signs, even for the same region and the same variable. Moreover, climate sensitivity also differs substantially among models. The MME averaging approach can be very useful in reducing the uncertainties related to model dependence. Although this method is widely used to investigate the future climate, only a single study (Wang, 2005) has attempted to evaluate the likelihood of future drought. That study used 15 state-of-the-art GCMs. Furthermore, there is another limitation. Most of them derived their results by comparing the climatologically averaged values of present-day and future. This simple comparison can only measure the climatological dryness and/or wetness but fail to catch the actual drought change, which is an extreme natural phenomenon with very irregular time scales.

This study examined the impact of greenhouse gas warming on East Asian drought by comparing the projected climate (2081$\sim$2100) in the SRES A1B experiment with the present-day control climate (1981$\sim$2000). The projected daily precipitation data were translated into drought climatology by using the effective drought index (EDI; Byun and Wilhite, 1999), which quantifies the drought intensity in daily time steps. We used the MME average from 14 GCMs and assessed its roughness.

2. Models and Methodology

2.1. *Models*

The 14 GCMs used this study are a part of the IPCC's data archives at the Lawrence Livermore National Laboratory. All of the models begin their integration from the "20th Century Climate in Coupled Model" run, in which the level of anthropogenic forcing is based on historical data from the late 19th century through the 20th century. From the end of the 20C3M run, SRES A1B conditions were imposed and integrated through the year 2100. The SRES A1B assumes rapid economic and population growths that peak mid-century and decline thereafter. Two time periods of twenty years each were chosen for analysis: the late 20th century (1981–2000; hereafter 20C3M) and the late 21st century (2081–2100; hereafter A1B). The analysis based on the 14 GCMs and their MME average (average of 14 GCMs). The following model data were used in this study: CCSM3, CGCM3.1 (T47), CGCM3.1 (T63), CNRM-CM3, CSIRO-Mk3.0, ECHAM5/MPI-OM, FGOALS-g1.0, GFDL-CM2.0, GFDL-CM2.1, GISS-AOM, INM-CM3.0, MIROC3.2 (hires), MIROC3.2 (medres), and MRI-CGCM2.3.2 In addition to the differences in the parametrization of the physical and dynamical processes, the models also differ from each other in their spatial resolution. The resolutions of the 14 models range from coarse (e.g., $4° \times 5°$ for the GISS-AOM) to fine (e.g., $1.4° \times 1.4°$ in CCSM3). To obtain the MME pattern, the original model outputs were converted to the same resolution (2.5° longitude/latitude) by employing the bi-linear interpolation technique.

Several studies have demonstrated that these models are capable of reproducing the temporal and spatial features of the East Asian precipitation climate (e.g., Min *et al.*, 2004; Kitoh and Uchiyama, 2006; Kripalani *et al.*, 2007).

2.2. *Effective drought index*

The effective drought index (EDI) was applied to measure the drought. Unlike many other drought indices, the EDI is calculated with a daily time step.

$$\begin{aligned} \mathrm{EP}_i &= \sum_{n=1}^{i}\left[\left(\sum_{m=1}^{i} P_m\right)\Big/ n\right] \\ \mathrm{DEP} &= \mathrm{EP} - \mathrm{MEP} \end{aligned}, \qquad (1)$$

$$\mathrm{EDI} = \mathrm{DEP}/\mathrm{ST}(\mathrm{DEP}),$$

where P_m is the precipitation m days before and the index i represents the duration of summations in days. Here $i = 365$ is used; that is, the summation is equal to a year which is the most dominant precipitation cycle worldwide. EP is the summed value of daily precipitation with a time dependant reduction function. DEP represents the deviation of EP from MEP (30-year average EP for the calendar date). ST(DEP) denotes the standard deviation of each day's DEP. EDI expresses the standardized deficit or surplus of stored water on a daily basis. It enables one location's drought severity to be compared to that of another location, regardless of climatic differences. The "drought range" of EDI indicates extreme drought at EDI < -2.5, severe drought at $-1.5 >$ EDI > -249 and moderate drought at $-0.7 >$ EDI > -1.49. Near normal conditions are indicated by $0.69 >$ EDI > -0.69. The use of EDI has been tested in several drought studies (e.g., Yamaguchi and Shinoda, 2002; Kang and Byun, 2004; Kim and Byun, 2006; Morid *et al.*, 2006; Smakhtin and Hughes, 2007; Akhtari *et al.*, 2008). When using EDI to explore the changes in drought as a result of future climate scenarios, the calibration factors were set at present-day values.

2.3. *Study area*

The East Asian region is defined as the domain of 10°–50°N and 100°–140°E; it includes eastern China, eastern Mongolia, western Japan, Korea, Vietnam, Laos, Cambodia, eastern Thailand, and northern part of Philippines (Fig. 1). With approximately two billion people, a quarter

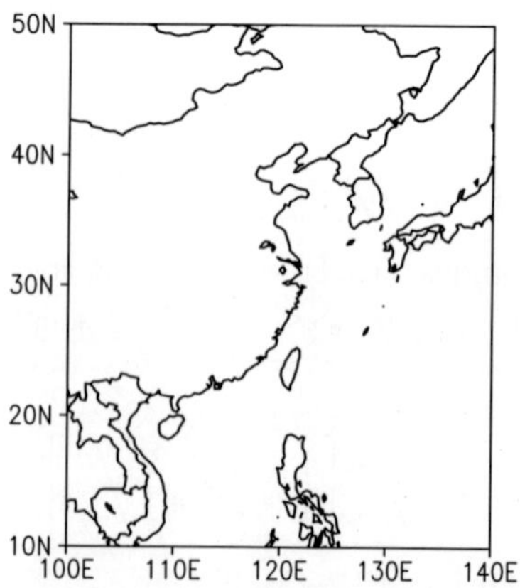

Fig. 1. The region of analysis in this study.

of all the people in the world, East Asia is one of the most heavily populated areas of the world. Exploitation of natural resources associated with rapid urbanization, industrialization, and economic development has led to increasing water demand and pollution. Hence, the water sectors are likely to be most sensitive to climate change-induced impacts in East Asia (IPCC, 2001).

3. Results

3.1. *Future precipitation climatology*

Figure 2 shows the area-averaged (10°–50°N, 100°–140°E) percentage changes (Eq. (2)) of the mean and the standard deviation of precipitation data set for 14 GCMs.

$$(\mathrm{A1B} - \mathrm{20C3M})/\mathrm{20C3M} \times 100. \quad (2)$$

It is confirmed that the increase of precipitations in East Asia, mentioned in the introduction is shown in all the 14 models used in this study. While the minimum increase is projected by GISS-AOM (3.56%), the maximum increase is projected by CCSM3 (13.06%). The increase of the MME average is 7.41%. The increases in the standard deviation vary from 5.26% (INM-CM3.0) to 22.73% (MIROC3.2 (hires)), and the MME average is 12.28%. In 12 out of 14 GCMs, the increase rate in the standard deviation is higher than that in the mean value. In the 4 GCMs (CNRM-CM3

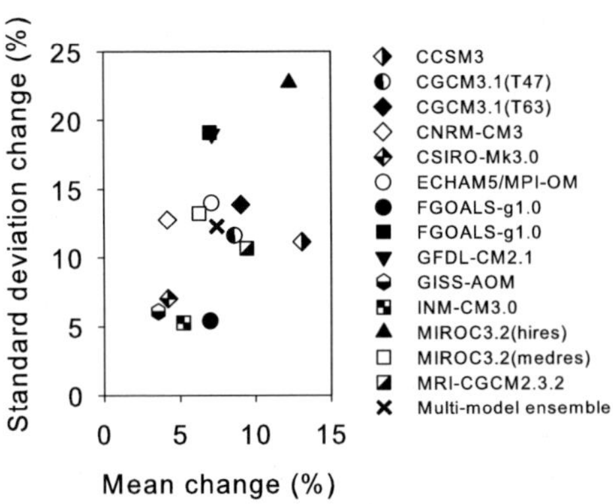

Fig. 2. Scatter plot of the area-averaged (East Asia; 10°–50°N, 100°–140°E) percentage change in the mean and the standard deviation of the precipitation data for 14 GCMs. The percentage change is defined as 100×(A1B-20C3M)/20C3M. A multi-model ensemble value is denoted by the X symbol.

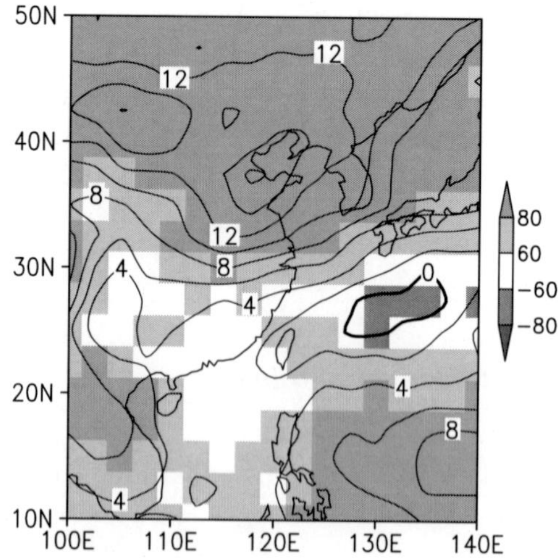

Fig. 3. Multi-model ensemble percentage change in mean precipitation from the 20C3M to the A1B experiments. Shading denotes the consistency level (%) of the 14 models in predicting the direction of change in mean precipitation.

MIROC3.2 (medres) GFDL-CM2.1 GFDL-CM2.0), the increase rate in the standard deviation is over two times that in the mean. These mean that the increase in changeability of precipitation is more distinct than the increase in the average precipitation.

Figure 3 shows the spatial distribution of the percentage changes of the MME average precipitation from the 20C3M to the A1B experiments. To investigate its robustness, consistency level among models is calculated (shading). Here the consistency is defined as a fraction of the number of models with either positive or negative change: that is, the value is +100% if all models have projected an increase, and is −100% if all models have projected a decrease in the future compared to the present. The absolute value of consistency level is always larger than or equal to 50%. In Northeast Asia above 30°N, the increase rate of precipitation is 8∼12%, and the model consistency level is high. On the other hand, in the Southeast Asia below 30°N, the increase rate is relatively low (0∼8%), and their model consistency level is low too

Figure 4(a) shows the time-latitude cross-section of monthly mean precipitation averaged for 100°–140°E for the MME of 20C3M experiments. Figure 4(b) shows the MME percentage changes from the 20C3M to the A1B experiments and the corresponding consistency index (shading). The comparison of these two figures shows that the increase rate of precipitation

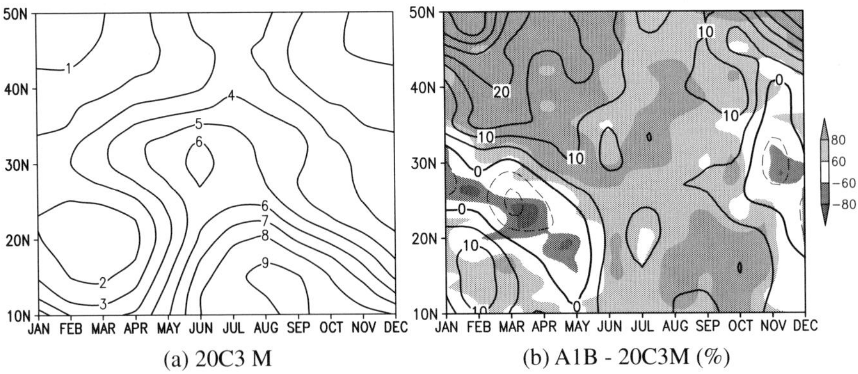

Fig. 4. Latitude-time cross section of monthly mean precipitation averaged for 100° E–140° E. (a) Multi-model ensemble for the 20C3M experiments (unit: mm/day). (b) Multi-model ensemble percentage change from the 20C3M to the A1B experiments. Shading denotes the consistency level (%) of the 14 models in predicting the direction of change in mean precipitation.

is bigger as the latitude is higher and the season is colder. It is confirmed that the precipitation in the entire East Asia increases during summer, implying the strengthening of the East Asian summer monsoon. From the region around 30°N during winter to Southeast Asia during spring, the precipitation decreases a little.

3.2. *Future drought climatology*

Figure 5 shows the difference in the frequency distribution of the EDI values over East Asia for 14 GCMs between the A1B and 20C3M experiments EDI, originally, is a standardized index, follows normal distribution with

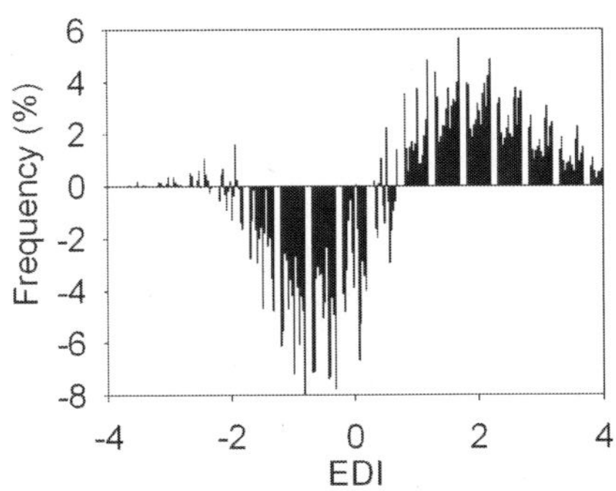

Fig. 5. Difference in frequency distribution of EDI over East Asia for 14 GCMs between the A1B and 20C3M experiments.

zero mean. The frequency decrease of the negative values which means dryness and the frequency increase of positive values which means wetness are definitely shown in Fig. 5. Hence, the center of EDI value frequency distribution moved toward wetness in the A1B experiments. In other words, frequency of wetness increases according to the overall precipitation increase in East Asia, and the frequency of dryness decreases. However, there is seen a special feature here. That is the increase of frequency of extreme values: below -2 and over $+2$. This means that the hydrologic variability increases greatly in East Asia. That is, the frequencies of extreme flood increase at the same time, and the frequencies of extreme drought increase as well.

Figure 6 shows the relationships between the intensity and duration of drought simulated by 14 GCMs in the 20C3M and the A1B experiments. The drought duration herein is the consecutive days of negative EDI, and the drought intensity, the minimum EDI during the duration. The regression coefficients (c) which show the relationship of the two variables are indicated in the bottom of each panel. The percentage change of the number of total drought events from the 20C3M to the A1B experiment is indicated in the top of the A1B panel. As analyzed above, the frequency of drought decreases from 14.0% (GFDL-CM2.1) to 42.6% (CCSM3) according to the increased precipitation in East Asia. In 13 out of 14 models, on the other hand, the slope of the linear regression line is steeper in the A1B experiment than that was in the 20C3M experiment. This means that the droughts in the future have the tendency of intensive precipitation lack although the frequency and duration of drought decrease.

3.2.1. *Drought intensity*

Figure 7 displays the difference in MME total drought days (extreme $< -2.5 <$ severe $< -1.5 <$ moderate < -0.7) between the A1B and the 20C3M experiments. The panels in the left are the spatial distribution (days/decade), and the panels in the right show the latitude-time cross sections of monthly total drought days (days/decade) averaged for 100°E–140°E. In the horizontal spatial distribution, moderate drought days decrease in the entire East Asia, corresponding to the increased mean precipitation. The magnitude of the decrease is large in the Northeast Asia where the model consistency exceeds 80%. Both the magnitude of the decrease and the model consistency level are relatively small in Southeast Asia. The drought days in Northeast Asia show a noticeable decrease during spring while in the Southeast Asia, the decrease is weak. At the latitude of 25°N, there is a weak increase of drought days in April and May.

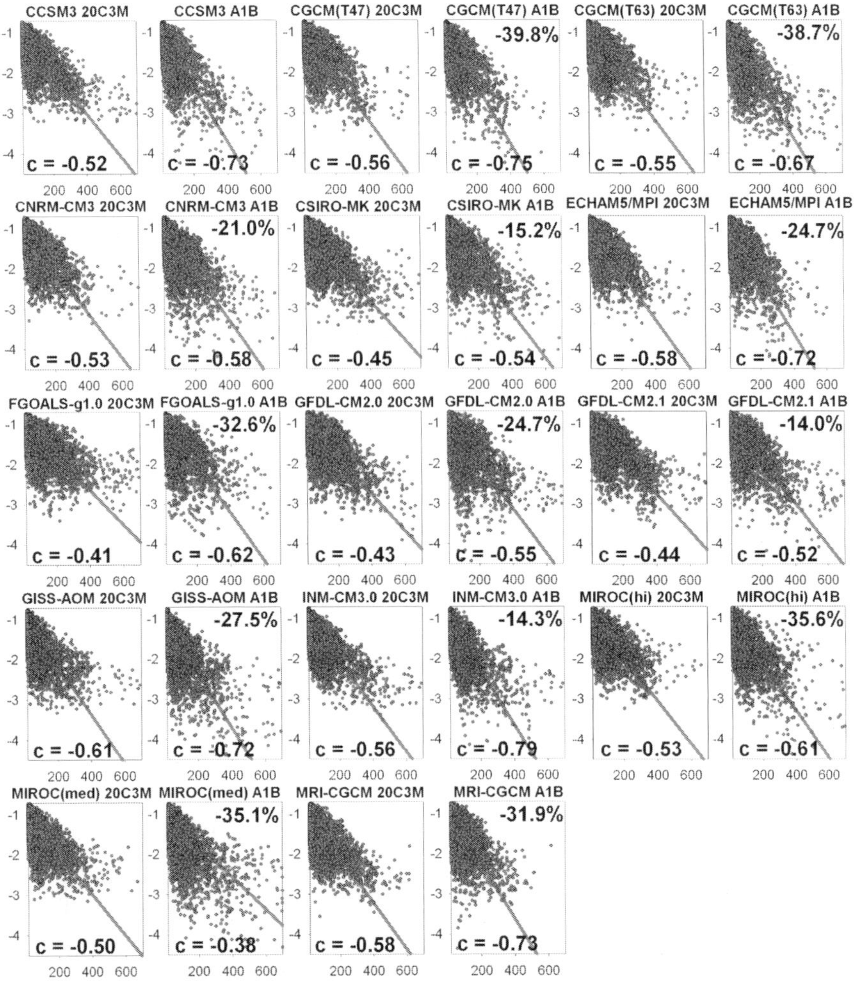

Fig. 6. Scatter diagrams showing the relationship between the intensity and the duration of drought. The left scale shows the EDI values. The bottom scale shows the duration of drought (unit: days). The percentage change of the number of total drought events from the 20C3M to the A1B experiment is shown in the upper part of the A1B panels. The regression coefficient (C) is shown in the bottom part of the panels.

In the difference map of severe drought days, although, decreases in the drought days are still dominant over Northeast Asia, the model consistency levels become lower compared with that in the moderate case. In the some regions of the Southeast Asia, there is shown an increase of drought days, which are predicted by approximately half of the models. These increases

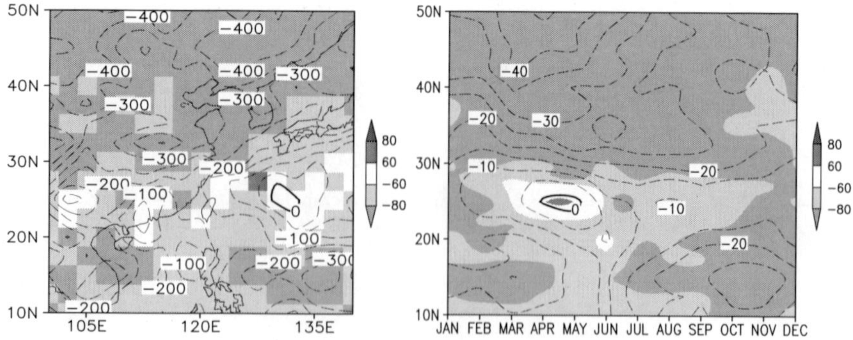

(a) Moderate drought

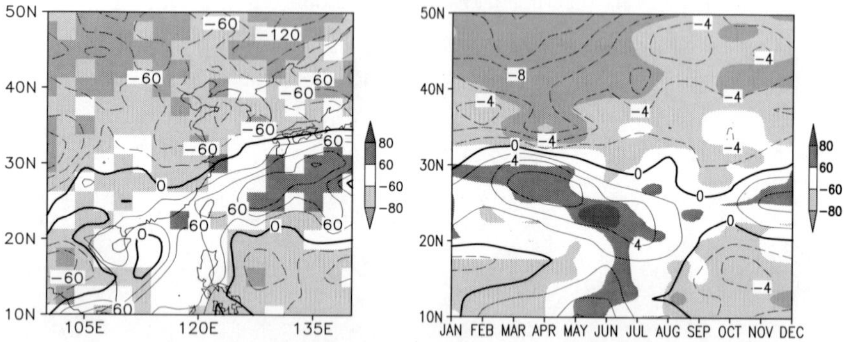

(b) Severe drought

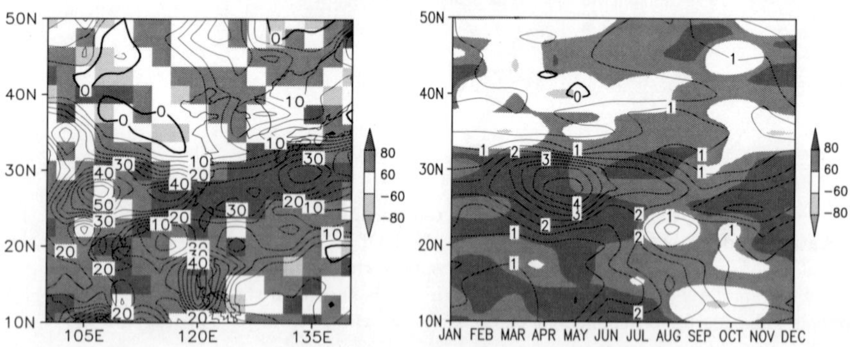

(c) Extreme drought

Fig. 7. The multi-model ensemble differences in (a) moderate, (b) severe, and (c) extreme drought days between the A1B and the 20C3M experiments. The left panels show the horizontal distribution of the difference in the number of total drought days (days/decade). The right panels show the latitude-time cross-section of the difference in the number of monthly total drought days (days/decade) averaged for 100°E–140°E. Shading denotes the consistency level (%) of the 14 models in predicting the direction of change in the number of total drought days.

are noteworthy in the areas of 25°–30°N during winter to spring, and in all Southeast Asia during summer.

Extreme drought days are predicted to increase in all Southeast Asia by the majority of the models. In Northeast Asia, the same reactions appear in some regions. Such an increase of drought days is shown in all seasons in Southeast Asia, especially the regions nearby of 25°–30°N in spring.

4. Conclusions

We analyzed the changes in precipitation and drought climatology over East Asia by global warming using daily precipitation data from 14 coupled atmosphere-ocean general circulation model (GCM) simulations under the SRES A1B scenario at the end of the twenty-first century The effective drought index (EDI) was applied to measure the drought.

The increase in the mean precipitation was predicted in East Asia by the majority of the models. This is outstanding in Northeast Asia, and in Southeast Asia, the magnitudes of the increase and model consistency levels are relatively small. These increases in the mean precipitation are accompanied by a bigger increase in the precipitation variability. The seasonal precipitation increase is the biggest in Northeast Asia during winter, and shown commonly in East Asia during summer. From the regions around 30°N during winter to Southeast Asia during spring, there are weak precipitation decreases.

All GCMs predicted that the frequency of droughts decreases in East Asia according to the increased mean precipitation and the frequency of wetness increases greatly. However, the frequency of extreme drought also projected to increase due to the greatly increased precipitation variability. That is, the frequency and duration of droughts showed a tendency of decrease, but the precipitation shortage is greater during the drought period.

We analyzed the spatial and seasonal changes in the three categories of drought according to the intensity (moderate, severe, and extreme drought). Moderate drought days are predicted to decrease in all East Asia except for regions of 25°–30°N during short period in spring. This weakening of drought is especially noteworthy in Northeast Asia during winter and spring when precipitation increases greatly. However, severe droughts show almost no change in Southeast Asia, but an increasing tendency during spring and summer in those regions. Extreme droughts projected to increase in all

Southeast Asia during all seasons. This strengthening of extreme drought intensity is also shown in some regions of Northeast Asia.

This study found that drought can become severe due to the greatly increased precipitation variability despite the increased mean precipitation in East Asia. According to an IPCC report (2001), it is expected that the average temperature in East Asia will increase more than 4°C in the late 21st century. When considering the increased evaporation in warmer climate, the severity of the future drought in East Asia is deemed to be more severe than the outcome of this study.

Acknowledgments

We acknowledge the modeling groups, the Program for Climate Model Diagnosis and Intercomparison (PCMDI) and the WCRP's Working Group on Coupled Modelling (WGCM) for their roles in making available the WCRP CMIP3 multi-model dataset. Support of this dataset is provided by the Office of Science, U.S. Department of Energy. This work was funded by the Korea Meteorological Administration Research and Development Program under Grant CATER 2006-2306.

References

1. R. Akhtari, S. Morid, M. H. Mahdian and V. Smakhin, Assessment of real interpolation methods for spatial analysis of SPI and EDI drought indices, *Int. J. Climatol.* (2008), doi:10.1002/joc.1691.
2. E. J. Burke, S. J. Brown and N. Christidis, Modeling the recent evolution of global drought and projections for the twenty-first century with the Hadley Centre climate model, *J. Hydrometeor.* **7** (2006) 1113–1125.
3. H. R. Byun and D. A. Wilhite, Objective quantification of drought severity and duration, *J. Climate* **12** (1999) 2747–2756.
4. Z. Z. Hu, M. Latif, E. Roeckner and L. Bengtsson, Intensified Asian summer monsoon and its variability in a coupled model forced by increasing greenhouse gas concentrations, *Geophys. Res. Lett.* **27** (2000) 2681–2684.
5. J. J. McCarthy, O. F. Canziani, N. A. Leary, D. J. Dokken and K. S. White (eds.), *Climate Change 2001: Impacts, Adaptation and Vulnerability* (Cambridge University Press, Cambridge, UK, 2001).
6. K. A. Kang and H. R. Byun, On the developing processes of the climatological drought over the East Asia in 1982, *J. Korean Meteorol. Soc.* **40** (2004) 467–483.
7. Y. W. Kim and H. R. Byun, On the causes of summer droughts in Korea and their return to normal, *J. Korean Meteorol. Soc.* **42** (2006) 237–251.

8. M. Kimoto, Simulated change of the East Asian circulation under global warming scenario, *Geophys. Res. Lett.*, Vol. 32 (2005), doi:10.1029/2005GL023383.
9. M. Kimoto, N. Yasutomi, C. Yokyama and S. Emori, Projected changes in precipitation characteristics around Japan under the global warming, *SOLA* **1** (2005) 85–88.
10. A. Kitoh, S. Yukimoto, A. Noda and T. Montoi, Simulated changes in the Asian summer monsoon at times of increased atmospheric CO_2, *J. Meteorol. Soc. Jap.* **75** (1997) 1019–1031.
11. A. Kitoh and T. Uchiyama, Changes in onset and withdrawal of the East Asian summer rainy season by multi-model global warming experiments, *J. Meteorol. Soc. Jap.* **84** (2006) 247–258.
12. R. H. Kripalani, J. H. Oh, A. Kulkarni, S. S. Sabade and H. S. Chaudhari, South Asian summer monsoon precipitation variability: Coupled climate model simulations and projections under IPCC AR4, *Theor. Appl. Climatol.* **90** (2006) 133–159.
13. R. H. Kripalani, J. H. Oh and H. S. Chaudhari, Response of the East Asian summer monsoon to doubled atmospheric CO_2: Coupled climate models simulations and projections under IPCC AR4, *Theor. Appl. Climatol.* **87** (2007) 1–28.
14. S. Manabe, R. T. Wetherald, P. C. D. Milly, T. L. Delworth and R. J. Stouffer, Century-scale change in water availability: CO_2-quadrupling experiment, *Climatic Change* **64** (2004) 59–76.
15. S. K. Min, E. H. Park and W. T. Kwon, Future projections of East Asian climate change from multi-AOGCM ensembles of IPCC SRES scenario simulations, *J. Meteorol. Soc. Jap.* **82** (2004) 1187–1211.
16. S. Morid, V. Smakhtin and M. Moghaddasi, Comparison of seven meteorological indices for drought monitoring in Iran, *Int. J. Climatol.* **26** (2006) 971–985.
17. M. Lal, G. A. Meehl and J. M. Arblaster, Simulation of Indian summer monsoon rainfall and its intraseasonal variability in the NCAR climate system model, *Reg. Environ. Change* **1** (2000) 163–179.
18. V. U. Smakhtin and D. A. Hughes, Automated estimation and analyses of meteorological drought characteristics from monthly rainfall data, *Environ. Model. Software* **22** (2007) 880–890.
19. G. Wang, Agricultural drought in a future climate: Results from 15 global climate models participating in the IPCC 4th assessment, *Clim. Dyn.* **25** (2005) 739–753.
20. R. T. Wetherald and S. Manabe, Detectability of summer dryness caused by greenhouse warming, *Climatic Change* **43** (1999) 495–511.
21. R. T. Wetherald and S. Manabe, Simulation of hydrologic changes associated with global warming, *J. Geophys. Res.* **107** (2002) 4379–4394.
22. Y. Yamaguchi and M. Shinoda, Soil moisture modeling based on multiyear observations in the Sahel, *J. Appl. Meteorol.* **41** (2002) 1140–1146.

Advances in Geosciences
Vol. 16: Atmospheric Science (2008)
Eds. Jai Ho Oh *et al.*

RELATIONSHIP BETWEEN SUMMER TROPICAL CYCLONE GENESIS FREQUENCY AND WINTER ALEUTIAN LOW OSCILLATION*

KI-SEON CHOI
Korea Meteorological Administration,
45 Gisangcheong-gil, Dongjak-gu,
Seoul 156-720, Republic of Korea

DO-WOO KIM, SU-BIN OH, JI-SUN LEE and HI-RYONG BYUN†
Department of Environmental Atmospheric Sciences,
Pukyong National University,
599-1 Daeyeon 3-dong, Nam-gu,
Busan 608-737, Republic of Korea
† *hrbyvn@pknv.ac.kr*

This research found a high positive correlation between tropical cyclone (TC) genesis frequency during the summer (July–September) and Aleutian low oscillation during the previous winter (November–March) for the last 26 years (1982–2007). In the years of high Aleutian low oscillation, the following characteristics were analyzed. From the previous winter, anomalous pressure pattern like south-low and north-high at the low-level was formed as the center for the regions of near 20°N in the north Western Pacific. And sea ice concentration has been less than the average around the Sea of Okhotsk and the Bering Sea has weakened the Aleutian low in this area. This became one of the causes that reinforced anomalous pressure pattern like south-low and north-high. This anomalous pressure pattern was continued until the summer, and especially reinforced the anomalous easterlies at the mid-latitudes (20°–40°N) in East Asia and contributed to the high TC passage frequency in the East Asian continent.

1. Introduction

Gray (1975) identified six physical parameters that have influence on tropical cyclone (TC) genesis: (i) low-level relative vorticity, (ii) local or planetary vorticity (Coriolis parameter), (iii) inverse of the vertical shear

*This work is supported the Korea Meteorological Administration Research and Development Program under Grant CATER 2006–2306.

of the horizontal wind between the lower and upper troposphere, (iv) ocean thermal energy due to temperatures above 26.8°C to a depth of 60 m, (v) vertical gradient of equivalent potential temperature between the surface and 500 mb (hPa), and (vi) middle-troposphere relative humidity. These gray parameters have been used as common large-scale predictors of statistical models to predict the seasonal genesis of TC, because of their merit that reflects the seasonal characteristics for the activity of TC well (Elsner and Schmertmann, 1993; Solow and Nicholls, 1990; Ward, 1995; DeMaria *et al.*, 2001; Mcdonnell and Holbrook, 2004). In particular, when combined, it has been shown that these parameters broadly identify the geographical and seasonal distribution of tropical cyclogenesis in each of the major ocean basins. This combination of parameters is known as the *seasonal genesis parameter* (e.g., Ryan *et al.*, 1992; Watterson *et al.*, 1995; Royer *et al.*, 1998). As well as statistical models to predict the genesis of TC, these Gray parameters were used mainly in the development of an index (Sall *et al.*, 2006; Venkatesh and Mathew, 2004).

As seen in the above research, the prediction of genesis of TC that used the atmosphere and ocean parameters (Gray parameters) in the main area of the genesis of TC was performed successfully in many researches. However, in addition to TC geneses due to environmental factors of tropical regions, TCs also often occurred due to interactions with many different teleconnection patterns that exist in areas other than the tropical regions. Therefore, it is most important to search for the signal of teleconnection patterns to have an influence on the genesis of TC and to find a clear relationship with the genesis of TC. Chan *et al.* (2001) used the predictors from the teleconnection patterns rather than Gray parameters. In other words, they showed that it's been successful in predicting TC genesis frequency (TCGF) during warm and cold events of the ENSO in 1997 and 1998 when TCGF was abnormal using the predictors such as (i) sea surface temperature anomalies over the central and eastern Pacific, (ii) indices that represent the characteristics of the circulation over Asia and the western Pacific from April of the previous year to March of the current year, (iii) trend of the interannual variations in TC activity (climatology and persistence), and indices that represent the characteristics of the circulation in the Australian region and South Pacific. However, this study has failed to find out what impact the diverse predictors have on the TC genesis. In a statistical model, the accurate prediction of TCGF is important. However, it's more important to have complete understanding of how the predictors are related to TCGF. The accurate prediction about TCGF plays an important role in the statistical model; however, a definite understanding concerning the relationship between the TCGF and the predictors to make up this model is more important. Although having not a high correlation for the TCGF, this study suggests the teleconnection pattern of the previous winter to be able to have an influence on the genesis of TC. In addition, this teleconnection pattern is analyzed which has an influence on the genesis of TC by each process.

Even though the western North Pacific is the ocean basin where TCs are most active in the whole world, the seasonal prediction problem in this area is relatively unexplored. Two well-known typhoon centers in this basin, namely, the Regional Specialized Meteorological Center (RSMC) Tokyo and the Joint Typhoon Warning Center (JTWC) provide only track and intensity forecasts for an individual TC but do not issue the seasonal prediction. That is, in the actual condition that the seasonal prediction of TC is conducted separately in each country in the western North Pacific and the method of the accurate seasonal prediction of TC genesis can become an important issue in these countries. Therefore, to judge whether the teleconnection pattern of the previous winter that was suggested in this study became a good prediction factor in prediction of the summer TCGF would be the ultimate end of this research.

2. Data and Methods

The information about TC activity is obtained from the best track archives of the RSMC, Tokyo Typhoon Center. The data sets consist of names, longitude and latitude positions, minimum surface central pressures, and maximum sustained wind speeds measured every 6 hour for TCs for 1982–2007 (26 years). TCs are generally divided into four grades based on their maximum sustained wind speed (MSWS): tropical depression (TD, MSWS $<$ 34 kts), tropical storm (TS, 34 kts $\leq$ MSWS $\leq$ 47 kts), severe tropical storm (STS, 48 kts $\leq$ MSWS $\leq$ 63 kts), and typhoon (TY, MSWS $\geq$ 64 kts). The present study focuses on these four grades TCs, as well as on extratropical cyclones (ECs) that transitioned from TCs. These are included because ECs do damage in the mid latitudes of East Asia.

We use the geopotential height (gpm) and horizontal wind (m s^{-1}) data reanalyzed by National Centers for Environmental Prediction–National Center for Atmospheric Research (NCEP-NCAR) (Kalnay *et al.*, 1996; Kistler *et al.*, 2001) for 26 years. These NCEP-NCAR reanalysis data have a horizontal and vertical resolutions of 2.5° × 2.5° latitude-longitude and

17 standard pressure levels (hPa), and available for the period of 1948 to the present. The NOAA interpolated Outgoing Longwave Radiation (OLR) data having the same horizontal resolution as NCEP-NCAR reanalysis data, retrieved from the NOAA satellite series, is available from June 1974. However, it includes a missing period from March to December of 1978. Sea ice concentration (hereafter, sea ice) data is also used. The data is divided at intervals of grid of $1° \times 1°$ latitude-longitude, and is made of a percentage of sea ice in each grid box. The data is available from 1982 to the present, and this is the reason that this study has been analyzing the variation of TC from 1982. All of the above reparse data is available in NOAA-CIRES Climate Diagnostics Center (CDC) website (http://www.cdc.noaa.gov).

3. Relationship between Summer TCGF and Winter ALI

Figure 1 represents the genesis frequency of TC that occurred in the western North Pacific in July, August, and September, and for these 3 months collectively and for a time series of Aleutian Low index (ALI) in the previous winter. On each of the three months or the three months collectively, a positive correlation between two variables can be found. The positive correlation has been reinforced by the increase of the moon that indicates a correlation of 0.55 at a 95% confidence level in September. In particular, between the TCGF and ALI for three months, it has a

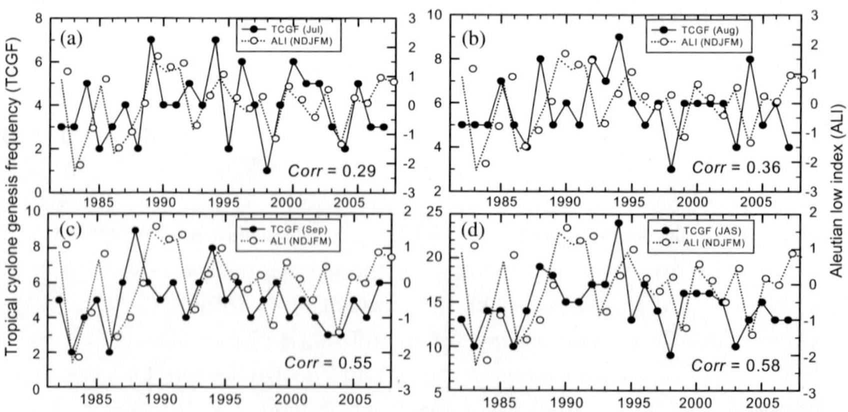

Fig. 1. Relationships between Aleutian low index (ALI) for preceding November–March (NDJFM) and tropical cyclone genesis frequency (TCGF) in the western North Pacific for (a) July, (b) August, (c) September, and (d) all three months. The correlation coefficients are significant at a 95% confidence level.

correlation coefficient of 0.58 (at a 95% confidence level) that is higher than the correlation coefficient in September. This means that the TCGF in the summer increases (decreases) if the Aleutian low in the previous winter is weak (strong).

4. Differences between High ALI Years and Low ALI Years

The TC genesis location makes an effect on TC track. Wang and Chan (2002) and Chen *et al.* (2006) noted that the TC passage tends to draw a curve along the western periphery of the western North Pacific high when more TCs occur at the southeast in the tropical western Pacific. In addition, Ho *et al.* (2005) stressed that more TCs show a tendency to move toward the west or the westnorth without turning when they occurred near the Philippine (in other words, when they occurred in the west of the tropical Western Pacific). Therefore, this study analyzed the difference for average TCPF between the two phases per grid box of $5° \times 5°$ latitude-longitude (Fig. 3(b)). The west (east) side based in about 150°E shows a higher frequency of TC for high (low) ALI years. During high ALI years in particular, it is apparent the passage moved to Korea and Japan through the East China Sea from the east sea of Philippines. In the area of Vietnam and South China, TCGF is not small. Therefore, it is judged that countries located on the coast in the East Asia continent should give a higher attention to the damage caused by the TC occurrences during the high ALI years.

4.1. *Tropical cyclone genesis frequency (TCGF)*

Figure 2 shows monthly differences in TCGF between the two phases. The differences between the two phases totaled 25 for the three months. These differences mean that about 3 TCs averagely occurred more during each summer of high ALI years than during the low ALI years. In the differences between the two phases for each month, as months increased, it can be seen that the differences also is increasing (July: 4 TCs, August: 7 TCs, September: 14 TCs). September in particular, accounted for more than half of the differences for the sum of three months. This had been previously determined because the correlation between the TCGF and ALI had been most highly seen in September. This had been previously determined because the correlation between the TCGF and ALI had been most highly seen in September.

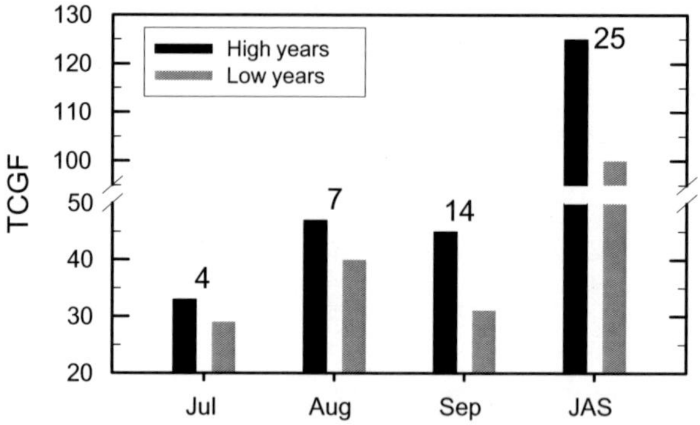

Fig. 2. Monthly variation of TCGF in eight highest ALI years and eight lowest ALI years that selected from ALI in Fig. 1. The number denotes a difference in TCGF between the two phases.

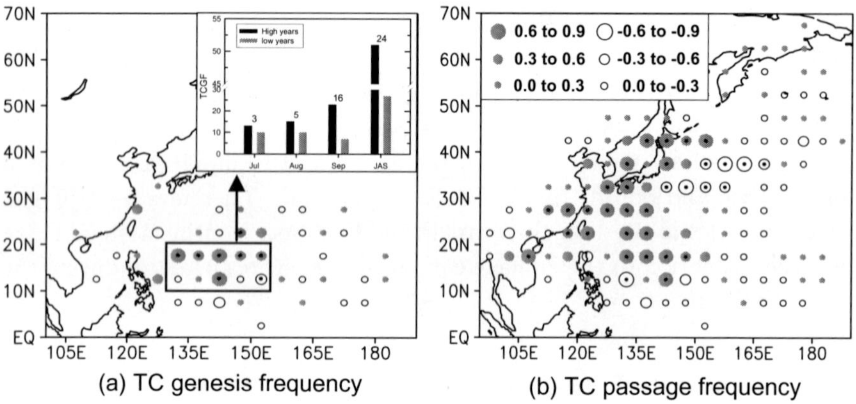

Fig. 3. Differences in (a) TC genesis frequency and (b) TC passage frequency between positive ALI years and negative ALI years. Small solid circle indicates that the differences are significant at a 95% confidence level.

Figure 3(a) shows the characteristics of spatial distribution for TCGF in the summer between the two phases. The largest difference between the two phases appears in the east sea (10°–20°N, 130°–155°E) of the Philippines. In other words, in this area the TCGF during the high ALI years is higher. The open ocean east of the Philippines is known as the western Pacific warm pool (WPWP), which has a dominant influence on the climate or interannual (interdecadal) variabilities of TCGF. Thus, the difference of TCGF in this region between the two phases is examined in Fig. 3(a) (bar graph). In the total of three months, it seems the TCGF of high ALI years is about two times higher than the low ALI years (high ALI years: 51 TCs, low ALI years: 27 TCs). In each month, it shows also the largest difference in September (July: 3 TCs, August: 5 TCs, September: 16 TCs). The differences between the two phases shown in September account for about 2/3 of the difference (24 TCs) in the total of three months in this area.

4.2. *TC passage frequency (TCPF)*

The TC genesis location makes an effect on TC track. Wang and Chan (2002) and Chen *et al.* (2006) noted that the TC passage tends to draw a curve along the western periphery of the western North Pacific high when more TCs occur at the southeast in the tropical western Pacific. In addition, Ho *et al.* (2005) stressed that more TCs show a tendency to move toward the west or the westnorth without turning when they occurred near the Philippine (in other words, when they occurred in the west of the tropical Western Pacific). Therefore, this study analyzed the difference for average TCPF between the two phases per grid box of $5° \times 5°$ latitude-longitude (Fig. 3(b)). The west (east) side based in about 150°E shows a higher frequency of TC for high (low) ALI years. During high ALI years in particular, it is apparent the passage moved to Korea and Japan through the East China Sea from the east sea of Philippines. In the area of Vietnam and South China, TCGF is not small. Therefore, it is judged that countries located on the coast in the East Asia continent should give a higher attention to the damage caused by the TC occurrences during the high ALI years.

4.3. *Environmental conditions*

As shown above, the characteristics of the environmental conditions that brought out the differences of TCPF and TCGF between the two phases were analyzed with regards to the previous winter and summer.

Figure 4 shows the differences between the two phases for 850 hPa geopotential heights and horizontal winds during the previous winter. In the high ALI years, in the north based on about 25°N, the anomalous anticyclonic circulations and in the south, the anomalous cyclonic circulations are strengthened. This is a positive North Pacific Oscillation (NPO), which is similar to the pattern analyzed by Walker and Bliss (1932). They discovered that there is a seesaw pattern in sea level pressure between

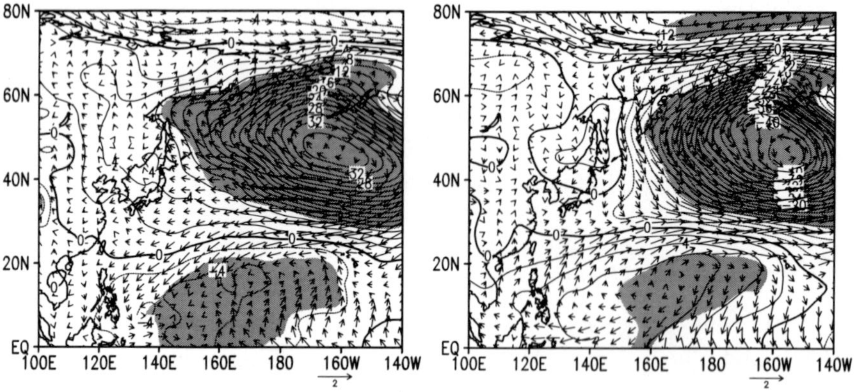

(a) High ALI years (b) Low ALI years

Fig. 4. Composites of anomalies of 850 hPa geopotential heights and horizontal wind for (a) positive ALI years and (b) negative ALI years for NDJFM. Shaded areas are significant at a 95% confidence level. Contour interval is 2 gpm.

high-latitude from eastern Siberia to western Canada during the winter and subtropical low-latitude below 40°N in the Pacific sector like NAO (North Atlantic Oscillation). They defined NPO as this oscillation between north and south regions in the North Pacific. Afterwards, Wang *et al.* (2007) showed that the summer TCGF in the western North Pacific is higher at the time of positive NPO phase than at the time of the negative NPO phase during the previous spring. Therefore, when the anomalous pressure pattern like south-low and north-high (north-low and the south-high) strengthens in the North Pacific during the previous winter and spring, we can know whether summer TCGF becomes high (low). These anomalous pressure patterns were also shown in 500 hPa geopotential height field (not shown). In particular, Aleutian low from the region near the Aleutian in the winter and spring develops climatologically. Thus, the anomalous anticyclonic (cyclonic) circulations that based in this region on high (low) ALI years means that the Aleutian low was weakened (strengthened). Eventually, when the Aleutian low weakens during the previous winter and spring, it can be determined that TCGF in the summer increases.

Meanwhile, Aleutian low can be associated with the sea ice condition in this region. Therefore, in the area near the Sea of Okhotsk and the Bering Sea, the difference in sea ice concentration between the two phases during the winter was analyzed in Fig. 5. In high ALI years, it appears that there was a negative anomaly in most areas except for the north areas of the Aleutian, but it appears the reverse pattern in low ALI years. This means

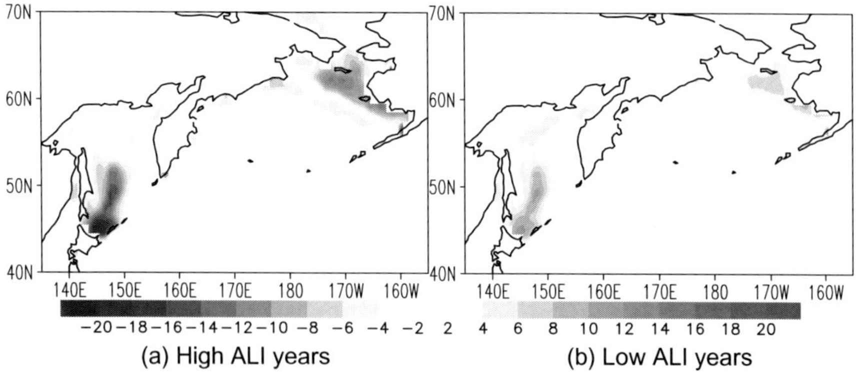

Fig. 5. Same as Fig. 4, but for sea ice concentration.

that the TCGF became high in the summer when sea ice concentration in the area becomes low in the previous winter. Cavalieri and Parkinson (1987), Parkinson (1990), Fang and Wallace (1994) showed that a higher than average existence of sea ice in the North Pacific during the winter and spring can help strengthen the Aleutian low. Fan (2007) also showed that sea ice being more than average in the North Pacific reinforces the Aleutian low in the North Pacific and then forms the anomalous pressure pattern like south-high and north-low, finally plays a negative role in the summer TCGF.

The difference in the environmental conditions between these two phases that appears during the winter can be seen to continue until the summer (Fig. 6). Although the anomalous cyclonic circulations are located near the Sea of Okhotsk and the Bering Sea, based at 25°N in the south of 45°N in high ALI years, the anomalous pressure pattern like south-low and north-high have been still maintained. The opposite pressure pattern shows in low ALI years. These anomalous pressure patterns between the two phases can be more obvious in OLR analysis (Fig. 7). In high ALI years, the OLR anomaly of each negative and positive is formed as the standard about 25°N in the region of South and North, respectively. This means that convection is much more active in the subtropical western Pacific in high ALI years. On the contrary, the positive OLR formed in the direction of northeast to southeast from the southeast region in the tropical western Pacific to the south region in China in low ALI years. In conclusion, it is shown that the anomalous pressure pattern like the south-low and north-high was reinforced since the previous winter in high

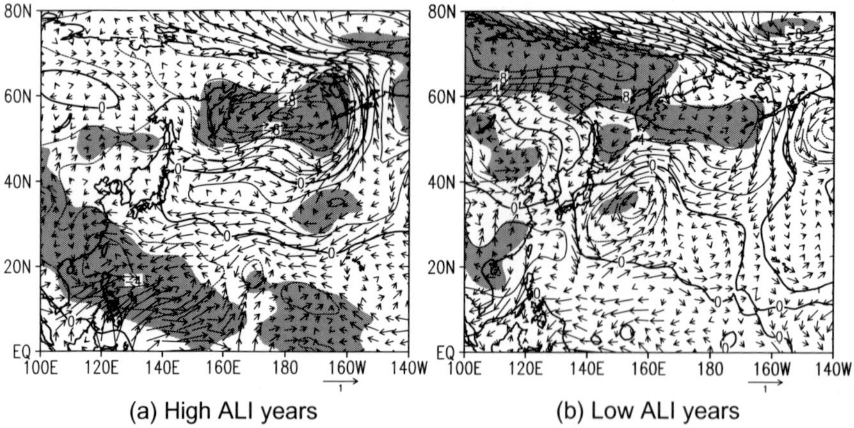

Fig. 6. Same as Fig. 4, but for summer (July–September).

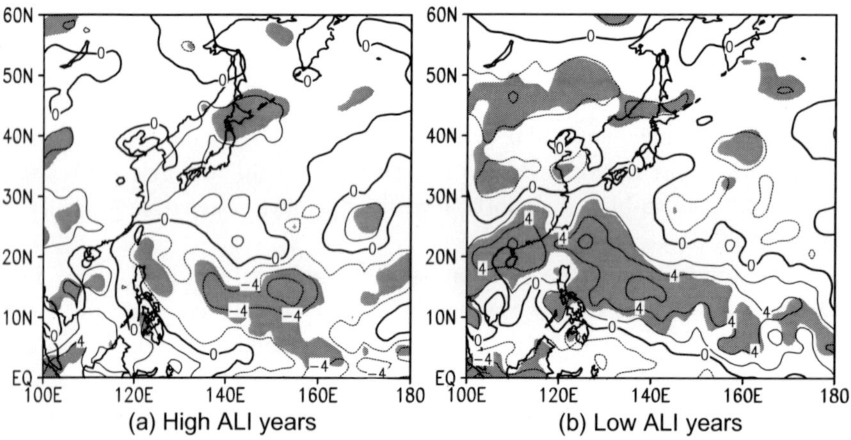

Fig. 7. Same as Fig. 6, but for OLR. Contour interval is 2 Wm^{-2}.

ALI years were given a good environment for the increase of TCGF in the summer.

On the other hand, we can notice that each of the anomalous pressure patterns which were reinforced from the previous winter to the summer during high ALI years and low ALI years (see Fig. 3(b)) caused the difference in TCPF between the two phases. In the case of high ALI years, it can be seen that the anomalous easterlies are especially notable on the latitude of 20°N–40°N from the previous winter to the summer. The steering flows play a role that TC can move easily toward East Asia. However, anomalous westerlies are reinforced near 20°N in the low ALI years, the steering flows can play the role which interfere with TC to move to the East Asian region.

5. Summary and Conclusions

This research has analyzed the relationship between TCGF in the summer and ALI in the winter during last 26 years (1982–2007). The high positive correlation (corr = 0.58) were formed between the two variables, and it showed the highest correlation (0.55) in September among the three months (JAS) in the summer especially.

In high ALI years, it has occurred in many more TCs from eastern sea areas in the Philippines (western Pacific warm pool region). These TCs have mainly gone to Korea and Japan through the East China Sea. Therefore, it has been determined that countries located on the coast of East Asian must give more attention to TCs that occur in high ALI years.

The anomalous pressure pattern like south-high and north-low formed in the western North Pacific since the previous winter continued until the summer and sea ice condition less than the average in the Sea of Okhotsk and the Bering Sea during the previous winter helped to reinforce this pressure pattern. Anomalous easterlies to East Asian region by the anomalous pressure pattern like south-high and north-low become a more frequent cause of the movement of TC to this area. Therefore, the anomalous pressure pattern formed in the western North Pacific and sea ice condition in the Sea of Okhotsk and the Bering Sea during the previous winter will become good predictors of the summer forecast of TCGF in the near future.

This research was analyzed not to consider the effects of ENSO. The condition of sea surface temperature also has considerable affects on the TC genesis. Therefore, the future research is also to be done that considers the effects of El Niño and La Niña as well as the role of the Aleutian low oscillation with regard to the TC genesis.

References

1. D. J. Cavalieri and C. L. Parkinson, On the relationship between atmospheric circulation and fluctuations in the sea ice extents of the Bering and Okhotsk Seas, *J. Geophys. Res.* **92** (1987) 7141–7162.

2. L. T. Chan, J. E. SHI and K. S. Liu, Improvements in the seasonal forecasting of tropical cyclone activity over the western North Pacific, *Weather Forecast.* **16** (2001) 491–498.
3. T. C. Chen, S. Y. Wang and M. C. Yen, Interannual variation of the tropical cyclone activity over the western North Pacific, *J. Climate* **19** (2006) 5709–5720.
4. M. DeMaria, J. A. Knaff and B. H. Connell, A tropical cyclone genesis parameter for the tropical Atlantic, *Weather Forecast.* **16** (2001) 219–233.
5. J. B. Elsner and C. P. Schmertmann, Improving extended range seasonal predictions of intense Atlantic hurricane activity, *Weather Forecast.* **8** (1993) 345–351.
6. K. Fan, North Pacific sea ice cover, a predictor for the western North Pacific typhoon frequency? *Sci. China Ser. D-Earth Sci.* **8** (2007) 1251–1257.
7. Z. Fang and J. M. Wallace, Arctic sea ice variability on a timescale of weeks and its relation to atmospheric forcing, *J. Climate* **7** (1994) 1897–1914.
8. W. M. Gray, Tropical cyclone genesis. Dept. of Atmospheric Science Paper 234, Colorado State University, Fort Collins, CO (1975), p. 121.
9. C. H. Ho, J. H. Kim, H. S. Kim, C. H. Sui and D. Y. Gong, Possible influence of the Antarctic Oscillation on tropical cyclone activity in the western North Pacific, *J. Geophys. Res.* **110** (2005), D19104, doi:10.1029/2005JD005766.
10. E. Kalnay, *et al.*, The NCEP/NCAR 40-Year reanalysis project, *Bull. Am. Meteorol. Soc.* **77** (1996) 437–471.
11. R. Kistler, *et al.*, The NCEP/NCAR 50-year reanalysis, *Bull. Am. Meteorol. Soc.* **82** (2001) 247–267.
12. K. A. McDonnell, and N. J. Holbrook, A Poisson regression model of tropical cyclogenesis for the Australian–Southwest Pacific Ocean Region, *Weather Forecast.* **19** (2004) 440–455.
13. C. L. Parkinson, The impacts of the Siberian high and Aleutian low on the sea-ice cover of the Sea of Okhotsk, *Ann. Glaciol.* **14** (1990) 226–229.
14. J. F. Royer, F. Chauvin, B. Timbal, P. Araspin and D. Grimal, A GCM study of the impact of greenhouse gas increase on the frequency of occurrence of tropical cyclones, *Climatic Change* **38** (1998) 307–343.
15. B. F. Ryan, I. G. Watterson and J. L. Evans, Tropical cyclone frequencies inferred from Gray's yearly genesis parameter: Validation of GCM tropical climates, *Geophys. Res. Lett.* **19** (1992) 1831–1834.
16. S. M. Sall, H. Sauvageot, A. T. Gaye, A. Viltard and P. Felice, A cyclogenesis index for tropical Atlantic off the African coast, *Atmos. Res.* **79** (2006) 123–147.
17. A. Solow and N. Nicholls, The relationship between the Southern Oscillation and tropical cyclone frequency in the Australian region, *J. Climate* **3** (1990) 1097–1101.
18. T. N. Venkatesh and J. Mathew, Prediction of tropical cyclone genesis using a vortex merger index, *Geophys. Res. Lett.* **31** (2004) L04105, doi:10.1029/2003GL019005.

19. G. T. Walker and E. W. Bliss, World Weather, *Mem. Roy. Meteorol. Soc.* 4 (1932) 53–84.
20. B. Wang, and J. C. L. Chan, How strong ENSO events affect tropical storm activity over the western North Pacific, *J. Climate* **15** (2002) 1643–1658.
21. H. J. Wang, J. Q. Sun and K. Fan, Relationship between North Pacific Oscillation and the typhoon and hurricane frequency, *Sci. China Ser. D-Earth Sci.* **50** (2007) 1409–1416.
22. G. F. A. Ward, Prediction of tropical cyclone formation in terms of sea-surface temperature, vorticity and vertical wind shear, *Aust. Meteorol. Mag.* **44** (1995) 61–70.
23. I. G. Watterson, J. L. Evans and B. F. Ryan, Seasonal and interannual variability of tropical cyclogenesis: Diagnostics from large-scale fields, *J. Climate* **8** (1995) 3052–3066.

Advances in Geosciences
Vol. 16: Atmospheric Science (2008)
Eds. Jai Ho Oh *et al.*

GEOMORPHIC INFLUENCES ON THE VARIABILITY OF COASTAL EROSION AND DEPOSITION ON AMBAE ISLAND, VANUATU, CAUSED BY CYCLONE FUNA IN JANUARY 2008

JAMES P. TERRY
Department of Geography, National University of Singapore,
AS2, 1 Arts Link, Kent Ridge, 117570, Singapore
geojpt@nvs.edu.sg

The coastline of north east Ambae Island in Vanuatu experienced both erosion and deposition during the passage of Cyclone Funa through the western South Pacific in January 2008. At Saratamata, the government station for Penama Province located at the extreme easterly point of the island, degraded copra plantations have left coconut trees standing on the intertidal flat, indicating 45 m of recent shoreline retreat. Regional eustatic sea-level rise, localised exposure to easterly swells, storm-wave scouring and a limited sediment supply from narrow fringing reefs combine to explain the erosion problem here. In contrast, volcanic geomorphology influenced the deposition of a large sand body at Torgil to the north. Here a breached volcanic crater provides a relatively sheltered embayment with a flat back-beach environment, features that encouraged sand accumulation during the cyclone.

1. Introduction and Aims

Tropical Cyclone Funa was the second cyclone to form within the South Pacific Ocean in early 2008, affecting northern parts of the island nation of Vanuatu. The storm began life as a tropical depression to the west of Vanuatu, but intensified to cyclone status at 6 am on 16th January as it approached the country on an easterly track. The cyclone intensified to category 2 as it traversed the large island of Santo, then at midnight local time[1] on 17th January it passed approximately 30 km to the north of smaller Ambae Island. At this time 10 minute average winds were 55 knots (102 km/hr). TC Funa's path then turned south easterly as the cyclone passed between the Vanuatu and Fiji archipelagoes.

[1]Vanuatu local time is GMT+11 hours.

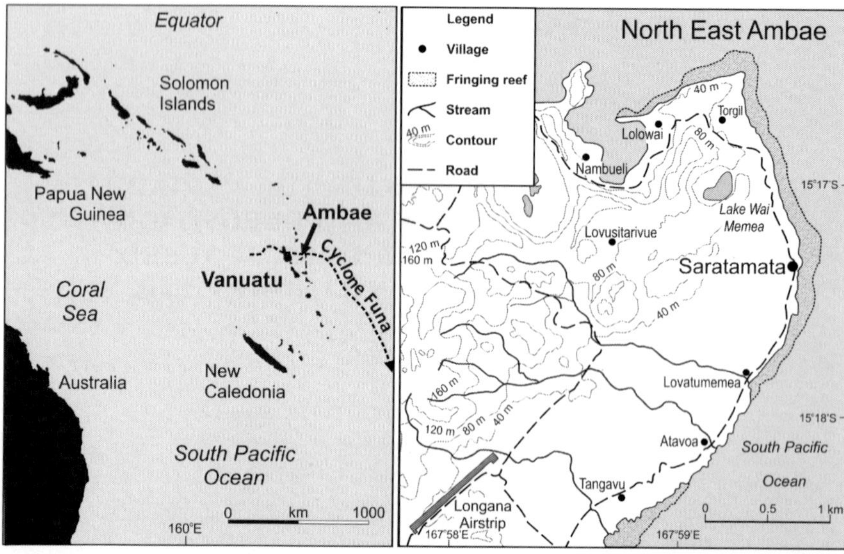

Fig. 1. *Left*: Location of Ambae Island in the south west Pacific and the track of Cyclone Funa over 16–19 January 2009 through the northern islands of the Vanuatu group. *Right*: far north east peninsula of Ambae, showing the drainage pattern, volcanic topography (crater lakes and breached craters at Torgil and Lolowai), the restricted extent of fringing reefs and Saratamata government station at the coast.

On Ambae Island, the close proximity of Cyclone Funa drove large waves ashore in the vicinity of the provincial government station for Penama Province, which is located at Saratamata on the windward north east coast of the island (Fig. 1). Partly in consequence, significant erosion was experienced at this location, although farther north along the coast a large sand body was deposited. The aim of this paper is to describe the variable nature of shoreline erosional/depositional effects felt in the vicinity of Saratamata after Cyclone Funa and to examine the geomorphological influences on this variability. Findings are based on ten days of field observations carried out in southern Santo and eastern Ambae islands in late January 2008, less than two weeks after the passage of the cyclone.

2. Background and Study Area

Ambae Island (15°30'S 167°30'E; 405 km^2) is a strato-volcano situated on the Vanuatu volcanic arc. The volcano is currently active with the last major eruption occurring in 2006. Much of the island is covered in Holocene-age ash and scoria deposits, with subordinate basaltic lava flows [1]. The geology is mostly obscured with either dense rainforest, coconut plantations or subsistence food gardens near villages. The highest elevation is 1,496 m at Mt Lombenben on the rim of the main caldera towards the geographical centre of the island. The structure of the north east peninsula of Ambae is characterised by a group of five small volcanic craters [2], some filled with freshwater lakes and others breached by marine erosion, for example at Torgil and Lolowai (Fig. 1).

The population of Ambae is approximately 10,000, who live subsistence lifestyles in traditional Melanesian villages along the coast as well as in more remote interior highlands. Saratamata is the government station for Penama Province, responsible for neighbouring Pentecost, Ambae and Maewo islands. Geomorphology at Saratamata is characterised by a gently-dipping coastal plain, fringed by narrow beaches comprising mixed terrigenous and coralline sediments. A fringing reef generally less than 100 m wide follows the coast north of Saratamata, but is absent 1.5 km to the south at Atavoa where freshwater discharge from several streams inhibits coral growth. The island has a tropical wet/dry maritime climate, with a hot wet season from November to April and a drier warm season from May to October. Average annual rainfall at Saratamata climate station is 2,406 mm. During the hot season, tropical cyclones may strike. Some 20 to 30 cyclones per decade pass through Vanuatu waters, with 3 to 5 storms causing severe damage [3].

3. Field Observations and Interpretation

Initial observations to the immediate south of Saratamata in January 2008, shortly after Cyclone Funa, indicated that approximately 0.5 km of coastline had experienced recent shoreline retreat. Removal of fine sands was a feature of the beach erosion, leaving a lag deposit of coarse gravels comprising black clasts of volcanic origin (65%) and coral fragments (35%). A number of dead coconut trees, normally growing in copra plantations in the back-beach area, were seen stranded on the intertidal flat (Fig. 2). Vertical erosion scarps with exposed tree roots were also noted at the edge of the coastal plain behind the current high-tide level. Behind this, the coastal section of the road to the airstrip at Longana was partly damaged and strewn with coarse rock debris.

To estimate the maximum amount of shoreline retreat at this location, the distance was measured at low tide from the farthest seaward-standing coconut trees (i.e., dead but *in situ* on the intertidal flat), back to the high-water mark. Trunks of additional coconut trees were seen lying

Fig. 2. *Below left*: Beach erosion and shoreline retreat at Saratamata, observed after Cyclone Funa in January 2008 (**A** on map). *Above left*: Deposition of a large body of fine sand in the sheltered embayment at Torgil by the same storm (**B** on map); the rocky promontory in the distance is the rim of a breached volcanic crater that forms the north east point of Ambae Island. *Right*: Locations of erosional and depositional sites and photographs A and B.

horizontally farther out on the intertidal flat, but these were ignored as they may have been transported to their present positions subsequent to falling over and therefore could not be confirmed as still being *in situ*. Measurements thus taken indicated up to 45 m of shoreline retreat had occurred at Saratamata. Interviews with local villagers gave convincing accounts that Cyclone Funa was mainly responsible for the coastal erosion observed here, based on them having witnessed storm waves estimated to be up to two metres in height that were driven onshore by Funa's high winds. However, several individuals were more circumspect and were keen to point out that the farthest coconut trees on the intertidal flat had already been stranded seaward of the shoreline prior to Cyclone Funa, although it could not be established exactly when these trees had last been growing.

In complete contrast to the degraded coastline at Saratamata, field investigation 1.5 km north at Torgil revealed that a large body of fine sand had been deposited during Cyclone Funa (Fig. 2). The rather dramatic variation from beach scouring to depositional effects, seen over a relatively short distance (1 km) along the coast of north east Ambae, needs careful interpretation. Any explanation must take into account the local coastal geomorphology, regional trends in sea level change, as well as any localised effects produced by Cyclone Funa.

It is well known from coastal research elsewhere in the Pacific Islands that cyclone-generated wave action can be constructive as well as destructive. On the surface of wide and well-developed reefs, large storm waves are able to throw up gravel ramparts [4, 5], the subsequent landward migration of which during quiescent periods nourishes coastlines with biogenic sediments [6, 7]. On Santo Island lying just 50 km to the west of Ambae, field investigation of the geomorphic effects of Cyclone Funa showed that spits and ridges comprising coarse coral gravels had indeed been built up in several places along the south coast (Fig. 3), although these features were constructed only where broad fringing reefs exist. Compared to the wide fringing reefs in parts of southern Santo, the intermittent coral reefs around Ambae are rather narrow and poorly developed. As a result they were not able to supply sufficient coral detritus required for gravel ridge formation at Saratamata. Also, since Cyclone Funa was not a storm event of extreme magnitude, wave power was insufficient to dredge up coarse talus from fore-reef sources deeper offshore, which may occur during very violent storms, i.e., intensity categories 4–5.

Another important consideration is the direction in which Cyclone Funa tracked near Ambae. Since the storm passed in an easterly direction close to the north east point of the island, this means that wind circulation in a counter-clockwise direction around the centre of the storm would have driven waves towards the Saratamata coastline on the south side of the track. Waves therefore approached from the east and later the south east as the cyclone centre moved by. Saratamata is the most exposed location facing this direction of wave attack. The oblique south easterly wave advance with respect to the NNW-SSE orientation of the shoreline would also have set up strong northward littoral drift, mobilising available beach sand away from this exposed point.

Yet, why did the Saratamata coastline not benefit in a similar way by sediment supply from the area to the south? Two geomorphic factors are responsible. First, there is a 1 km long gap in the fringing reef at

Fig. 3. *Above*: New gravel spit comprising coarse coral fragments, thrown up during Cyclone Funa at Naoneban Point on the southern coast of Santo Island. *Below*: A typical steep stream channel on Ambae with many pools. The higher sections of the rocky bed, exposed here at low flow, were swept clean of sediments during Cyclone Funa.

Atavoa (Fig. 2), so reef-derived sediments were not available for transport northwards towards Saratamata. Secondly, although the heavy rainfall delivered by Cyclone Funa (76 mm and 35 mm on 16th and 17th January, respectively) did cause increased discharges in the island's streams, fluvial bedload materials are not readily delivered to the coast. This is owing to the steep step-and-pool configuration that is characteristic of the long-profiles of streams on Ambae, causing effective sediment entrapment in the sequence of deep bedrock pools along stream beds (Fig. 3).

The anecdotal accounts of the local people are also valuable, as their evidence suggests that although wave erosion did occur during Cyclone Funa, this storm actually exacerbated a pre-existing problem of shoreline retreat at Saratamata. In this context it is therefore necessary to consider the influence of Pacific-wide sea-level rise on Ambae Island. Decadal trends in eustatic sea levels are currently being measured at instrumented sites as part of the South Pacific Sea Level and Climate Monitoring Project (SPSLCMP), managed by the Australian Bureau of Meteorology. Monitoring stations with high-resolution SEAFRAME instrumentation (Sea Level Fine Resolution Acoustic Measuring Equipment) established throughout the Pacific include a station at Port Vila, the capital of Vanuatu on Efate Island. This station has been in operation since January 1993. Figure 4 shows the net eustatic sea-level trends in mm/yr for Port Vila and other island nations in the western South Pacific, based on data collected by SPSLCMP until the end of June 2008 [8]. Vanuatu is seen to be experiencing 3.3 mm/yr of sea-level rise.

Recent studies in the South Pacific Islands have shown that beach scour, shoreline retreat, exposure of cemented beachrock and loss of coastal vegetation are some of the physical responses of coastlines to modern eustatic sea-level rise [9–12]. In view of the fact that Saratamata is the most easterly point on the Ambae mainland, and is therefore particularly exposed to strong easterly swells generated by the persistent South East Trade Winds, it is reasonable to interpret the ongoing (i.e., pre-Cyclone Funa) shoreline retreat here as a consequence in part of the annual sea-level rise experienced by the region as a whole.

At Torgil, located at the north eastern extremity of Ambae Island, evidence points to geomorphic control being the principal factor in the accumulation of the large sand body here, instead of the erosional impact during Cyclone Funa felt elsewhere. A flat back-beach area and an embayment in the shape of the coastline are both notable characteristics of the denuded volcanic crater at this site. These features

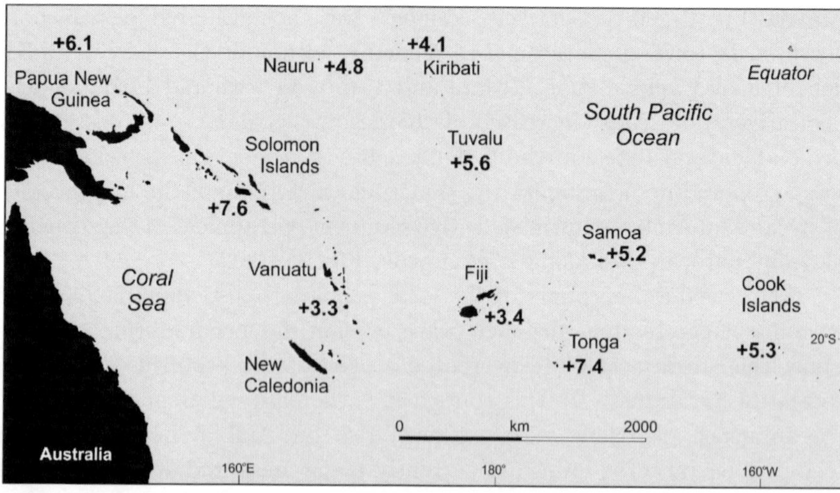

Fig. 4. The net relative sea-level trends in mm/year (bold numerals) for the island groups of the western South Pacific, measured by the South Pacific Sea Level and Climate Monitoring Project (SPSLCMP) of the Australian Bureau of Meteorology (BoM). Trends are based on data collected since the start of the project in the early 1990s (slightly different installation times for each station shown) up to the end of June 2008. Only eustatic sea-level changes are represented, since vertical movement of the instrumentation sites (wharf platforms) and atmospheric pressure effects have been subtracted. Redrawn from original data in BoM, 2009 [8].

of the breached crater provided a relatively sheltered environment that encouraged sediment deposition at Torgil during the storm. In addition, it is seen that Ambae Island's northernmost rocky promontory is one arm of the eroded rim of the volcanic crater. This jutting promontory marks the northward edge of the Torgil embayment and disrupts littoral current flow around the point. During Cyclone Funa, interference with the northward-oriented drift pattern would therefore have prevented some sediment from being swept beyond the point, again favouring deposition in the more sheltered crater embayment.

4. Conclusions

The most exposed area on the north east coast of Ambae Island at the provincial government settlement of Saratamata is suffering significant erosion, affecting coastal roads and coconut plantations and threatening some houses and other buildings. Evidence from dead coastal forest and the present intertidal position of stranded *in situ* coconut trees indicates that up to 45 m of shoreline retreat has occurred within recent years. Some of the erosion at Saratamata can be attributed to wave attack during the passage of tropical cyclone Funa in January 2008, owing to the close proximity of the storm centre to Ambae and the direction of associated storm waves on shore. Yet Cyclone Funa's effects were localised and variable.

The track orientation caused wind-driven waves from the south east to generate northward littoral drift that mobilised available beach sand at Saratamata, depositing this farther north as a large sand body near Torgil. At Torgil, a breached volcanic crater provided geomorphic control, encouraging sediment accumulation in a relatively sheltered coastal embayment bounded by a rocky promontory (denuded crater rim). However, replacement sand was not supplied by drift to Saratamata from the coastline immediately south owing to a lack of sediment sources. This is associated with both the 1 km gap in the fringing reef (limiting the supply of biogenic detritus) and the step-and-pool configuration of steep coastal streams that effectively traps coarse fluvial sediment along the stream bed. The frequent recurrence of tropical cyclones in Vanuatu and the modern eustatic sea-level rise of 3.3 mm/yr suggest that the outlook for the exposed coastline of Saratamata is continuing shoreline retreat, punctuated by episodes of more severe cyclone-accelerated erosion. Appropriate adaptation measures are probably needed to mitigate further damage to both copra plantations and existing infrastructure adjacent to the shoreline in north east Ambae.

Acknowledgments

The author extends his appreciation to the following individuals at the University of the South Pacific for their companionship and assistance with fieldwork: Nick Rollings (formerly USP Laucala Campus), Ketty Napwatt and Alfred Maoh (USP Santo Sub-centre) and Sandy Banga (USP Penama Sub-centre). Mr Augustine Garae (Secretary-General of the Penama Provincial Government at Saratamata) is especially thanked for his hospitality and for facilitating travel on Ambae Island.

References

1. D. I. J. Mallick, Some petrological and structural variations in the New Hebrides, in *The Western Pacific: Island Arcs, Marginal Seas, Geochemistry*, ed. P. J. Coleman (Univ. of Western Australia Press, Nedlands, 1973), pp. 193–211.

2. K. Németh and S. J. Cronin, Phreatomagmatic volcanic hazards where rift-systems meet the sea, a study from Ambae Island, Vanuatu, *J. Volcanol. Geoth. Res.* **180** (2009) 246–258.
3. VMS, Vanuatu Meteorological Services, Port Vila, *Climate of Vanuatu* (2009), http://www.meteo.gov.vu/VanuatuClimate/tabid/196/Default.aspx.
4. J. E. Maragos, G. B. K. Baines and P. J. Beveridge, Tropical Cyclone Bebe creates a new land formation on Funafuti Atoll, *Science* **181** (1973) 1161–1164.
5. D. M. Rearic, Survey of Cyclone Ofa damage to the northern coast of Upolu, Western Samoa, South Pacific Applied Geoscience Commission, Suva., SOPAC Technical Report No. 104 (1990), p. 37.
6. N. D. Newell and A. L. Bloom, The reef flat and "two-meter eustatic terrace" of some Pacific atolls, *Geo. Soc. Am. Bull.* **81** (1970) 1881–1894.
7. T. P. Bayliss-Smith, The role of hurricanes in the development of reef islands, Ontong Java Atoll, Solomon Islands, *Geo. J.* **154** (1988) 377–391.
8. BoM, Australian Bureau of Meteorology, National Tidal Centre, The South Pacific Sea level and climate monitoring project, sea level data summary report, July 2007–June 2008, http://www.bom.gov.au/ntc/IDO60102/IDO60102.2008_1.pdf, (2009), p. 37.
9. P. D. Nunn and N. Mimura, Vulnerability of South Pacific nations to sea-level rise and climate change, *J. Coast. Res.* **24** (1997) 133–151.
10. P. D. Nunn, Coastal changes over the past two hundred years around Ovalau and Moturiki Islands, Fiji: Implications for coastal-zone management, *Aust. Geo.* **31** (2000) 21–39.
11. P. S. Kench and P. J. Cowell, The impacts of sea-level rise on Pacific islands. Part 1: A case study of low-lying atoll islands, Tarawa, Kiribati, *Asia Pacific J. Environ. Dev.* **9** (2002) 43–68.
12. J. P. Terry, Shoreline erosion on a low coral island in Fiji — Causes and consequences, *South Pacific Studies* **24** (2004) 55–66.

Advances in Geosciences
Vol. 16: Atmospheric Science (2008)
Eds. Jai Ho Oh *et al.*

STRATOSPHERIC OZONE CLIMATOLOGY AND ITS VARIABILITY OVER ANKARA

DENIZ DEMIRHAN BARI*, CEYHAN KAHYA, SEMA TOPCU
and SELAHATTIN INCECIK
Faculty of Aeronautics and Astronautics,
Istanbul Technical University,
Meteorological Engineering Department,
Istanbul 34469, Turkey
**demirhand@itu.edu.tr*

The climatological characteristics of ozone based on Total Ozone Mapping Spectrometer (TOMS) and Ozone Monitoring Instrument (OMI) data, during 1979–2008 over central Anatolia, Ankara (32°E, 39°N) is analyzed. The location is highly effected by a number of atmospheric perturbing events such as strong jet winds. Main characteristics of the annual and interannual variability of total ozone over Turkey are presented. The maximum decreasing winter-spring decadal trend was found to be 0.16%. Furthermore seasonal variation of ozone is examined by calculating the deviations from the mean. The variations in low ozone days are 23.7%, 26.6% and 30% for spring, winter and summer, respectively.

1. Introduction

Ozone in the stratosphere acts as a shield protecting living organisms from biologically effective solar irradiance. Ozone layer absorbs 90% of harmful ultraviolet radiation that causes structural DNA deformation. Thus total ozone data has been critically examined by using the data from satellites or ground-based instruments all around the world.

Atmospheric ozone has been declining since the end of 1970's. The relevant studies indicate that ozone decline in the northern hemisphere is mainly due to the anthropogenic emissions. The ozone depletion in both hemispheres that has accelerated up to the 1990s is remarkably well-documented. This study focuses the total ozone variability with recent ozone data on one site in the central Anatolia as a representative of the whole of Turkey.

Total ozone over the globe is about 4% below the 1964–1980 average nowadays. Furthermore global ozone has decreased by 4% from 1997 to 2005 [10]. Indeed in mid-latitudes in the Northern Hemisphere a significant loss of ozone has been discovered. Observed trends at 30°N–60°N, from 1979 to 1993 is about −4% [5]. However a recovery is observed after the ozone control regulations. In the studies covering the last 10 years the sign of ozone recovery has been recognized [5, 9]. In the study of Hadjinicolaou *et al.* [5] at the mid-latitudes (30°–60°N) from 1994 to 2003 the trend shifts up to +3%.

1.1. *Location and data*

In this study Earth Probe TOMS Version-8 total column ozone data from 1979 to 2004 and OMI data from 2004 to 2007 for Ankara were used. Both instruments measure total column ozone on a daily basis. There is a missing period in TOMS total column data from 1994–1996 with 906 days.

Figure 1 indicates the selected station in Turkey. The selected region Ankara is the capital city of Turkey. As Ankara is located in the Central Anatolia of Turkey, the city could be assumed to be in the center of the country. Being measured in the center of Turkey, total ozone data for Ankara could be a good indicator for the rest of Turkey. The daily total column ozone data for the study area is obtained from National Aeronautics & Space Administration (NASA) via internet

Fig. 1. Map of the station in Ankara (32°E, 39°N), Turkey.

(http://jwocky.gsfc.nasa.gov/). Since TOMS and OMI data sets have been regularly measured, data for Ankara is fairly homogeneous.

2. Total Column Ozone Trends in the Region

In the latest report of WMO, comparing the 1964–1980 averages, total ozone averaged for 2002–2005 is about 3% lower in the mid-latitudes, 35°N–60°N. In the same report [10], the seasonal change is also analyzed in detail. In the summer/autumn periods, from 2002 to 2005, the decline is 2% and also in winter/spring, it is about 5–6%. Demirhan *et al.* [2] has examined the 7 stations in the South Eastern Europe region and Eastern Mediterranean from 1979 to 2003. From 1979 to 2003, a maximum ozone depletion yearly trend is found to be in S.Pietro, Italy (44°N, 11°E) with 0.25% and a minimum yearly trend in Ankara with 0.17% is declared.

In this study, TOMS-V8 and OMI data from 1979 to 2007 were examined to find out trends for Ankara. Figure 2 gives the monthly mean variation of total ozone from 1979 to 2007 for the study area. A decadal trend is found with a 2.7% decrease. Average total ozone value is 321 DU for the selected period.

According to Fig. 2, values exceeding the average are found in the winter/spring seasons. Summer and autumn seasons experience the ozone

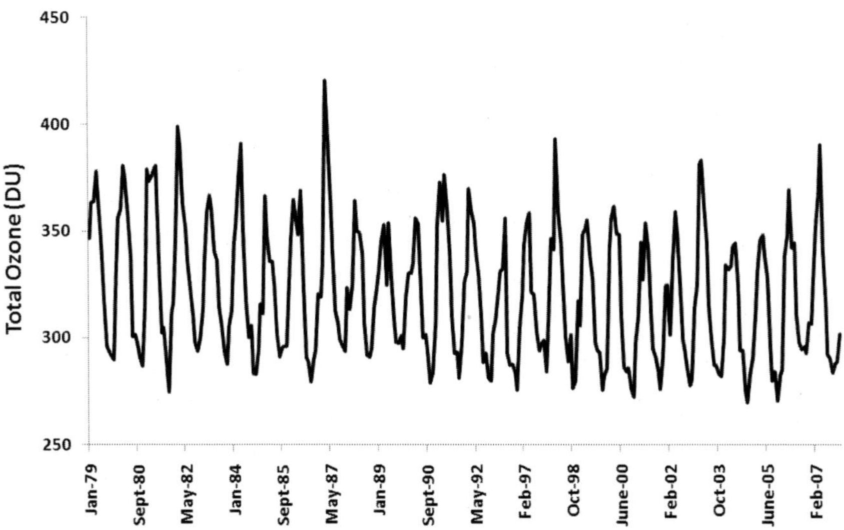

Fig. 2. Monthly mean total ozone variation in Ankara from 1979 to 2007.

Table 1. Minimum, maximum, mean values and trend of monthly averaged total ozone from 1979 to 2007.

	Total ozone	
Max	421 DU	Observed in Mar. 87
Min	270 DU	Observed in Oct. 04
Mean	321 DU	
Trend	2,7 % / decade	

values mostly below the average. A maximum value is noticed in March 1987 with 421 DU and a minimum value, 270 DU is found in October 2004 (Table 1).

It is well-defined that total ozone varies with seasons. In the winter/spring seasons, the observed ozone values are usually higher than summer/autumn seasons [10]. Ankara also shows a seasonal variability, with larger values in spring/winter and lower in autumn/summer [5]. This is a result of the increased dynamical activities in winter/spring season in the Northern Hemisphere.

Monthly mean cycle of total ozone over Ankara between 1979 and 2007 is presented in Fig. 3. Besides the monthly extreme values, standard deviation (σ) limits are shown. The missing values in the TOMS data set has been excluded.

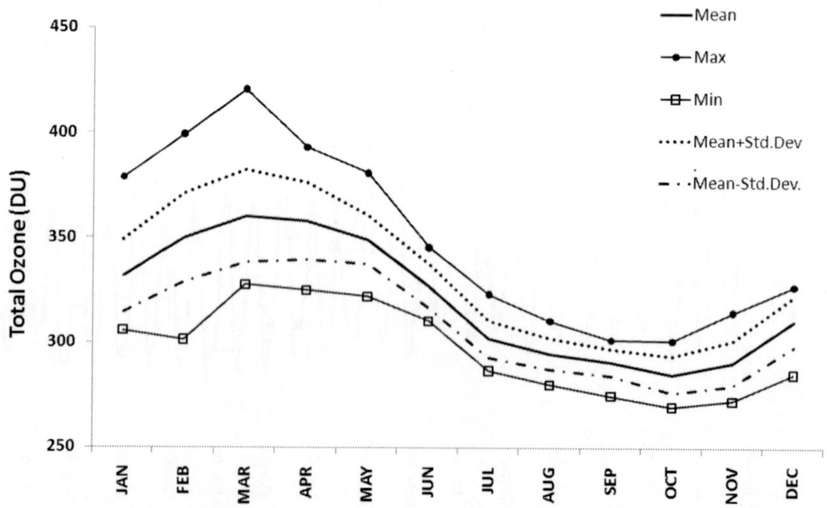

Fig. 3. Monthly mean cycle of total ozone over Ankara from 1979 to 2007.

Table 2. Positive and negative deviations in the monthly averaged total ozone data for Ankara from 1979 to 2007. Months in parenthesis indicate the month of occurrence.

	Max.	Min.
Mean+Std. Dev	382 DU (Mar.)	293 DU (Oct.)
Mean−Std. Dev	338 DU (Mar.)	276 DU (Oct.)

In Fig. 3 from December to March increase in the maximum, minimum, mean and standard deviation is clearly shown. "Mean+standard" deviation and "mean−standard" deviation indicates the positive and negative deviations from the mean respectively. Detailed explanation of Fig. 3 is given in Table 2.

Differences between the maximum and minimum monthly mean total column ozone is also analyzed in Fig. 3. Highest difference between the maximum and minimum monthly averaged total ozone is observed in February is 97 DU and lowest difference between the maximum and minimum ozone found in September which is 26 DU.

2.1. *Seasonal trends*

Seasonal trends for different time periods have been examined in several studies [4]. Total column ozone seasonal trends at northern mid-latitudes are highly influenced by interannual variability of dynamical perturbations [7].

On the seasonal basis, dynamics in fall and summer are less variable than in winter and spring [8]. Thus Total ozone data for winter-spring seasons are significantly larger than any other season throughout a year in northern mid-latitudes [7].

Figures 4(a) and 4(b). show the winter-spring and summer averages of total ozone over Ankara. Seasonal variability of total ozone is linearly decreasing from 1979 to 2007. In the seasonally averaged data winter-spring maximum, observed in 1981, is 359 DU. Summer maximum which is 340 DU measure in 1979 is much lower than winter-spring season maxima. Minimum ozone values from winter-spring and summer seasons are 307 DU in 2006 and 293 DU in 2000, respectively. It is remarkable that maximum values are declared before 1990s and minimum values are observed after 1990s. But it cannot be coincidence that after 2001, no distinguished minima is found. This could be sign of decreasing depletion.

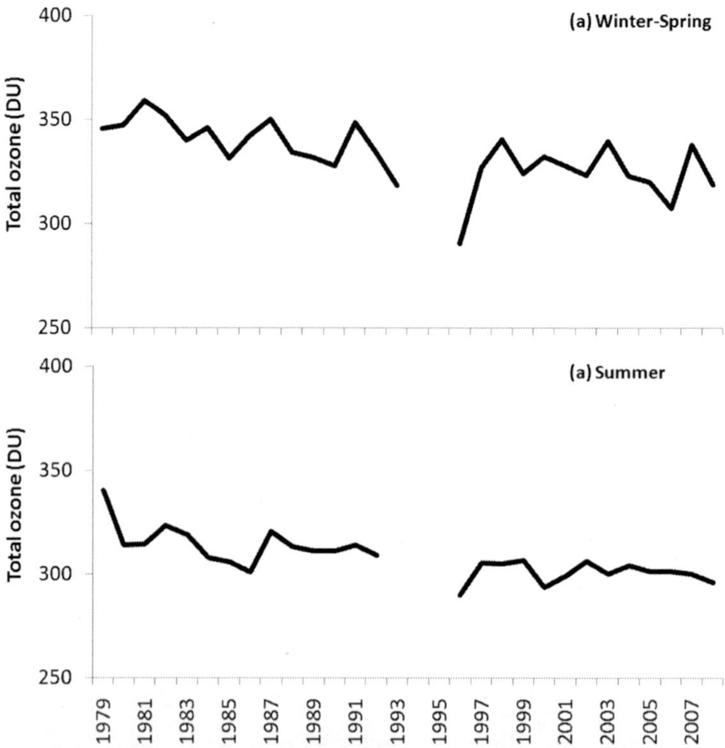

Fig. 4. Variability of total ozone (a) for winter/spring seasons (Dec–May.) (b) for summer season (Jun–Aug) from 1979 to 2007.

Table 3. Maximum, minimum, average values and yearly trend for seasonally averaged total ozone from 1979 to 2007. Dates in parenthesis indicate the date of occurrence. Units of maximum, minimum, average and standard deviation values are Dobson Units (DU).

	Max	Min	Average	Trend	Std. Dev
Winter-spring	359 (1981)	307 (2006)	335	−1.60%	12
Summer	340 (1979)	293 (2000)	309	−0.70%	9.7

Maximum, minimum, average values and trends for seasonally averaged total ozone from 1979 to 2007 given in Table 3. Winter-spring and summer averages are 335 DU and 309 DU, respectively. Yearly seasonal trends are also analyzed. Summer trend is almost half the winter trend. The winter-spring trend found to be −1.6% and summer trend is −0.7%.

3. Deviations from Mean

The tendency of increased number of low ozone days has been evident since 1979 [4]. In order to examine the change in the low and high ozone days, the yearly numbers of low and high ozone days and the extreme ozone days are obtained (Fig. 5). The analysis depends on deviations from mean. High ozone days have the ozone values between the selected month's mean and the standard deviation added to that month's mean (Mean $<$ value $<$ Mean $+ \sigma$) and the situation for the low ozone days is vice verse (Mean $>$ value $>$ Mean $- \sigma$). There are also extreme ozone days which have the ozone values between the month's mean and the twice the standard deviation plus or minus the mean value (Mean $<$ value $<$ Mean $+ 2\sigma$ and Mean $>$ value $>$ Mean $- 2\sigma$).

In Fig. 5, from 1979 to 2007, yearly numbers of low (a) and high (b) ozone days is given. According to the figure, within 28 years, the yearly

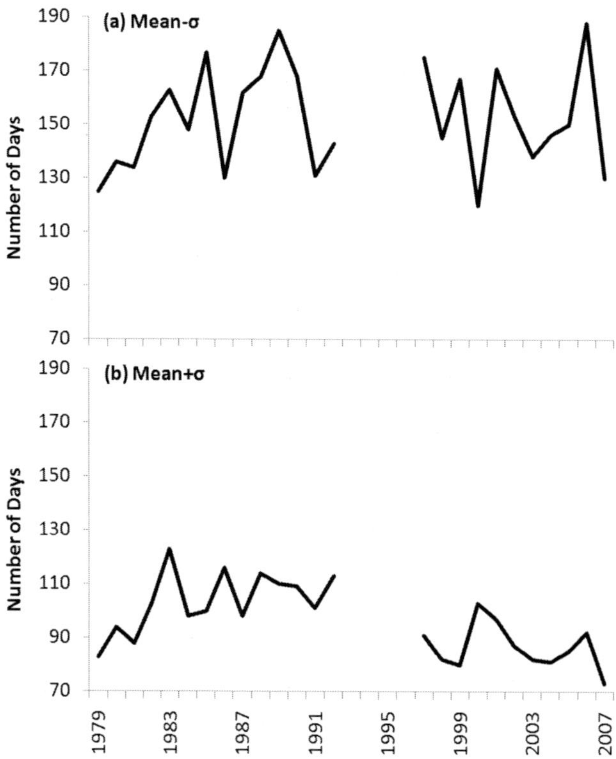

Fig. 5. Number of total ozone days with (a) low ozone values (Mean $- \sigma$) (b) high ozone days (Mean $+ \sigma$) from 1979 to 2007.

number of days with low ozone values has increased (a) and also the number of high ozone days has decreased (b).

In 2006, within 365 days, the 188 of the days were low ozone days. This is the maximum value since 1979. The minimum number of low ozone days observed in 2000 which is 120 days. The number of high ozone days was greatest in 1983 with 123 days. After the 1980s, the number of high ozone days has dropped instantly and in 2007, the minimum number of high ozone days is experienced with 73 days in 365 days.

Figure 6 shows the monthly averaged distribution of low ozone days' (a) and high ozone days' (b) percentages from 1979 to 2007. The percentages have been calculated by taking the ratio of number of low (high) ozone days

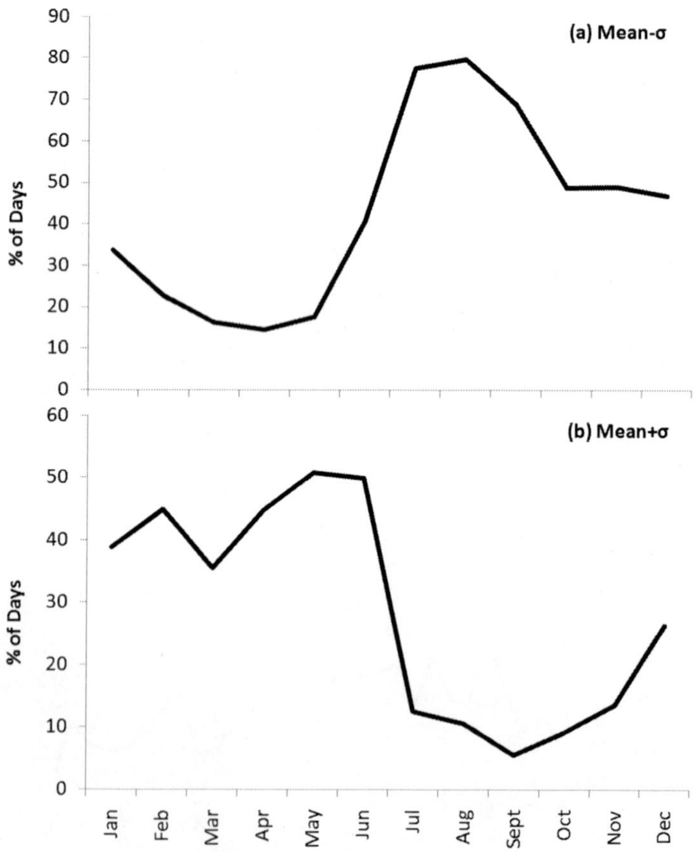

Fig. 6. Percent of days with (a) low ozone values (Mean $-\sigma$) (b) high ozone days (Mean $+\sigma$) from 1979 to 2007.

to total number of days in that month. In Fig. 6, from December to end of May, the percentage of low ozone days is decreasing. In the summer and autumn months from June to October, the percentage of low ozone days is increasing. The Percentage of low ozone days in Fig. 6(a), is minimum in April which is 15% and maximum in August with 80%. It is clear that there is a huge difference between the percentage of low and high ozone days.

Seasonal distribution of high ozone days (Mean + σ) from 1979 to 2007 is examined in Fig. 7. In summer season, number of high ozone days from 1999 to 2007 is remarkably less than the number of days from 1979 to 2003. It is apparent that high ozone days in summer are decreasing. Maximum and minimum number of high ozone days acquired in summer is 37 and 7 and they are found in 1983 and 2000, respectively. In winter-spring season, the maximum number of high ozone days occurred in 2000 is 96 and minimum is found in 1979 which is 55.

Summer and autumn seasons both experience low ozone days. In order to distinguish between them, summer and autumn seasons are examined separately and thus depletion is analyzed clearly. In Fig. 8, seasonal distribution of low ozone days (Mean − σ) from 1979 to 2007 for summer (a) and autumn (b) is shown. It is clear that a significant addition to large number of low ozone days comes from the summer and autumn months. In the study of Eerme *et al.* [4], the similar result explaining low ozone days are notably high in summer and autumn seasons is established.

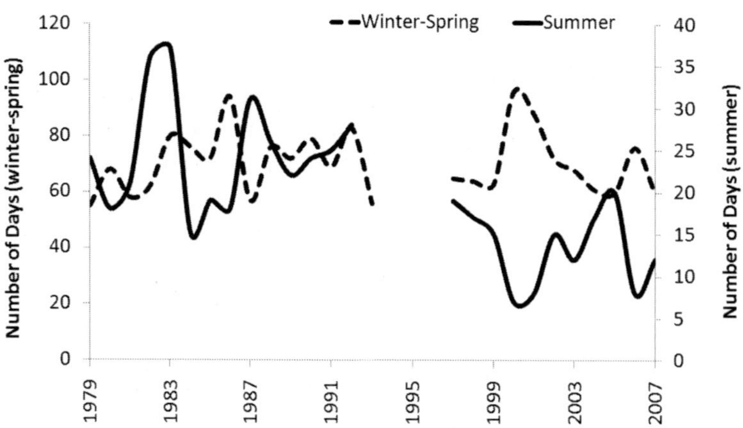

Fig. 7. Seasonal distribution of high ozone days (Mean + σ) from 1979 to 2007. Thick line indicates number of high ozone days in the winter-spring seasons and dashed lines gives the summer season numbers.

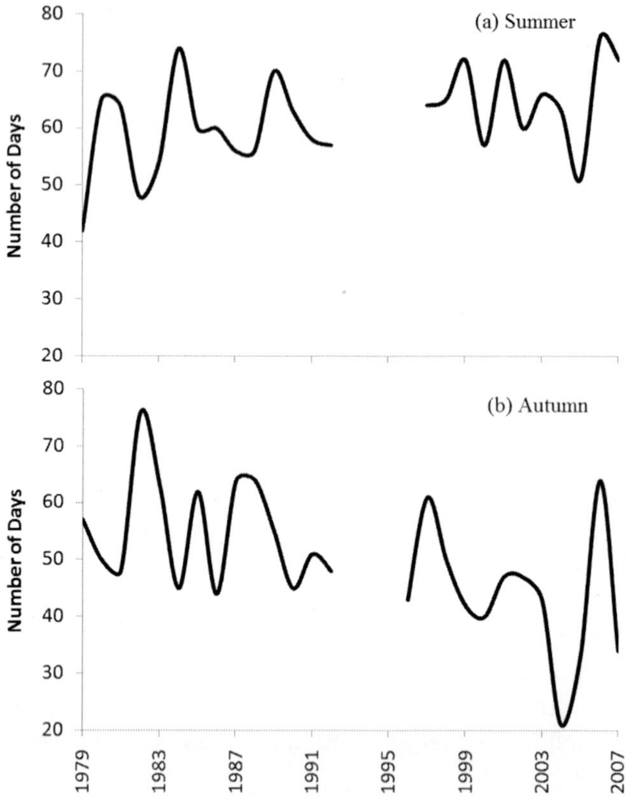

Fig. 8. Seasonal distribution of low ozone days (Mean $-\sigma$) from 1979 to 2007. (a) Summer distribution of low ozone days. (b) Autumn distribution of low ozone days.

In Fig. 8(a), increasing number of low ozone days is observed in summer. Minimum number of low ozone days in summer season is 42 and it is observed in 1979, maximum number is 76 and occurred in 2006. In Fig. 8(b), autumn number of low ozone days establishes the most interesting part because there is a decreasing trend of low ozone days. Fig. 8(b) shows the variability of low ozone days in autumn from 1979 to 2007.

Maximum and minimum number of low ozone days in autumn is 76 and 21 and they are occurred in 1982 and 2004, respectively. As mentioned above, in autumn the number of low ozone days is decreasing gradually. In further studies this could be a sign of recovery.

Ankara does not encounter extreme ozone days every year. Indeed there are no extreme high ozone days which correspond to total ozone values over 398DU since 1979. The extreme number low ozone days which correspond

Table 4. Extreme number low ozone days from 1979 to 2007.

	Total number of extreme ozone days (Mean -2σ)
1996	2
2000	1
2004	4
2005	5

to total column ozone values below 242 DU are given in Table 4. In the selected period lowest ozone events that exceeded mean -2σ were; 232 DU (in Nov. 2004), 233 DU (in Nov. 2005), 236 DU (Oct. 2004), 237 DU (Dec. 2005).

Before 1996, no extreme low ozone days are experienced. After 1996, the number of extreme low ozone days is increasing. By 2005, there are 5 extreme low ozone days in 365 days.

4. Conclusions

Monthly, yearly and seasonal deviations and trends of total ozone are examined in order to make a detailed climatology of ozone over Ankara (32°E, 39°N) as a representative of the entire country. Long term Earth Probe TOMS and OMI data from 1979 to 2007 is examined in seasonal and yearly basis. Both seasonal and yearly evaluations reveal a decreasing total ozone values.

The total ozone trend from 1979 to 2007 for Ankara is −2.7% per decade. The average is 321 DU. Seasonal trends of total ozone over Ankara are analyzed for two separate season groups that are dynamically related each other. These are winter/spring and summer/autumn. Trends for winter/spring and summer are −1.6% and −0.7%, respectively. Maximum values for winter/spring and summer are 359 (in 1981) and 340 (1979). Seasonal minimums for winter/spring and summer are 307 DU (in 2006), 293 (2000), respectively. Seasonal averages are 335 DU in winter/spring and 300 DU in summer.

The yearly numbers of low and high ozone days and the extreme ozone days are obtained. Analysis depends on deviations from mean. Low ozone days and high ozone days correspond to ozone values between mean $-\sigma$ and mean $+\sigma$ respectively. The extreme values are explained as mean $\pm 2\sigma$. From 1979 to 2007, the minimum number of low ozone days obtained in

2000 is 120. On the other hand maximum number of high ozone days was 123 days in 1983. After 1980s number of high ozone days dropped instantly and in 2007 minimum number of high ozone days is found to be 73 days.

In summer season, the number of high ozone days is remarkably dropped starting from 2003. But this is not the fact in the winter/spring period. Indeed in winter/spring seasons, the maximum number of high ozone days occurred in 2000 which is 96 days.

If the summer and autumn seasons inquired separately, it is interesting to find that only in autumn season number of low ozone days is decreasing. In further studies, this could be a sign of recovery.

Since 1979, there are no extreme high ozone days which correspond to the total ozone values over 398 DU, in Ankara. The first four of the extreme number low ozone days that exceeded mean -2σ were; 232 DU (in Nov 2004), 233 DU (in Nov 2005), 236 DU (Oct 2004), 237 DU (Dec 2005).

References

1. B. Aksoy, S. Incecik, S. Topcu, D. Demirhan Bari, C. Kahya, Y. Acar, M. Ozunlu and M. Ekici, *Int. J. Rem. Sens.* **30** (2009) 4387.
2. D. Demirhan, C. Kahya, S. Topcu and S. Incecik, *Int. J. Rem. Sens.* **26** (2005) 3479.
3. F. S. Rowland, *Ann. Rev. Phys. Chem.* **42** (1991) 731.
4. K. Eerme, U. Veismann and R. Koppel, *Annales. Geophys.* **20** (2002) 247.
5. P. Hadjinicolaou, J. Pyle and N. Harris, *Geophys. Res.Lett.* **32** (2005) 1–5.
6. www.esa.int (2009).
7. S. Chandra, C. Varotsos and L. E. Flynn, *Geo. Phys. Lett.* **23** (1996) 555.
8. S. R. Kawa, P. A. Newman, R. S. Stolarski and R. M. Bevilacqua, *Atmos. Chem. Phys.* **5** (2005) 1655.
9. E. C. Weatherhead and S. B. Andersen, *Nature* **441** (2006) 39.
10. WMO, *Global Ozone Research and Monitoring Project Report*, Vol. 50 (2006).

Advances in Geosciences
Vol. 16: Atmospheric Science (2008)
Eds. Jai Ho Oh *et al.*

NEAR SURFACE ATMOSPHERIC METHANE CONCENTRATION AT SUBURBAN ENVIRONMENT OF GUWAHATI BY DIFFERENTIAL ABSORPTION LIDAR SYSTEM

MANOJ SAIKIA*, MINAKSHI DEVI and ANANDA KUMAR BARBARA
Department of Physics, Gauhati University,
Guwahati 781014, India
**manojksaikia@gmail.com*

Starting from the basic design aspect of the Differential Absorption Lidar (DIAL) of Gauhati University, the paper presents the result of atmospheric methane measured by the system at the study area which is enriched with both natural and anthropogenic sources of methane. The paper describes the approaches adopted for selection of absorption line, λ_{On} and off-absorption line, λ_{Off} for methane at $3\,\mu$m range. A brief discussion on laboratory set up of the DIAL is presented where methane measurements is done in artificial environment. To measure the atmospheric methane concentration, λ_{On} and λ_{Off} lines are alternately transmitted and the medium is probed horizontally at a height of around 5 m from ground. Here we have adopted both range resolution and integrated measurement techniques and average concentration comes to be 1.8 ppm.

1. Introduction

Methane is an abundant greenhouse gas in the troposphere after water vapour and carbon dioxide [1] and its green house effect is 21 times more than that of carbon dioxide [2]. It is also observed that even on a per molecule basis, it is much more effective as greenhouse gas than additional CO_2. At $\sim$1.75 ppm, current globally averaged tropospheric mixing ratio of methane is more than twice as high as in pre-industrial times [3]. While analyzing methane data for arctic and Antarctic environment, Khalil and Rasmussen [4] have also shown that methane concentration has increased at an average rate of 16.6 ± 0.4 ppbv/year.

Therefore, measurements of spatial and temporal variability of methane has gained importance for understanding its influence on atmospheric processes associated with global warming as well as green house effect. Chemical techniques based on sampling and successive laboratory analysis have no possibility of continuous working. Further, their integration time for analysis is long, ranging from 1 min to a few hours. After the advent of tunable solid state lasers use of LIDARs for this purpose has gained momentum of late. In fact, LIDAR has now become one of the most powerful tools for remote sensing of trace elements because of its sensitivity and capability of working round the clock. The Differential Absorption Lidar (DIAL) offers a unique opportunity for receiving reliable high resolution concentration measure of trace compounds without disturbing mixing dynamics of surroundings. Historically, the first DIAL measurements were confined to making humidity profiles and then on NO_2 detection. After these pioneering works, several DIAL systems have been developed for many applications [5–7]. Simulated measurement on methane concentration with a mid-Infrared optical parametric oscillator (OPO) based DIAL system having frequency range of 1.45–4 micrometer was done by Lee *et al.* [8]. By adopting Injection-seeded OPO for range resolved DIAL measurements of atmospheric methane, Milton *et al.* [9] had obtained 1.75 ppm concentration over frequency range 1490.6–1580.3 cm^{-1}. In this work, result of methane concentration measured by Gauhati University DIAL system along with the system development, calibration and approach adopted are discussed.

2. Site Environment

Our study area is over Gauhati University (GU) which is a situated on south bank of river Brahmaputra about 12 km away from the main traffic and business zone (marked as A in Fig. 1) of the Guwahati city (91.47°E, 26.11°N). The site is surrounded by paddy fields, forests and wetlands. The residential area is mostly built up over marshy fields and filled up wetlands.

3. DIAL Principle

The technique is based on the spectral absorption properties of gases and requires a tunable laser which can generate two different but closely spaced very stable wavelengths [7]. One of the wavelengths matches with an absorption frequency (λ_{On}) of the target molecule, while the other off-resonance wavelength (λ_{Off}) remains unabsorbed as the two waves travel through the atmosphere. It is necessary that the difference of scattering

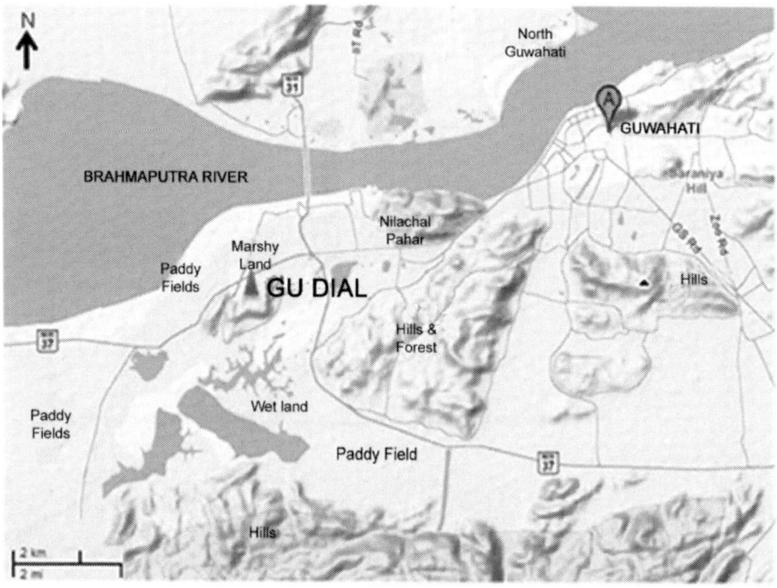

Fig. 1. Study area over Gauhati University.

coefficient of On and Off line must be very small. A receiver collects the backscattered energy as a function of distance. The λ_{On} and λ_{Off} lines are made close enough so that they exhibit the same scattering properties. Backscattered power ratio (at the two wavelengths) is used to extract gas concentration as function of distance [5–7] given by Eq. (1).

$$N(R) = \frac{1}{2\Delta\sigma\Delta R} \ln \left[\frac{P(\lambda_{\text{Off}}, R_2) P(\lambda_{\text{On}} R_1)}{P(\lambda_{\text{Off}}, R_1) P(\lambda_{\text{On}} R_2)} \right], \tag{1}$$

where

$N(R)$ = Number density of the target molecule.
ΔR = Range of the target cell.
$\Delta\sigma$ = Differential absorption Cross-section. $P(\lambda_{\text{On}}, R1)$, $P(\lambda_{\text{On}}, R2)$ and $P(\lambda_{\text{Off}}, R1)$, $P(\lambda_{\text{Off}}, R2)$ are backscattered laser power for On line beam at distances $R1$, $R2$ and that of Off line at $R1$, $R2$, respectively.

Along with the basic DIAL range resolved approach, integrated measurement of methane concentration through the entire probing path of transmitted wavelengths can also be made. Here a tropographic target is generally used to scatter the probing waves and backscattered signal

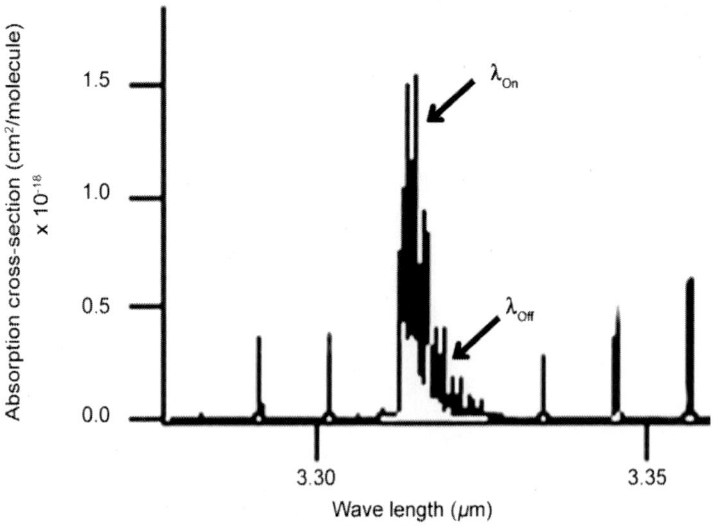

Fig. 2. Absorption spectra of methane at 3.3 μm range.

powers from the target are used to calculate concentration. For this purpose the On and Off line should be very close to each other. If the online and offline wavelengths are so close to each other that the atmosphere behaves identically for both the lines, one can assume the reasonable approximations, *viz.* $P(\lambda_{\text{On}}, R1) \approx P(\lambda_{\text{Off}}, R1)$ in Eq. (1). One then gets

$$N(R) = \frac{1}{2\Delta\sigma R_T} \ln\left[\frac{P(\lambda_{\text{Off}}, R_2)}{P(\lambda_{\text{On}} R_2)}\right]. \tag{2}$$

In DIAL measurements, selection of λ_{On} and λ_{Off} lines is very crucial. The first important aspect is that the lines should be transparent to the other molecules present in the atmosphere so that absorption/scattering magnitude received through the λ_{On} line is only for the selected molecule. After a thorough study and analysis of methane absorption spectra (source: HITRAN and PNNL IR database), 3,316 nm and 3,320 nm lines are respectively selected as λ_{On} and λ_{Off} for our system (Fig. 2).

4. System Description and Specifications

A block diagram of the Gauhati University DIAL system is given Fig. 3. The transmitter is a tunable laser source along with the integrated electronics, power supply and cooling unit. The receiving system comprises of a telescope, InSb photovoltaic detector, triggering and data processing unit.

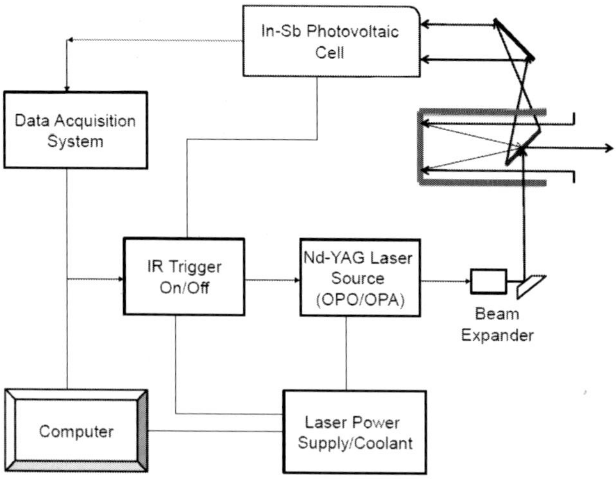

Fig. 3. Block diagram of G U DIAL system.

The tunable laser source consists of Nd YAG pumped OPO and optical parametric amplifier (OPA). The transmitted energy is in 5 ns pulse in the range of 2.8 to 3.45 micron wavelength with pulse repetition frequency (PRF) of 20 Hz. The coolant is ion free distilled water and it is pumped by a motor with a rate of 0.25 gallons/min.

The detailed specifications are given below:

Transmitter:
Laser source: Nd YAG pumped OPO/OPA, Wavelength range: 2.8–3.45 μm, Energy $>$ 1 mJ across the tuning range, Pulse length ~5 ns, PRF: 20 Hz, External trigger: +5 V, 100 μs, Lamp volt: 120 μs pulse, 1000 V at 300 W, Coolant: Distilled water (1–5 MΩ-cm resistivity).

Receiver:
Telescope: 10 inch Schmidt Cassegrain type, Detector element: InSb, Maximum Input voltage: $\pm$15 V, Maximum output voltage: $\pm$10 V, Element Temperature: $-196°$C, Cooling: Liquid N_2, Peak response at λ: 5.3 μm, Cut off λ: 5.5 μm, Photosensitivity: 3×10^8 V/W.

5. Calibration

Both Eqs. (1) and (2) indicate that the power received for the λ_{On} and λ_{Off} lines is important. Therefore, the selection of relevant wave pairs is most important for translation of backscatter power to concentration. For this

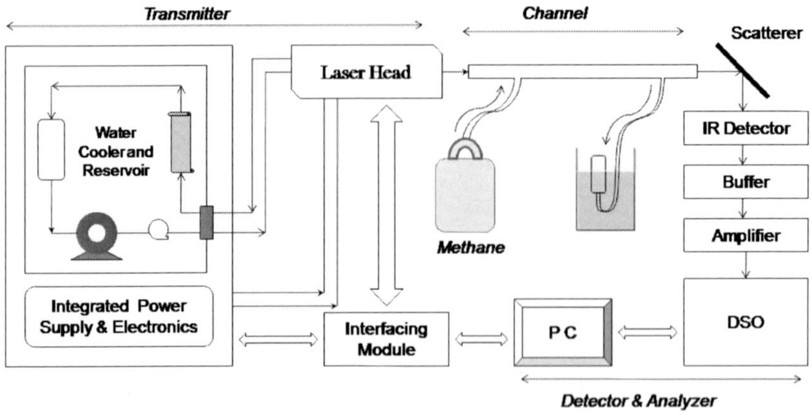

Fig. 4. Laboratory set up of DIAL system.

purpose we have established a laboratory set up of DIAL system (Fig. 4) for identifying the probing lines. The system consists of a transmitter, channel and receiver. The transmitter is a tunable laser source along with the integrated electronics, power supply and cooling unit. The receiving system comprises of InSb photovoltaic detector, filter, triggering and data processing unit. The channel consists of glass tubes of different lengths filled with *gobar gas*. This bio gas used is locally produced from cow dung and contains methane: 55 to 75%, carbon dioxide: 25 to 45% and other minor constituents.

The IR radiations is passed through the channel of glass tubes of different lengths filled with *gobar* gas at atmospheric pressure and the outgoing scattered energy is detected and processed for estimating the absorption of different wave component by methane molecules. The process is repeated for different lengths of methane columns. Some results of absorption at our interested IR range are shown in Fig. 5. These data clearly indicates that maximum absorption takes place at 3,316 nm as supported by methane absorption spectra given in Fig. 2 and justifies our selection of λ_{On} line.

6. Methane Measurement and Results

To measure the atmospheric concentration of methane we have adopted two approaches: one is through range resolved differential absorption and the other is by using a tropospheric target commonly known as differential optical absorption spectroscopy (DOAS).

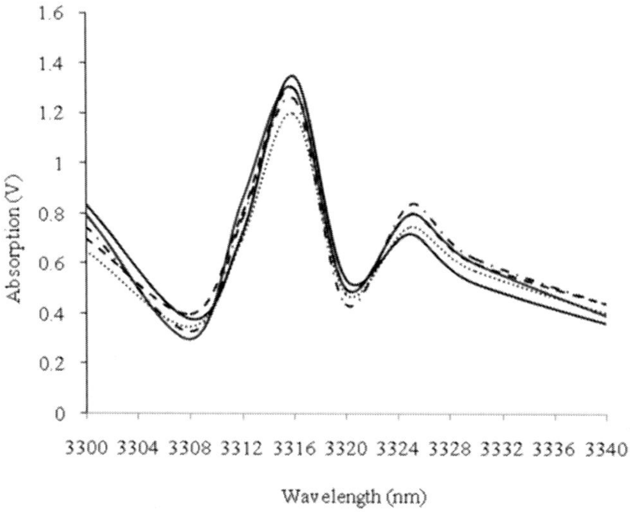

Fig. 5. Methane absorption spectra at wavelengths 3,300 nm to 3,340 nm (five different events as marked with different symbols) observed through laboratory set up of GU.

In the first approach, to measure the atmospheric methane concentration, λ_{On} and λ_{Off} lines are alternately transmitted and the medium is probed horizontally at a height of around 5 m from ground. This height is selected mainly because of clear view and safety reason. The distance between the source and the detector is adjusted by placing a reflector at one end and the receiver and the transmitter in the same platform at the other end. For varying atmospheric column length from 30 to 92 m the reflector position is altered. Each observation is an average of at least 50 to 100 returned pulse-pairs with transmitter energy less than 1 mJ. The detected signal is plotted against distance (Fig. 6) both for Off and On lines and the gradient for each selected cell are taken for calculation of methane concentration by using Eq. (1). For this purpose, cells are taken with width of 5 m to 15 m and the average value of methane concentration comes to be 1.8 ppm. The $\Delta\sigma$, value is taken from PNNL IR database. The experiment was conducted only during night hours of pre-monsoon (mid March to May) period of 2007, 2008 and repeated in 2009. From the observations, it is noted that the significant attenuation of transmitted pulses are obtained only after the pulse travels a distance of minimum 50 to 70 m from the source, which we call as "no attenuation range". Similar observations on range associated attenuation through DIAL set up was also seen in the measurement of Milton *et al.* [9].

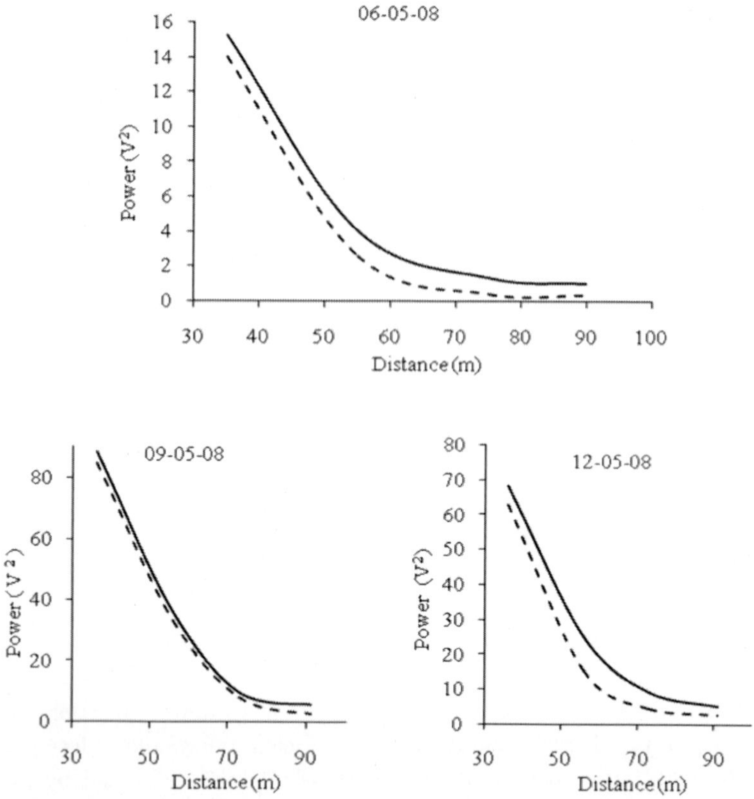

Fig. 6. Typical examples of On and Off line slope as detected by our system. Solid line represents the On line and dashed line represents the Off line.

In the second approach, we measure the methane concentration by using a tropospheric target where integrated measurement of concentration over the probing path is done, range resolved situation does not arise. This technique is adopted to examine the reliability of the method by comparing methane concentration obtained from Eq. (2) with those obtained by Eq. (1). Table 1 shows a comparison of results obtained by the two approaches and it is noted that Eq. (2) can well be adopted for methane concentration measurement.

On examining the reliability of Eq. (2), a few case studies are conducted for tropospheric methane measurement by taking natural hard targets at a range varying from 50 to 250 m. The on and off lines are transmitted at a height of around 5 m and backscattered signals from such tropospheric targets are detected. Figures 7(a) and 7(b) show some detected signal from

Table 1. A few case studies on Methane concentration at different seasons.

	Average Methane concentration (ppm)	
Year	Range resolution	Integrated measurement
2007	1.74, monsoon	1.5–1.74, monsoon
2008	1.77, pre-monsoon	1.2–1.72, monsoon
2009	1.8, pre-monsoon	1.2–1.76, monsoon

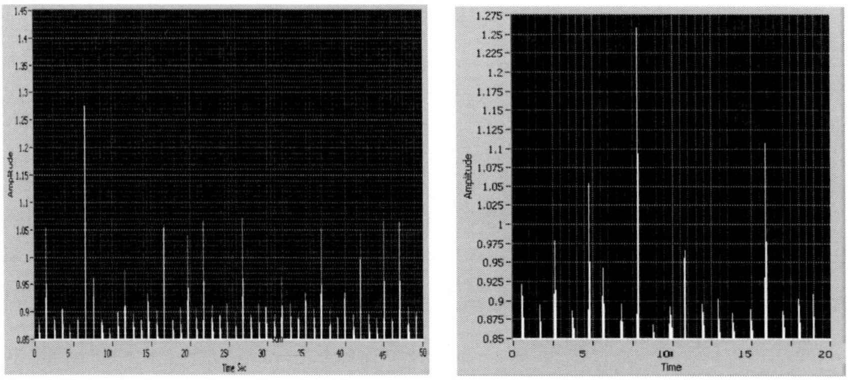

(a)

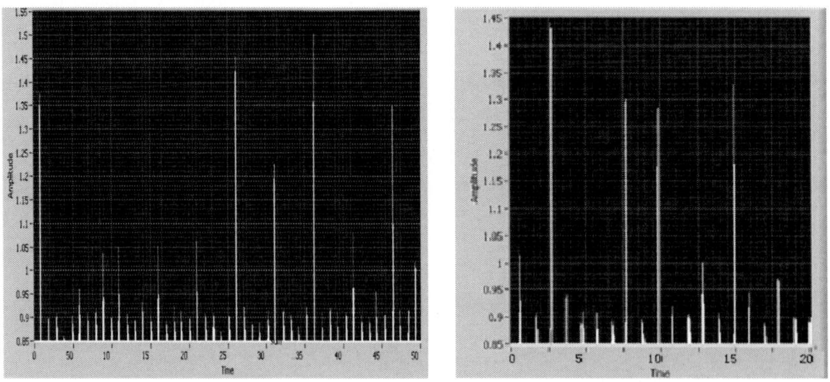

(b)

Fig. 7. (a) Shows backscattered power from the hard target at On line and (b) shows the same for Off line.

the hard target both for On and Off line. The readings are taken for different time frame ranging from 1 sec to 5 min and average powers from all detected pulses are used to calculate methane concentration as given by Eq. (2). The methane concentration given by Eq. (2) shows values lying within 1.2 to 1.76 ppm. It is to be noted that these case studies are conducted mainly in the monsoon (June to August) period. Such situation perhaps inhibits emission and flow of methane to the atmosphere resulting in lower methane concentration in this season compared what we have obtained prior to rainy season.

7. Conclusions and Discussion

The result of methane concentration obtained by the DIAL is compared with reported results [8, 9] and global average value which is 1.75 ppm. Our measured value of 1.8 ppm during the pre-monsoon period indicates a slightly increased concentration. This is likely due to the period of observation which is basically warm and humid after a cool and dry winter season. Therefore advent of rain in this period enhances the anaerobic decomposition of biomaterials; thereby increases methane emission. This statement is supported by the observation of *will-o'-the-wisp* incidences at this season which is a spontaneous blue flame of methane that floats in air. Because, a considerable portion of the study area is wet and marshy; a high emission rate of methane from these wetlands is expected during this period [10]. Forest fires for shifting cultivation in the nearby hilly area also contribute to the high concentration of methane during this period [11].

References

1. D. J. Wuebbles and K. Hayhoe, *Earth Sci. Rev.* **57** (2002) 177–210.
2. IPCC, *Climate Change 1995: The Science of Climate Change* (Cambridge Univ. Press, 1996).
3. D. Etheridge, L. Steele, R. Francey and R. Langenfelds, *J. Geophys. Res.* **103** (1998) 15979–15993.
4. M. A. K. Khalil, and R. A. Rasmussen, *Environ. Sci. Technol.* **24** (2002) 549–553.
5. R. L. Byer, *Opt. Quant. Electron.* **7** (1975) 147–77.
6. T. Kobayashi, *Rem. Sens. Rev.* **3** (1987) 1–56.
7. R. M. Measures, *Laser Remote Sensing* (Wiley, New York, 1984).
8. S. W. Lee, J. McNeil, T. Zenker and T. H. Chyba, *Proc. ILRC* (1998).

9. M. J. T. Milton, T. D. Gardiner, F. Molero and J. Galech, *Optics Comm.* **142** (1998) 153–160.
10. S. Mallick and V. Dutta, *J. Sustain. Dev.* **2** (2009) 2.
11. T. R. Krishnachand and K. V. S. Badarinath, *Cur. Sci.* **92** (2007) 10.

Advances in Geosciences
Vol. 16: Atmospheric Science (2008)
Eds. Jai Ho Oh *et al.*

SOURCE APPORTIONMENT ANALYSIS OF MEASURED FINE PARTICULATE MATTER IN A SEMI-ARID URBAN AIRSHED IN CORPUS CHRISTI, U.S.A

SARITHA KARNAE
Department of Environmental Engineering,
Texas A&M University-Kingsville,
Kingsville, TX-78363, U.S.A

KURUVILLA JOHN
Frank H. Dotterweich College of Engineering,
Texas A&M University-Kingsville,
Kingsville, TX-78363, U.S.A
k-john@tamuk.edu

Corpus Christi is an industrialized urban area of South Texas that is currently in compliance with the National Ambient Air Quality Standards (NAAQS) for $PM_{2.5}$ as set by the United States Environmental Protection Agency (U.S EPA). However a gradual increase in the annual and 24-hour $PM_{2.5}$ concentrations was noted since 2001. In this study, principal component analysis/absolute principal component scores (PCA/APCS) was used as a source apportionment technique to identify key source categories that affected the measured $PM_{2.5}$ concentrations at a continuous ambient monitoring station (CAMS) 04 maintained and operated by Texas Commission on Environmental Quality (TCEQ) during 2000 through 2007. Cluster analysis using computed backward trajectories was performed on days with high $PM_{2.5}$ concentrations. The elevated PM days were heavily influenced by transported levels of PM during three types of episodic events including smoke plumes due to biomass burning in Mexico and Central America during April and May, sub-Saharan dust transport from Africa during June and July, and regional haze transport from highly industrialized areas of Texas and surrounding Midwestern states during September. Pyrotechnic emissions during local firework events during the New Year day celebrations under stagnant meteorological conditions also resulted in elevated $PM_{2.5}$ concentrations. PCA/APCS identified five key source categories that accounted for 78% of the variance in the $PM_{2.5}$ concentrations measured within the urban airshed. Secondary sulphates were identified to be the major contributor accounting for 46% of the apportioned mass. This was followed by mobile sources which accounted for 26%. The other sources that were identified by PCA/APCS included crustal dust, a commingled source of biomass burning and sea salt, and secondary nitrates. Increase in secondary sulphates was observed during August and

September typically associated with the long range transport of continental haze from industrialized areas in Texas and surrounding states. Mobile source contributions increased during the winter months due to an increase in tourism related activities in the area. Biomass burning in Mexico and Central America during April and May contributed to elevated $PM_{2.5}$ concentrations observed in the Corpus Christi urban airshed.

1. Introduction

Several epidemiological studies have related aerosol composition in ambient air with adverse health effects on respiratory and cardiovascular systems causing increased mortality and morbidity [1–4]. Based on the physical properties, fine particulate matter were observed to have longer residence time than coarse particulate matter and are transported over larger distances under favourable meteorological conditions causing acute long term health effects. As noted by various recent scientific studies, long range transport of $PM_{2.5}$ from regional and local sources that are either natural or anthropogenic is critical in the understanding of air pollution events affecting most urban areas [3–5].

Corpus Christi, located in South Texas, is a growing industrialized area and is designated as the eighth largest metropolitan statistical area. It has the fifth largest port in U.S.A. with several major petrochemical refineries located along the port. As per the National Ambient Air Quality Standards (NAAQS) for PM set by the United States Environmental Protection Agency (U.S. EPA), the region is currently in compliance. However, a preliminary trend analysis performed showed a gradual increase in the $PM_{2.5}$ concentrations since 2001, possibly due to an increase in the influence of local and/or regional sources. Thus, the primary objective of this study was to identify these local and regional sources of $PM_{2.5}$ a source apportionment analysis technique.

During a $PM_{2.5}$ characterization study conducted by Kleanthous *et al.* [6], the air quality in Cyprus, a dry Mediterranean island, was observed to be influenced by approximately 6–8 long-range transported events of Sahara dust storms. Galvas *et al.* [7] identified long-range transport to be a dominant factor influencing $PM_{2.5}$ concentrations measured at a rural coastal Mediterranean site. According to a $PM_{2.5}$ source apportionment study conducted by Lall *et al.* [8] for the New York City area, regional sources were observed to contribute approximately 90% of the measured sulphate mass. Sillanpaa *et al.* [9] used backward trajectories for a $PM_{2.5}$ source identification study and they identified biomass combustion events to be the major regional source impacting the measured $PM_{2.5}$ mass at an urban background site in Helsinki, Finland. Pongkiatkul *et al.* [10] using potential source contribution function (PCSF) analysis (trajectory-based model) identified long range transport from Southern China, the South China Sea and nearby countries in the South East Asia including northern Thailand as the major sources of elevated $PM_{2.5}$ concentrations measured in the Bangkok Metropolitan Region (BMR).

Unlike trajectory based models, source apportionment models including Principal Component Analysis/Absolute Principal Component Scores (PCA/APCS), UNMIX, Positive Matrix Factorization (PMF) and Chemical Mass Balance (CMB) have been widely used to identify and quantify local anthropogenic sources of $PM_{2.5}$ [11]. Kim and Hopke [12] used PMF for a source apportionment study using the measured $PM_{2.5}$ mass concentrations observed at four Speciation Trends Network (STN) monitoring sites in Seattle, U.S.A. during 2000–2005. They concluded that secondary aerosols including sulphates, nitrates and primary emissions from gasoline-fueled vehicles were identified as the major contributors of the PM observed at the urban sites. However, at a residential site, wood smoke was identified to be the major contributor in their study. Using PMF, Birnur *et al.* [13] identified mobile sources (55–76%) to be the major contributors of $PM_{2.5}$ mass concentrations measured at four STN sites in the midwestern states of U.S.A. and at a rural monitoring site belonging to the Interagency Monitoring of Protected Visual Environments (IMPROVE). The other major sources identified in this study included secondary sulphates and crustal soil sources [13]. CMB identified vehicular sources along with road dust as the major contributor of PM mass measured at seven different sites located in the Los Angeles–Long Beach harbour area, which is one of the busiest ports in U.S.A. [14].

PCA is a simple multivariate technique and has been applied to identify sources of various pollutants including PM and VOC [11, 15–17]. Garcia *et al.* [18] used PCA/APCS for a source apportionment study and concluded that burning activities were a major contributor accounting for about 68% of the measured $PM_{2.5}$ mass concentrations at a monitoring site located in Sunland Park, New Mexico, U.S.A. In a separate source apportionment study conducted by John *et al.* [19], PCA identified industrial, fossil fuel combustions and geological sources to be the major factors affecting $PM_{2.5}$ concentrations measured at three elementary schools in Central and Southeast Ohio.

In the present study, a detailed source apportionment analysis was performed using PCA/APCS technique on the speciated $PM_{2.5}$ concentration measured at an industrialized urban site in Corpus Christi, Texas.

2. Methodology

2.1. *Data acquisition*

The Texas Commission on Environmental Quality (TCEQ) is responsible for maintaining and operating continuous ambient monitoring stations (CAMS) measuring ambient air quality in Texas. CAMS 04 and CAMS 635 are two of the 18 monitoring sites maintained by TCEQ in the Corpus Christi region. Figure 1 shows the location of $PM_{2.5}$ monitoring sites within the study region. These sites provide measurements of $PM_{2.5}$ concentrations. CAMS 04 located in the core of the urban airshed at 902 Airport Boulevard (27°45′N, 97°26′W) is equipped with a $PM_{2.5}$ analyzer measuring hourly $PM_{2.5}$ concentrations and it has been in operation since 2000. Since 2001, twenty-four hour averaged filter samples were also collected once every sixth day at CAMS 635, located at 5,707 Up River road (27°48′N, 97°27′W) and the samples were analyzed further to characterize individual chemical components of $PM_{2.5}$. Chemical analysis of the filter samples was performed by Research Triangle Institute (RTI), in North Carolina to identify the concentrations of elements, ions, elemental carbon and organic carbon. Quality assurance and quality control checks were performed on the data acquired and reported to TCEQ by RTI. Speciation data measured at CAMS 635 and continuous $PM_{2.5}$ concentrations measured at CAMS 04 were acquired from TCEQ for this source apportionment study.

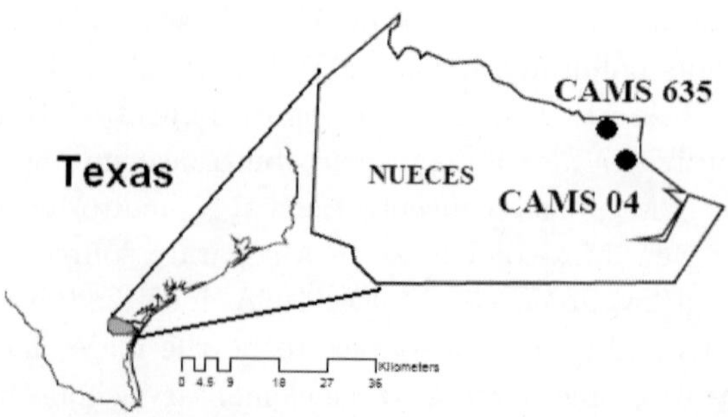

Fig. 1. Geographical map of $PM_{2.5}$ monitoring sites within the Corpus Christi region.

2.2. *Cluster analysis of backward trajectories*

Backward trajectories indicate the historical pathway of air parcel approaching the receptor site and thus can be used to identify regional sources impacting the monitoring site. The National Oceanic and Atmospheric Administration (NOAA) has developed HYbrid Single-Particle Lagrangian Integrated Trajectory (HYSPLIT) model to compute air parcel trajectories from any given location. HYSPLIT assumes constant pressure surfaces and neglects vertical transport leading to uncertainties in the trajectories computed. Based on the uncertainties involved in the computation, a single trajectory cannot be used to represent the pathway of an air parcel [20]. Thus cluster analysis, a multivariate statistical technique, has been used to track the air parcels and identify source-rich areas that could potentially impact the receptor site. A cluster can be defined as a group of trajectories with similar meteorological traits. Hence in the present study, cluster analysis was applied to identify the probable upwind source regions influencing the $PM_{2.5}$ concentrations measured in Corpus Christi.

2.3. *Principal component analysis/absolute principal component scores (PCA/APCS)*

PCA is one of the simplest source apportionment models used to identify the probable sources of a pollutant. In the current study varimax rotation with Kaiser normalization was applied to identify principal components with eigen values greater than 1.0 explaining maximum variance in the data. The identified principal components were then characterized based on the species with factor loadings greater than 0.5. Factor scores computed by PCA were used for further quantification using multiple linear regression analysis (MLRA). As the factor scores were estimated using the normalized data, they were then rescored using true zero factor score. Regression analysis was performed on the absolute principal component scores and the total $PM_{2.5}$ concentrations measured during each sampling day to estimate source contributions as shown in Eq. (1).

$$M_k = \zeta_0 + \sum_{j=1}^{p} \zeta_j APCS_{jk}, \tag{1}$$

where M_k is the measured concentration in sample j. The $\zeta_j APCS_{jk}$ term is the rescored absolute component score for the jth source in the kth sample. ζ_0 represents the mass contributions of unaccounted sources. The $PM_{2.5}$ mass contributions of each factor were then used to estimate the average concentrations of species contributed by each source.

3. Results and Discussion

Hourly $PM_{2.5}$ concentrations measured at CAMS 04 during 2000–2007 were used to calculate the three-year annual arithmetic mean concentrations as per the NAAQS for $PM_{2.5}$ as set by the United States Environmental Protection Agency (EPA). The three-year averaged annual concentrations were calculated to be 7.4, 7.7, 8.1, 8.9, 9.4 and 10.3 μg.m^{-3}, respectively. The values were certainly below 15.0 μg.m^{-3} and the region was identified to be presently in attainment with the NAAQS for $PM_{2.5}$. However, a gradual increase was observed from 2000 through 2007 with an overall 39% increase was noted possibly due to possibly an increase in local emissions or due to the increased influence of regional sources under favourable meteorological conditions. Days with 24-hour averaged $PM_{2.5}$ concentrations above 35.0 μg.m^{-3} were identified as episode days in this study. A temporal analysis showed elevated $PM_{2.5}$ concentrations during spring and summer months, and low concentrations during winter months.

3.1. *Analysis of $PM_{2.5}$ episodes*

HYSPLIT4 developed by NOAA was used to estimate the hourly backward trajectories starting from CAMS 04 at a height of 500 m during the episode days. Cluster analysis was performed using backward trajectories for the three major types of $PM_{2.5}$ episodes that affected the Corpus Christi urban airshed. Stagnant meteorological conditions during the midnight hours on New Year's Eve of 2003 along with emissions from pyrotechnic during the celebration resulted in very high levels of $PM_{2.5}$ concentrations measured within the urban area. For instance, hourly $PM_{2.5}$ concentrations of 771 μg.m^{-3} were measured at CAMS 04 at 1:00 a.m. on January 2003. These types of local emissions were major contributors to the elevated $PM_{2.5}$ concentrations observed within this urban airshed.

$PM_{2.5}$ concentrations ranging from 37 μg.m^{-3} to 68 μg.m^{-3} were observed during April and May months and the typical winds were blowing from south-southeast at a speed of 5 to 10 m.s^{-1}. Selected episode days identified for the cluster analysis included April 23rd through 26th, 2003;

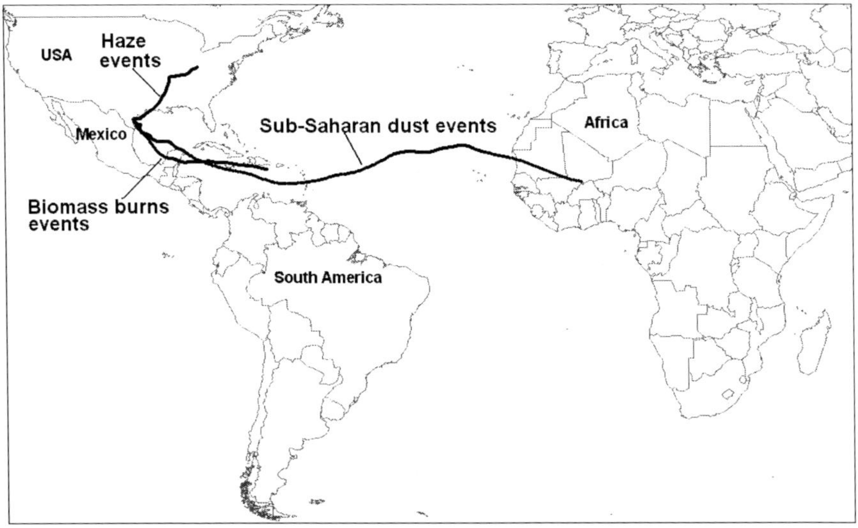

Fig. 2. Cluster analysis of the episode days observed at CAMS 04 during 2001–2007.

May 3rd through 7th, 2003; and May 7th through 11th, 2006. Figure 2 shows the clusters estimated based on the computed backward trajectories using HYSPLIT4. The air parcels during these episode days seem to originate from Mexico and Central America. Biomass burning events in Mexico and Central America occurred during April and May and this was identified to be the possible source of elevated $PM_{2.5}$ concentrations observed in the study region.

Days with elevated $PM_{2.5}$ concentrations ranging from $36\,\mu g.m^{-3}$ to $48\,\mu g.m^{-3}$ were observed during June and July also with predominant winds blowing from the southeast at about 5 to $10\,m.s^{-1}$. The $PM_{2.5}$ episodes identified included June 17th through 20th, 2004; July 28th through August 1st, 2005; June 21st through 24th, 2007; and July 28th through August 1st, 2007. Cluster analysis conducted with fifteen day backward trajectories originating from the sub-Saharan region of Africa is shown in Fig. 2. The sub-Saharan dust storms were identified as a possible source of elevated $PM_{2.5}$ concentrations that affected Corpus Christi during June and July.

Elevated $PM_{2.5}$ concentrations were observed during August and September with predominant winds blowing from the northeast. Episode days identified for further cluster analysis included September 11th through 15th, 2002; September 25th through October 1st, 2004; September 3rd through 6th, 2005; and August 26th through 31st, 2006. Figure 2 shows the cluster generated using three-day backward trajectories and the air parcels were observed to originate from highly industrialized areas of southern and Midwestern U.S.A. Continental scale haze transport from the industrialized areas of Texas, Louisiana, middle Mississippi River Valley, Tennessee and the lower Ohio River Valley was identified to impact the elevated $PM_{2.5}$ concentrations observed during August and September.

3.2. *Principal component analysis/absolute principal component scores (PCA/APCS)*

The speciation dataset measured at CAMS 635 during 2001 through 2007 consisted of 377 observations with concentrations provided for 55 species. Species with over 25% of missing values were excluded from further source apportionment analysis. Thus, 17 key species were identified which included elements (Ca, Fe, Ti, Si, Zn, K, V, S, Cl and Na), water soluble ions (Na^+, K^+, NH_4^+, NO_3^-, and SO_4^{2-}), elemental carbon and organic carbon. Observations with missing and incomplete concentrations (over 25%) were excluded from the analysis to reduce the uncertainty in source contribution estimates, leading to a dataset with 352 observations, which was approximately 93% of the initial dataset.

By applying PCA with varimax normalized rotation, five possible sources were identified that explained 78% of the variance in the $PM_{2.5}$ concentrations measured within the Corpus Christi region. Factor loadings greater than 0.5 were used to characterize the PCA identified sources and are shown in Table 1. Factor 1 explained 26% of the variance and was observed to have high factor loadings of Ca, Fe, Ti, Si, and K. In a source apportionment study conducted by Gildemeister *et al.* [21] using speciation data collected at two monitoring sites in Detroit, a source profile classified as soil with elevated concentrations of Si, Fe, Al, and Ca was observed. As the monitoring site in the present study was located in an urban region, factor 1 was classified as crustal dust from paved and unpaved roads and was also influenced by transported Si concentrations from Africa during sub-Saharan dust storm events.

High factor loadings of S, SO_4^{2-} and NH_4^+ were observed in factor 2. In a $PM_{2.5}$ source apportionment study conducted by Lee *et al.* [22], source profile rich in SO_4^{2-} and NH_4^+ was characterized as secondary sulphates that was formed during photochemical reactions. Thus, factor 2 accounting for 21% of the variance was classified as secondary sulphates as a result of

Table 1. Factor loadings greater than 0.5 of the PCA identified sources.

Species	Factor 1	Factor 2	Factor 3	Factor 4	Factor 5
Ca	0.8313				
Cl			0.8432		
Fe	0.9814				
Ti	0.9498				
V				0.5775	
Si	0.9801				
Zn				0.5209	
S		−0.9729			
K	0.6415		0.5096		
Na			0.9120		
NH_4^+		−0.9448			
Na^+			0.7904		
K^+			0.6741		
OC				0.6851	
EC				0.7851	
NO_3^-					0.7447
SO_4^{2-}		0.9586			
% Source contribution	15	46	10	26	3
Probable source category	Crustal dust	Secondary sulphates	Biomass Burns & Sea salt	Mobile sources	Secondary nitrates

atmospheric transformation of sulphur emissions from local anthropogenic sources including refineries. This factor also accounted for transported levels of SO_4^{2-} from industrialized urban areas of Texas and surrounding states during regional haze events. Several source apportionment studies have characterized fresh sea salt source with the presence of Na, Na^+ and Cl^- and have associated biomass burning with K, K^+ and OC. Thus factor 3 with high factor loadings of Na, Cl, Na^+, K and K^+ was classified as a combined source of sea salt and biomass burns and this accounted for 15% of the variance [22, 23]. The proximity of the monitoring site to the coast along suggests the influence of commingled sources such as sea salt with $PM_{2.5}$ transported from biomass burning sources. In the source apportionment study conducted by Gildemeister *et al.* [21] gasoline and diesel emissions were characterized by the presence of OC, EC and S. Vanadium and Zn are the byproducts from catalytic converters. Thus factor 4 accounting for 10% of the variance was identified as mobile sources. Factor 5 rich in nitrates was classified as secondary nitrates and this accounted for 6% of the variance. The possible sources of nitrates were photochemical reactions, emissions from secondary industrial processes and aged sea salt.

$PM_{2.5}$ mass contributions of the PCA identified sources were estimated by applying multiple linear regression to the absolute principal component scores and the observed $PM_{2.5}$ concentrations. The coefficient of correlation (R^2) between the observed and predicted $PM_{2.5}$ mass was identified to be 0.92, indicating good model performance. The mass contributions estimated were then used to identify the percentage contribution of each source and are shown in Table 1. Secondary sulphates were identified to be the largest source contributor accounting for 46% of the apportioned $PM_{2.5}$ mass concentrations. Mobile sources including gasoline and diesel vehicles were identified to be the second-largest contributor and this accounted for 26%. The other source contributions included crustal dust from local sources and sub-Saharan dust storm events and this accounted for 15% of the apportioned mass. A commingled source of plumes from biomass burning and sea salt accounted for 10%, while secondary nitrates accounted for 3% of the apportioned mass. Seasonal variations in the percentage contributions of the PCA identified sources were calculated and are shown in Fig. 3. Higher contributions of crustal dust with increased Si concentrations were observed in June and July primarily influenced by sub-Saharan dust storms in Africa. Secondary sulphate contribution was observed to be high in August and September (60%) due to an increase in photochemically

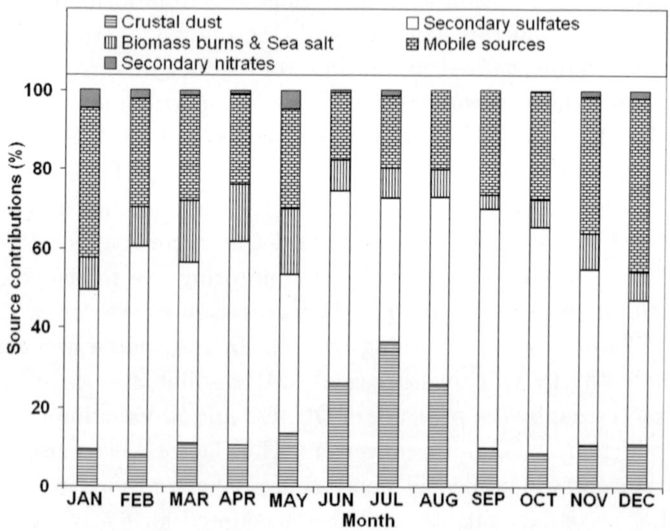

Fig. 3. Seasonal variations in percent source contributions observed at CAMS 635 during 2001 through 2007.

formed secondary aerosols during long range transport of haze from highly industrialized areas of Texas and surrounding states as highlighted by Kim and Hopke [12]. An increase in K^+ and OC concentrations of $PM_{2.5}$ was observed during April and May primarily due to an increase in the transported levels of $PM_{2.5}$ from areas with large biomass burning including Mexico and Central America. An increase in the contribution of mobile sources was observed during the winter months of December and January possibly due to the influence of increased local tourism related activities within the study region. In addition, higher contribution of secondary nitrates was observed during the winter months due to lower temperatures and higher relative humidity favouring the formation of secondary particles [12].

4. Conclusions

Corpus Christi region is currently in attainment of the NAAQS for $PM_{2.5}$, however a 39% increase was observed in the three-year annual averaged $PM_{2.5}$ concentrations since 2001. The region was observed to be affected by three major types of episodes with high $PM_{2.5}$ influenced by biomass burning in Mexico and Central America during April and May, transport of sub-Saharan dust storm events from Africa during June and July and regional haze transport from highly industrialized areas of Texas and surrounding states during September. Pyrotechnic emissions from fireworks during New Year's day contributed to a unique $PM_{2.5}$ episode. On applying PCA/APCS to the $PM_{2.5}$ speciation data, five possible sources accounting for 78% of the variance were identified. Secondary sulphates were identified to be the major contributor and this accounted for 46% of the apportioned $PM_{2.5}$ mass. Mobile sources were the second-largest contributor accounting for 26% followed by crustal dust, which accounted for 15%. The other source contributions included 10% from a combination of biomass burning and sea salt and 3% due to secondary nitrates. An increase in the contribution of secondary sulphates was observed in September that was influenced by continental haze transport. Transported levels of Si from sub-Saharan dust storms of Africa during July months caused an increase in the contribution of crustal dust sources. Secondary nitrates were observed to be high during the winter months due to lower temperatures and higher relative humidity favouring. Thus, the gradual increase in the observed $PM_{2.5}$ concentrations in Corpus Christi, U.S.A. can be attributed to recent $PM_{2.5}$ episodes that are influenced by both local and regional sources.

Acknowledgments

This material is based upon work supported by the Center for Research Excellence in Science and Technology — Research on Environmental Sustainability of Semiarid coastal arid (CREST-RESSACA) at Texas A&M University at Kingsville through funding from the National Science Foundation (NSF) under Cooperative Agreement No. HRD-0734850 from the National Science Foundation, Washington, D.C. Any opinions, findings, and conclusions or recommendations expressed in this material are those of the author and do not necessarily reflect the views of the National Science Foundation.

References

1. K. Ramgolam, S. Chevaillier, F. Marano, A. Baeza-Squiban and L. Martinon, *Chemosphere* **72** (2008) 1340.
2. G. T. O'Connor, L. Neas, B. Vaughn, M. Kattan, H. Mitchell, E. F. Crain, R. Evans III, R. Gruchalla, W. Morgan, J. Stout, G. K. Adams and M. Lippmann, *J. Allergy Clin. Immunol.* **121** (2008) 1133.
3. L. H. Chen, S. F. Knutsen, L. Beeson, M. Ghamsary, D. Shavlik, F. Petersen and D. Abbey, *Ann. Epidemiol.* **15** (2005) 642.
4. M. Kumpa and E. Castanas, *Environ. Pollut.* **151** (2008) 362.
5. M. Zhang, Y. Song and X. Cai, *Sci. Total Environ.* **376** (2007) 100.
6. S. Kleanthous, M. A. Bari, G. Baumbach and L. Sarachage-Ruiz, *Atmos. Environ.* (2008), doi: 10.1016/j.atmosenv.2008.06.025.
7. S. D. Glavas, P. Nikolakis, D. Ambatzoglouo and N. Mihalopoulos, *Atmos. Environ.* **42** (2008) 5365.
8. R. Lall and G. D. Thurston, *Atmos. Environ.* **40** (2006) S333.
9. M. Sillanpaa, S. Saarikoski, R. Hillamo, A. Pennanen, U. Makkonen, Z. Spolnik, R. V. Grieken, T. Koskentalo and R. O. Salonen, *Sci. Total Environ.* **350** (2005) 119.
10. P. Pongkiatkul and N. T. K. Oanh, *Atmos. Res.* **85** (2007) 3.
11. M. Sanchez, S. Karnae and K. John, *Int. J. Environ. Res. Publ. Health* **5** (2008) 130.
12. E. Kim and P. K. Hopke, *Atmos. Environ.* **42** (2008) 6047.
13. B. B. Guven, S. G. Brown, A. Frankel, H. R. Hafner and P. T. Roberts, *J. Air Waste Manag. Assoc.* **57** (2007) 606.
14. M. C. Minguillon, M. Arhami, J. J. Schauer and C. Sioutas, *Atmos. Environ.* **42** (2008) 7317.
15. G. Thurston and J. Spengler, *Atmos. Environ.* **19** (1985) 9.
16. K. F. Ho, J. J. Cao, S. C. Lee and C. K. Chan, *J. Hazard. Mater.* **138** (2006) 73.
17. G. Dogan, G. Gullu and G. Tuncel, *Microchem. J.* **88** (2008) 142.
18. J. H. Garcia, W.-W. Li, N. Cardenas, R. Arimoto, J. Walton and D. Trujillo, *Chemosphere* **65** (2006) 11.

19. K. John, M. Kim, K. Crist, S. Karnae and A. Kulkarni, *J. Air & Waste Management Association* **57** (2007) 394.
20. S. R. Dorling, T. D. Davies and C. E. Pierce, *Atmos. Environ.* **26A** (1992) 2583.
21. A. E. Gildemeister, P. K. Hopke and E. Kim, *Chemosphere* **69** (2007) 1064.
22. J. H. Lee and P. K. Hopke, *Atmos. Environ.* **40** (2006) S360.
23. K. J. Moom, J. S. Han, Y. S. Ghim and Y. J. Kim, *Environ. Int.* **34** (2008) 654.

Advances in Geosciences
Vol. 16: Atmospheric Science (2008)
Eds. Jai Ho Oh *et al.*

MULTI-SCALE ORGANIZATION OF WATER VAPOR OVER LOW AND MID-TROPICAL AFRICA

ONDEGO JOEL BOTAI
Department of Geography, Geoinformatics and Meteorology, University of Pretoria, Pretoria 0002, South Africa
joel.botai@up.ac.za

VENKATARAMAN SIVAKUMAR
National Laser Centre, Council for Scientific and Industrial Research, P.O. Box 395, Pretoria, South Africa
Department of Geography, Geoinformatics and Meteorology, University of Pretoria, Pretoria 0002, South Africa
svenkataraman@csir.co.za

WILLEM LUDWIG COMBRINCK
Hartebeesthoek Radio Astronomy Observatory, P.O. Box 443, 1740, Krugersdorp, South Africa
Department of Geography, Geoinformatics and Meteorology, University of Pretoria, Lynnwood Road 0002, South Africa

CORNELIS JOHANNES DE WET RAUTENBACH
Department of Geography, Geoinformatics and Meteorology, University of Pretoria, Pretoria 0002, South Africa

In this study, data from the Southern Hemisphere ADditional OZonesondes (SHADOZ) are used to compute the spatially averaged water vapor (WV_g) over low and mid-tropical Africa. Based on the SHADOZ stations, reanalysis data from the National Centres for Environmental Prediction and Atmospheric Research (NCEP/NCAR) were partitioned into four grid cells: (1) over Ascension [10.0°S to 7.5°S, 12.5°W to 15.0°W]; (2) Nairobi [0.0°N to 2.5°S, 35.0°E to 37.5°E]; (3) Irene [27.5°S to 25.0°S, 27.5°E to 30.0°E] and (4) Reunion [22.5°S to 20.0°S, 55.0°E to 57.5°E]. The mean Water Vapor (WV) was computed over these grid cells to obtain (WV_g). The temporal scales of WV_g fluctuations have been analyzed using wavelet transforms and Principal Components. Analysis (PCA) of the WV_g wavelet-transformed coefficients was used to demonstrate spatial organization of WV. The results indicate that WV exhibits localized spatial coherence patterns. Further, the analysis show strong seasonal dependence of WV which is associated with global and local atmospheric circulation.

1. Introduction

Generally, Water Vapor (WV) plays a key role in both the radiative and dynamic processes of the climate system (Zveryaev *et al.*, 2007). It is a major greenhouse gas which predominantly absorbs the Earth's outgoing thermal energy while, part of this energy re-radiates back to the surface. It is the most variable atmospheric constituent (Bevis *et al.*, 1992; Sun and Lindzen, 1993) which has applications in both short-term Numerical Weather Prediction (NWP) (Cucurell *et al.*, 2000) and in Earth's climate changes and hydrological cycles (Johnsen, 2003). Global distribution and variability of atmospheric WV has also been documented in the literature (see Dai, 2006) and there is also documented literature on WV variability over regional scales (e.g., Trenberth *et al.*, 2005).

A number of ground based and space-borne remote sensors are available that measure vertical and horizontal profiles of water vapour: e.g., radiosondes, light detection and Ranging (LiDAR-Raman), Global Navigation Satellite Systems (GNSS) receivers, Very Long Baseline Interferometry (VLBI) and Water Vapor Radiometers (WVR) (Raschke, 2002). One important consideration in WV analysis is to show how the WV fluctuations are organized into diurnal, synoptic, seasonal and climatic categories and if WV variability is associated with spatial structure and dominant modes of the rotated Principal Component Analysis (PCA) of the WV energy spectra. The rotated PCA component of WV would capture the dominant modes of the WV in temporal scales with similar spatial organization (Petr, 2005). This linear transformation of PCA allows for easy interpretation of the strongest spatial relationships of WV features that drive atmospheric weather systems, such as clouds, over a particular region. Although, the global spatial distribution and trends in WV are dominated by large-scale dynamics, such as the El Nino-Southern Oscillation (ENSO) rather than the thermodynamics, (see Zveryaev and Allan, 2005), the linkage between WV anomalies and atmospheric circulation processes is difficult to establish due to the complexity of the spatial-temporal structures of WV. The spatial and temporal variability of WV in the mid and low-tropical Africa ranges from a few kilometers to thousands of kilometers, and from a few minutes to several days, similar to meso or synoptic-scale processes, respectively (Husak, 2005). Therefore, analyses of correlations of WV between the spatial grids are of great practical importance for a better understanding of the background processes that lead to the development of hazardous weather systems (for example floods, thunderstorms, tropical cyclones).

In order to understand the feedback processes in mid- and low-tropical Africa, a robust methodology of examining the spatial-temporal structure of WV is required. In this study, the spatial and temporal organization of WV is analyzed simultaneously using orthogonal wavelet transform which allows for calculating the total energy of WV by accumulating individual scales of the spatial or temporal scale. Results obtained would form the basis for future comprehensive analysis to establish the relation between WV variability and the associated atmospheric weather systems, as well as any other forcing mechanisms observed in low- and mid-tropical Africa.

2. Data and Methodology

The main data source used in this study is about 8 years worth of upper air radiosonde/ozonesonde data archived at the Southern Hemisphere ADditional OZonesondes (SHADOZ) station network of four stations: Ascension, Irene, Reunion and Nairobi (see Thompson *et al.*, 2003 for further details about the SHADOZ network). The geographical locations of the SHADOZ stations and details about data periods considered in this paper are tabulated in Table 1.

The SHADOZ stations were configured in order to obtain spatially averaged WV values (WV_g) over four grid boxes and time series from reanalysis data of the National Centres for Environmental Prediction and Atmospheric Research (NCEP/NCAR) (Kalnay *et al.*, 1996). The time series of integrated WV (hereafter, WV_{ncep}) was constructed from the average of the closest four NCEP/NCAR reanalysis data grid points at each one of the SHADOZ stations (see Fig. 1) using NCEP/NCAR reanalysis data. Take note that the NCEP/NCAR reanalysis data has a latitude and longitude resolution of 2.5°.

For each NCEP/NCAR grid point, the temporal series of WV_g is tested manually for inherent normal distribution and then transformed into Box–Cox (Box and Cox, 1964) which ensures a normal distribution. Prior to the

Table 1. SHADOZ stations used in the current study and the time period considered for analyses.

Station	Longitude[°]	Latitude [°]	Elevation [m]	Launch [No]	Time period
Nairobi	36.80 E	1.27 S	1795.00	370	Jan 1998 to Aug 2007
Reunion	55.48 E	21.06 S	24.00	293	Jan 1998 to Oct 2006
Irene	28.22 E	25.90 S	15.24	232	Nov 1998 to Dec 2006
Ascension	14.42 W	7.98 S	91.00	397	Jan 1998 to Dec 2006

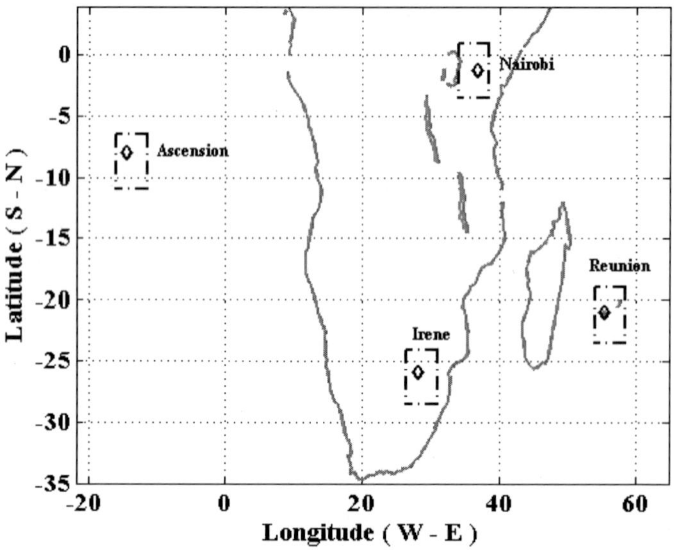

Fig. 1. The SHADOZ stations with the corresponding grid boxes formed by the closest four grid points of reanalysis data from the National Centers for Environmental Prediction and Atmospheric Research (NCEP/NCAR).

Box–Cox transformation, the data have been detrended. Further, in order to account for latitudinal distortions, each point of WV_g anomalies was weighted by the square root of the cosine of latitude (North *et al.*, 1982). The resulting time series has been linearly detrended and subjected to non-decimal Haar wavelet (Lindsay *et al.*, 1996) transform to capture localized temporal fluctuations.

Comparable to the Fourier Transform (FT), the wavelet power spectrum (absolute value squared of the wavelet transform) provides the total energy of the WV_g time series at a given scale. FT gives information about what frequencies are present in the signal, but lacks the ability to correlate the frequencies with the time of their presence. In general, the difference between Fourier and wavelet coefficients is that the former is influenced by a function on its entire domain (global measure), while the latter is influenced by local features. The wavelet power spectrum is therefore chosen in this paper as a better measure of variance attributed to localized events.

The wavelet coefficients at each time scale were used to compute the energy spectrum per spatial scale to form a temporal scale series (S) over the grid points (G): i.e. a matrix D with dimensions of S $\times$ G.

3. Results

A daily vertically integrated WV_g time series in mid- and low-tropical Africa is computed from the SHADOZ network points and from gridded NCEP/NCAR reanalysis data that extent over a period of 8 years. The SHADOZ point data were integrated over the height column from 2 km to 13 km and whereas, the NCEP/NCAR water vapor data integrated between the 700 hPa and 200 hPa pressure levels. The upper and lower limits chosen were, respectively, based on the sensitivity of the balloon-borne measurements and mean-sea level of the SHADOZ stations. In order to compare the corresponding WV measurements from NCEP/NCAR reanalysis data, corresponding SHADOZ dates are considered. The NCEP/NCAR reanalysis data are available four times a day (every six hours) and the daily mean was computed for comparison with SHADOZ observations which occur once a day.

The calculated WV_g values from radiosonde measurements at the SHADOZ stations and from the NCEP/NCAR reanalysis data are plotted in Fig. 2. It is clear from the figure that the NCEP/NCAR reanalysis data

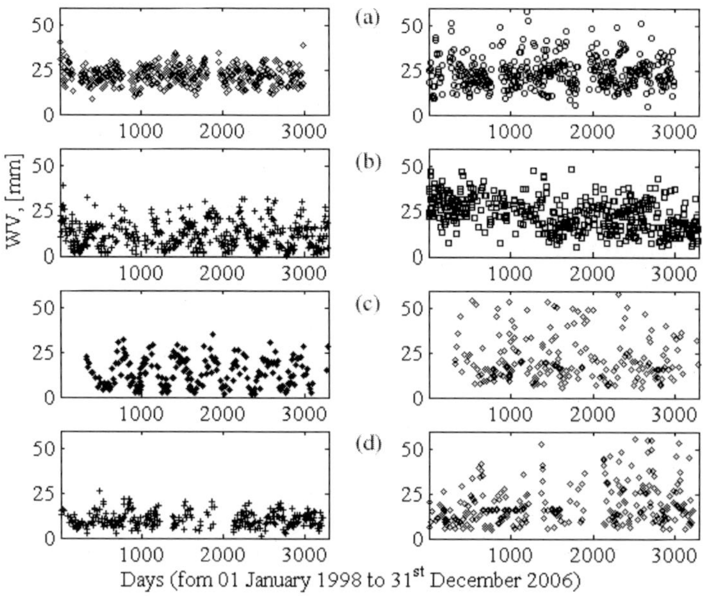

Fig. 2. Daily integrated spatially averaged Water Vapor (WV_g) in mm, calculated from four SHADOZ stations (right) and the gridded NCEP/NCAR reanalysis data (left) in the lower and mid-tropical Africa region over the 8-year period 1998–2007 for the SHADOZ stations (a) Nairobi (b) Ascension (c) Irene and (d) Reunion.

exhibit a cyclic trend over the period of observations, whereas such cycles are not evident in the SHADOZ observations. The difference might be due to the coarse latitude and longitude resolution of NCEP/NCAR data that were averaged over the station grid box, while each SHADOZ station corresponds to a particular location. In addition, sensitivity of the balloon measurements may have contributed to the differences in WV_g from the two measurements. Further to this NCEP/NCAR reanalysis data are based upon simulation with possible inherent biases. The differences between the NCEP/NCAR reanalysis data and SHADOZ station data were calculated for each station. Results concluded that the Irene and Reunion stations have higher mean deviations ($\sim$ 40 mm) while the Nairobi and Ascension stations show a mean WV_g deviation of $\sim$ 30 mm (figures are not shown).

It is understandable from Fig. 2 that WV_g fluctuations are difficult to discern from the time series. The excursions from the mean denote the presence of exogenous processes that play a significant role in WV_g fluctuations. These stochastic processes are manifestations of local weather system processes (eg. convection, precipitations, etc.). In order to better understand these fluctuations, the nature of distribution needs to be known. The standard probability distributions of WV_g are used and are compared to the normal Guassian distribution. The normal Guassian distribution has been generated by selecting random data sets. In order to assess the normal (Gaussian) distributions of WV_g, the Quartile–Quartile (QQ-plot were drawn between the Guassian generated and the probability distribution of WV_g). A linear variation in the QQ plot could signify a normally distributed time series. The distribution has been tested, individually for each station (Fig. 3). The regression co-efficients obtained illustrate that the SHADOZ station Ascension has high linearity in comparison to that of Nairobi, Irene and Reunion. A maximum non-linear fluctuation component of $\sim$10% was obtained for Reunion. On the other hand, Irene, Nairobi and Ascension have values of $\sim$8%, 5% and 1%, respectively. It implies that Ascension results follow a normal distribution and appear not to be affected by non-linear local weather conditions.

In order to study the local temporal fluctuations of WV, the Haar wavelet transform of Maximum overlap discrete technique has been applied. The wavelet coefficients derived from the wavelet transform capture local fluctuations in time series in both time and frequency. Each SHADOZ station's data were grouped in terms of month and year, following by the calculation of the corresponding mean. The corresponding monthly mean over the 8 year period of data is subjected to the wavelet transform after

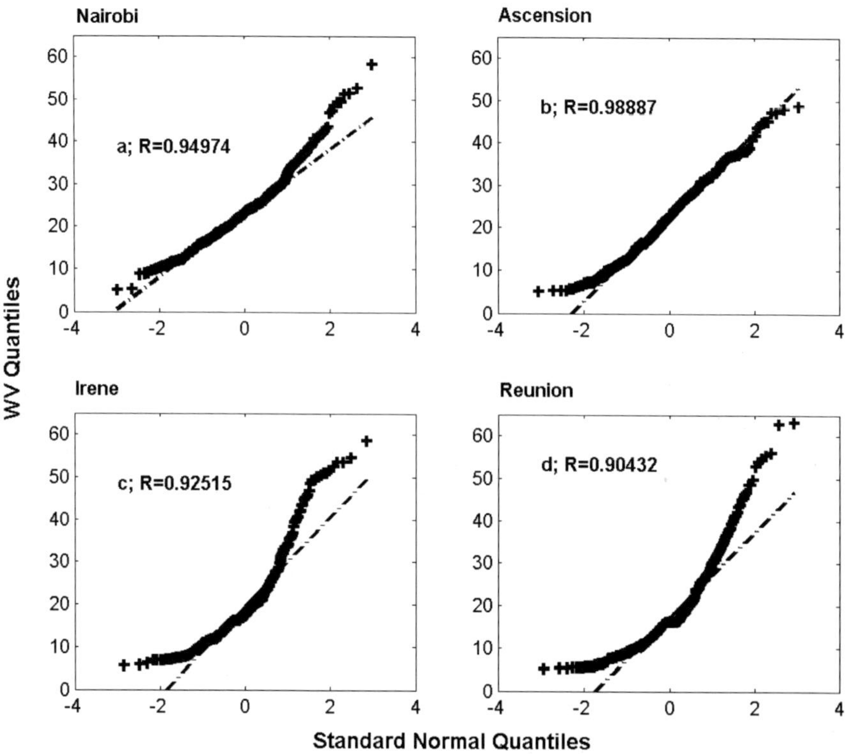

Fig. 3. Quartile–quartile (QQ) plot of a Gaussian distribution, and the probability distribution of WV_g at the four SHADOZ stations under consideration.

performing de-trending. Figure 4 depicts the obtained wavelet co-efficient (amplitude) at different temporal scales of 3, 8, 12 and 36 months (from bottom to top) or the time period of oscillation of WV_g at a given location. The relation between period of oscillation of WV_g fluctuations and the wavelet scale index is based on the relation $s = 2^{j-1}$, where the jth index denotes the period. The method of deducing the wavelet co-efficient is documented in Percival and Walden (2000). Although, scale-1 ($\sim$3 months) does not offer any clear information on the fluctuations, other higher order scales show a significant oscillation at all the stations. Notably, the annual oscillation (scale-3) is clearly distinguishable at all the stations. If compared to the other stations, Nairobi exhibits a clear cyclic variation. For almost all the stations, the scale-4 (3-year) component does not complete one period of a cycle, inferring that the periodicity is more than 12 years. It is noted here that the maximum possible number of scales obtained depends on the length

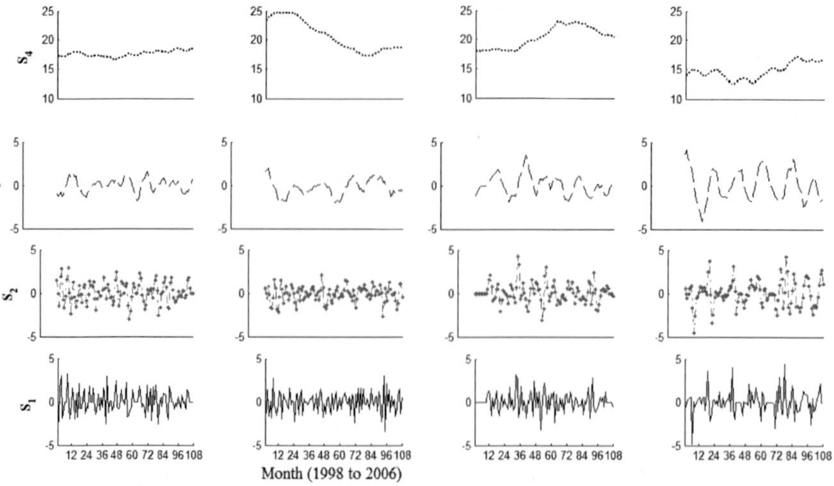

Fig. 4. Haar wavelet spectra at different scales (n = 1, 2, 3, 4) and at different station locations (Ascension, Reunion, Irene and Nairobi) — from left to right, respectivily. See Table 1 for the corresponding geographic coordinates.

of data period used. The log–log plot of the wavelet energy (not shown here) reveal an approximate power law scaling at lower time scales, which breaks down at high time scales. These results are consistent with those of Lay (1997). At high time scales, the break down in the linear relationship is associated with response of WV_g fluctuations to tele-connection patterns such as the influence of ENSO in the low- and mid-tropical Africa, see for example Trenberth *et al.* (2005).

PCA has been determined for the wavelet coefficients of all the four stations, and the calculated variance is presented in Fig. 5. The first three variance components account for 98% of the WV_g variations. The first component represents high frequency temporal fluctuations (monthly time scales), and accounts for 67% of the variability. Component two

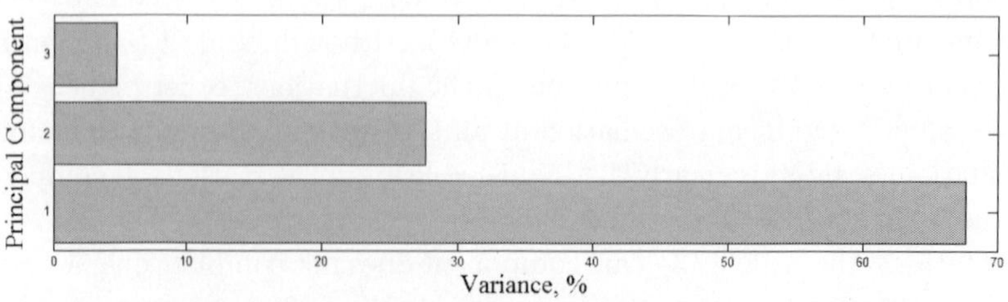

Fig. 5. Co-variance of the Principle Component Analysis (PCA) components obtained from the four stations under consideration.

represents the variance associated with annual fluctuations, and accounts for about 27% of the WV_g fluctuations. About 4% of WV_g variability is associated with low frequency fluctuations (1 < timescales < 9 years). Decadal fluctuations cannot be inferred convincingly due to the short time-span of the data (8 years: 1998 to 2006. These results show that there is a distinct spatial structure for each short term temporal WV_g variation in the low and mid-tropical Africa region that could be attributed to synoptic/seasonal-scale weather systems, which is consistent with the findings from Husak (2005) who reported that seasonal weather systems, topography, the Inter-tropical Convergence Zone (ITCZ) and monsoon winds affect WV distribution and fluctuation. Jin *et al.* (2008) also reported that the variability of water vapor in China is dominated by seasonal variations. In addition, the spatial distribution of WV dependence on the thermodynamic relationship between WV and temperature has been reported in Zveryaev and Allan (2005). The marked differences between WV fluctuations at longer timescales could be attributed to the WV response to tele-connection patterns such as ENSO in the low- and mid-tropical Africa, which is in line with the findings of Trenberth *et al.* (2005) who had indicated that the variability of WV is dominated by the evolution of ENSO. This link shows a strong relationship over the oceans between WV and Sea Surface Temperatures (SSTs). Further, the African low- and mid-latitude WV has a strong link to rainfall due to its close association with the mean wind flow, and convergence of moisture by trade winds as well as the links to SSTs. In addition, the correlation analyses performed between surface temperature and WV show that there exists a link between WV anomalies to regional air temperature variations with marked seasonal dependence (the results are not shown here) over all four SHADOZ stations.

4. Concluding Remarks

In an effort to analyze regional spatial and temporal features of WV variability over low- and mid-tropical Africa, NCEP/NCAR reanalysis data around the SHADOZ network of four stations were used to calculate spatially averaged WV (refer to VW_g) over the period 1998 to 2006. WV_g was calculated as the spatial average of the four closest NCEP/NCAR grid points around the SHADOZ stations to form grid cells. Based on these grid cells, data from NCEP/NCAR reanalysis data were also used to calculate the vertically integrated column of WV over the same time epoch for comparison. For the first time, the WV_g variability in the low- and mid-tropical Africa were analyzed using point data from the SHADOZ network indicating high frequency fluctuations in the wavelet space. Common to the entire SHADOZ network considered in this study is the pattern of temporal WV_g fluctuations with monthly time scales dominating. This dominant variance appears to be associated with locally driven WV variations such as the local weather systems. Our results show the power law scaling in the wavelet energy. The approximate log-log linear relationship at smaller temporal scales that breaks down at synoptic scales suggests that the energies at WV_g on different temporal scales are closely related. In addition, from the PCA, three dominant modes emerge that explains ~98% of the total spatial variance of the normalized energy. To validate the current findings, future studies will involve the use of observations such as HALOE (ref. Russell *et al.*, 1993), ECMWF (ref. Bock *et al.*, 2007) and regional numerical simulation model data sets to determine the temporal and spatial organization of PWV data at finer spatial and temporal scales.

References

1. M. Bevis, S. Businger, T. A. Herring, C. Rocken, R. A. Anthes and R. H. Ware, GPS meteorology: Remote sensing of the atmospheric water vapor using the global positioning system, *J. Geophys. Res.* **97** (1992) 15787–15801.
2. G. E. Box, and D. R. Cox, An analysis of transformed data, *J. Roy. Statist. Soc. B.* (1964) 211–252.
3. O. Bock, M.-N. Bouin, A. Walpersdorf, J. P. Lafore, S. Janicot, F. Guichard and A. Agusti-Panareda, Comparison of ground-based GPS precipitable water vapor to independent observations and NWP model reanalyses over Africa, *Q. J. R. Meteorol. Soc.* **133** (2007) 2011–2027.
4. L. Cucurull, B. Navascues, G. Ruffin, P. Elosegui, A. Rius and J. Vila, The use of GPS to validate NWP systems: The HIRLAM model, *J. Atmos. Oceanic Technol.* **17** (2000) 773–777.
5. A. Dai, Recent climatology, variability of and trends in global surface humidity, *J. Climate* **19** (2006) 3589–3606.
6. G. J. Husak, Methods for statistical evaluation of African precipitation, Ph.D. thesis, University of California, Santa Barbara (2005).
7. S. Jin, Z. Li and J. Cho, Integrated water vapor field and multiscale variations over China from GPS measurements, *J. Appl. Meteorol. Climatol.* **47** (2008) 3008–3015.
8. K. P. Johnsen, GPS atmosphere sounding project — An innovative approach for the recovery of atmospheric parameters (GKSS-Forschungzentrum Geesthacht, Geesthacht, 2003).

9. E. Kalkany, M. Kanmistu, R. Kistler, W. Collins, D. Deaven, L. Gandin, M. Iredell, S. Saha, G. White, J. Woollen, Y. Zhu, A. Leetmaa, B. Reynolds, M. Chelliah, W. Ebisuzaki, W. Higgins, J. Janowiak, K. Mo, C. Ropelewski, J. Wang, R. Jenne, and D. Joseph, The NCEP/NCAR 40-year reanalysis project, *Bull. Am. Meteorol. Soc.* **77** (1996) 437–471
10. O. P. Lay, The temporal power spectrum of atmospheric fluctuations due to water vapor, *Atron. Astrophys. Suppl. Ser.* **122** (1997) 535–545.
11. R. W. Lindsay, D. B. Percival and D. A. Rothrock, The discrete wavelet transform and the scale analysis of the surface properties of sea ice, *IEEE Trans. Geo. Rem. Sens.* **34** (1996) 771–787.
12. G. North, T. L. Bell and R. F. Calahan, Sampling errors in the estimation of empirical orthogonal functions, *Man. Weather Rev.* **110** (1982) 699–706.
13. D. B. Percival and A. T. Walden, *Wavelet Methods for Time Series Analysis* (Cambridge University Press, Cambridge, 2000).
14. P. Petr, Water quality assessment using SVD-based principal component analysis of hydrological data, *Water SA* **31** (2005) 417–422.
15. E. Raschke, Water vapor in the atmosphere, Technical report work package HCP of the CM-SAF, 3001393-RIN, Deutscher Wetterdienst, Offenbach, Germany (2002).
16. J. M. Russell III, L. L. Gordley, J. H. Park, S. R. Drayson, D. H. Hesketh, R. J. Cicerone, A. F. Tuck, J. E. Frederick, J. E. Harries and P. J. Crutzen, The halogen occultation experiment, *J. Geophys. Res.* **98** (1993) 10777–10797.
17. D. Sun and R. S. Lindzen, Distribution of tropical tropospheric water vapor, *AMS* **50** (1993) 1643–1660.
18. K. E. Trenberth, J. Fasullo and L. Smith, Trends and variability in column-integrated atmospheric water vapor, *Clim. Dyn.* **24** (2005) 741–758.
19. A. M. Thompson, J. C. Witte, S. J. Oltmans, F. J. Schmidlin, J. A. Logan, M. Fujiwara, V. W. J. H. Kirchhoff, F. Posny, G. J. R. Coetzee, B. Hoegger, S. Kawakami, T. Ogawa, J. P. F. Fortuin and H. M. Kelder, Southern Hemisphere Additional Ozonesondes (SHADOZ) 1998–2000 tropical ozone climatology. Tropospheric variability and the zonal wave-one, *J. Geophys. Res.* **108** (2003) 8241, doi:10.1029/2002JD002241.
20. I. I. Zveryaev and R. P. Allan, Water vapor variability and in the tropics and its links to dynamics and precipitation, *J. Geophys. Res.* **110** (2005) D21112, doi:10.1029/2005JD006033.
21. I. I. Zveryaev, J. Wibig and R. P. Allan, Contrasting interannual variability of atmospheric moisture over Europe during cold and warm seasons, *Tellus* **60A** (2007) 32–41, doi:10.1111/j.1600-0870.2007.00283.

Advances in Geosciences
Vol. 16: Atmospheric Science (2008)
Eds. Jai Ho Oh *et al.*

AEROSOL MEASUREMENTS OVER SOUTH AFRICA USING SATELLITE, SUN-PHOTOMETER AND LIDAR

VENKATARAMAN SIVAKUMAR*,†, MELAKU TESFAYE*, WONDIMU ALEMU**, AMEETH SHARMA, CHRISTOPH BOLLIG and GIZAW MENGISTU**

National Laser Centre, Council for Scientific and Industrial Research, P.O. Box 395, Pretoria, South Africa
†*svenkataraman@csir.co.za*

**Department of Geography, Geoinformatics and Meteorology, University of Pretoria, Lynnwood Road 0002, South Africa*

***Department of Physics, Addis Ababa University, P.O. Box 1176, Addis Ababa, Ethiopia*

In this study, we present the climatological picture of aerosols over South Africa using 20 years of Stratosphere Aerosol and Gas Experiment (SAGE-II) satellite, 6–10 years of AErosol RObatic NETwork (AERONET) and ground based mobile LIDAR datasets The climatological variation of aerosol concentration indicate minimum during winter and maximum over September months. The satellite and ground based measurements are found to be in good agreement with each other. The study affirms the presence of fine and accumulation mode aerosols over industrial areas of South Africa.

1. Introduction

Aerosols (solid particles suspended in the air) play an important role in the global climate, the radiative forcing of the climate, and the Earth's radiative balance. The interaction between atmospheric aerosol and incoming solar radiation influences the radiative forcing which in turn affects temperature. Aerosols can also have a significant impact on visibility and air quality, with potential economic consequences for tourism, cloud formation, meteorology, and climate. Aerosol particles are a largely natural, though highly variable component of our atmosphere. This is due to widely varying aerosols microphysical properties, such as their sources, sinks and loading, composition, size distribution, chemical interaction, life time and diurnal variation in space and time. Atmospheric aerosols originate from anthropogenic (human) activities such as biomass burning and industrial pollution as well as from natural processes, such as wind generated dust and sea spray, volcanic eruptions and smoke from natural forest fires, etc.

The knowledge of aerosol characteristics at a local and global scale, their temporal change interrelations with other atmospheric parameters and with solar radiation is of great importance for atmospheric research. Improved aerosol climatology may enable more accurate estimations of the direct and indirect aerosol forcing [1, 2]. The aerosol optical depth (AOD), which is an indicator of the aerosol loading in the vertical column of the atmosphere, constitutes the main parameter to assess the aerosol radiative forcing and its impact on the climate. The rationale for analyzing the Angstrom exponent (α) which is important in the interpretation of these datasets and in providing further information on the particle size. The values of both parameters exhibit strong depends on the amount of aerosols of different sizes and number concentration — their chemical composition and the wavelength of the incident radiation. It is critical that scientists continue to gather information about aerosols, especially in the Southern hemisphere regions [3]. In and around Southern Africa, major cities produce large amounts of aerosols as a result of industrial activities and automobile emissions, in addition to the natural bio-mass burning, volcanic eruptions and desert dust (Sahara in North Africa).

In this paper, we present our results over Southern Africa regions based on ground based and space borne datasets. It includes the data obtained from (a) Stratosphere Aerosol Gas Experiment (SAGEII), (b) Sun-Photometer measurements through AEROsol NETwork (AERONET) programme and (c) mobile LIDAR (LIght Detection And Ranging) system operational at the Council for Scientific and Industrial Research (CSIR) National Laser Centre (NLC), Pretoria (25°5°S; 28°2°E), South Africa.

2. Data

2.1. *SAGE-II data*

SAGE-II was launched into orbit aboard the Earth Radiation Budget Satellite in October 1984. The SAGE-II instrument vertically scans the limb of the atmosphere during sunsets and sunrises (15 observations) each day. During each sunrise and sunset encountered by the orbiting spacecraft, the instrument uses the solar occultation technique to measure attenuated solar radiation through the Earth's limb in seven channels centered at wavelengths ranging from 0.385 to 1.02 μm. The exo-atmospheric solar irradiance is also measured in each channel during each event for use as a reference in determining limb transmittances. The transmittance measurements are inverted using the "onion-peel" approach to yield 1 km vertical resolution profiles of aerosol extinction, ozone, nitrogen dioxide, and water vapour [4, 5]. The SAGE-II instrument has collected vertical profiles of stratospheric and troposphere aerosol extinction at four wavelengths (0.385, 0.453, 0.525, and 1.02 μm) with high resolution since the program's inception in October 1984. Near-global coverage 80°S to 80°N was achieved over time spans of about 1 month. The instrument mission was terminated on 8 September 2005. For clear geographical observation of the trend of aerosols, we have extracted the aerosol parameter; the aerosol extinction coefficients derived from version 6.20 series of 21 years (1984–2005) of data over southern Africa region (Latitude, 15°S to 40°S and 10°E to 40°E and Longitude).

2.2. *Aeronet*

Aerosol Robotic Network (AERONET) is a federation of ground-based remote sensing aerosol network which provides a long-term, continuous and readily accessible public domain database of aerosol optical, microphysical and radiative properties for aerosol research and characterization, validation of satellite retrievals, and synergism with other databases (www.aeronet.gsfc.nasa.gov). Basically, it uses the sun-photometer at different wavelengths to obtain aerosol microphysical and radiative properties. AERONET collaboration provides globally distributed observations of spectral AOD, Angstrom exponent, inversion products, and perceptible water in diverse aerosol regimes [6, 7]. The present study examines the climatology of aerosols over South Africa (three sites) for a longer period of data (see Table 1) over selected locations characterized by differing environments, and influenced by various air masses, anthropogenic activities and natural sources. From seven South African AERONET sites

Table 1. Quantity of data used in three selected South African AERONET sites.

Site and Location	Data used
Skukuza (24 S, 31 E)	January 1998–December 2008
Johannesburg (26 S, 28 E)	January 2002–December 2008
Bethlehem (28 S, 28 E)	January 1996–December 2001

selection of three stations were based on the criterion of the availability of enough (6–10 years) cloud-screened and quality-assured (level 2.0) aerosol data.

2.3. *LIDAR*

CSIR-NLC mobile LIDAR is also used for the present study. A neodymium-doped yttrium aluminium garnet (Nd:YAG) laser, which is presently employed at the second harmonic (532 nm), is used for transmission at a repetition rate of 10 Hz. The receiver system employs a Newtonian telescope configuration with a 406 mm diameter primary mirror. The backscattered signal is subjected to fall on the primary mirror of the telescope and is then focused toward a plane mirror kept at an angle of 45 degrees. It is detected by the Photo-Multiplier Tube (PMT) and the PMT output signal is transmitted to the transient digitizer and PC for analysis and archiving. The data acquisition is performed by a transient recorder which communicates with a host computer for storage and offline processing of data. More details about the system are available in the literature [3, 8, 9]. The LIDAR inversion technique is then applied to the signal returns in order to obtain the aerosol backscatter and extinction co-efficient. Retrieve aerosol extinction coefficients are used to determine the optical depth with SAGE-II satellite measurement and are being compared with AERONET results.

3. Results and Discussion

3.1. *SAGE-II aerosol extinction measurements*

The SAGE-II provides a height profile of aerosol extinction coefficient for the height region from 0.5 km to 40 km. We have grouped the SAGE-II retrieved profiles in terms of months to obtain the individual monthly mean aerosol-extinction profiles for the height region from 0.5 km to 40 km. The monthly mean aerosol extinction profile obtained for Southern Africa region is shown in Fig. 1. We have considered the SAGE-II profile as far as possible above 3–4 km, keeping in mind that the lower height region measurements are inaccurate due to a low signal to noise ratio (SNR) [10]. Figure 1 shows a larger extinction values during late summer periods but the accuracy of the satellite measurement is questionable at lower height regions. Relatively, it is found that the larger values below 5 km might be due to aerosol loading in the lower troposphere. The variations show a moderate value during winter

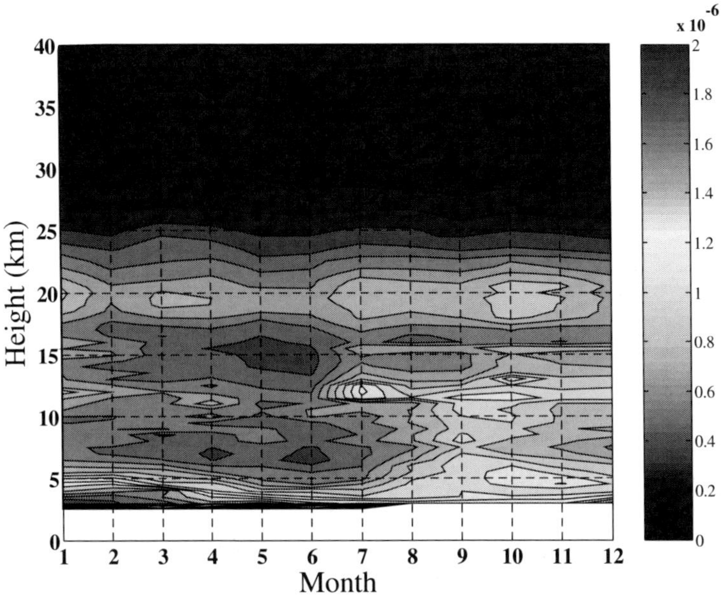

Fig. 1. Monthly variation of aerosol extinction coefficient for Southern Region of Africa, for = 525 μm.

period and enhanced values during the September month. In addition, the lower stratosphere at approximately 20 km, shows an aerosol layer (Junge layer) which is in general dominated by liquid sulfate aerosols and soot particles. This illustrates a semi-annual oscillation (SAO) with peaks during March and October months.

Aerosols at any given location in the tropics are subject to large variations by day, month and year. In order to observe the inter-annual variability of aerosols, we have studied the long-term variations in aerosol concentration based on the above 20 years of data. Figure 2 shows the lower stratosphere (18 km–21 km) aerosol extinction measured at 525 nm using SAGE-II data. Here, we have used the every-year-individual-monthly-mean datasets. Though, there are some data gaps (less than 2 months over year), we have plotted a running average on the measured data sets (solid-line). The data gaps are partly due to the satellite coverage periodicity which is generally about two months. The figure illustrates a clear annual oscillation of the aerosol concentration with a maximum during summer months (Nov–Jan) for most of the years. It also shows aerosol concentration

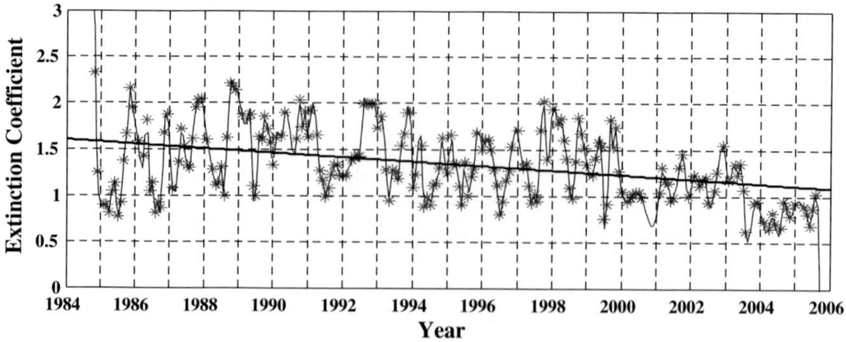

Fig. 2. Temporal evolution of lower stratosphere aerosol extinction at 525 nm over Southern Africa region and measured by SAGE-II.

has decreased in the lower stratosphere height region of ~15% over the 20-year period which indirectly reflects the loading of the aerosol concentration in the troposphere. The initial increase in aerosol concentration might due to the Pinatubo volcanic eruptions.

3.2. *AERONET — Sun photometer measurements*

In this study, we investigate the climatology of aerosols over South Africa using the monthly average AOD and Angstrom exponent (α). The AOD is obtained from direct-beam irradiance measurements measured by the AERONET sun photometer at 500 nm wavelength and the Angstrom exponent, determined from the spectral dependence of the measured optical depth, indicator of the aerosol size and its variations. Values of α approaches zero correspond to coarse-mode aerosols (sea spray and desert dust), while values of α above 1.5 indicate significant presence of fine-mode particles (mainly smoke or urban aerosols) and α is greater than 2, for very small particles in the Rayleigh limit [11].

The monthly average variation of the AOD_{500} (AOD at 500 nm) and $\alpha_{440-675}$ (Angstrom exponent (α) in the wavelength band 440 nm–675 nm) are presented in Figs. 3 and 4 for all three sites. Johannesburg and Skukuza show a clear annual oscillation (AO) with maximum values during summer (October–March) and minimum during the winter (May–July). On the other hand, the Bethlehem site illustrates a semiannual oscillation (SAO) with two maxima during March and October and minimum values during the winter period. In comparison to the three sites, Skukuza shows a larger

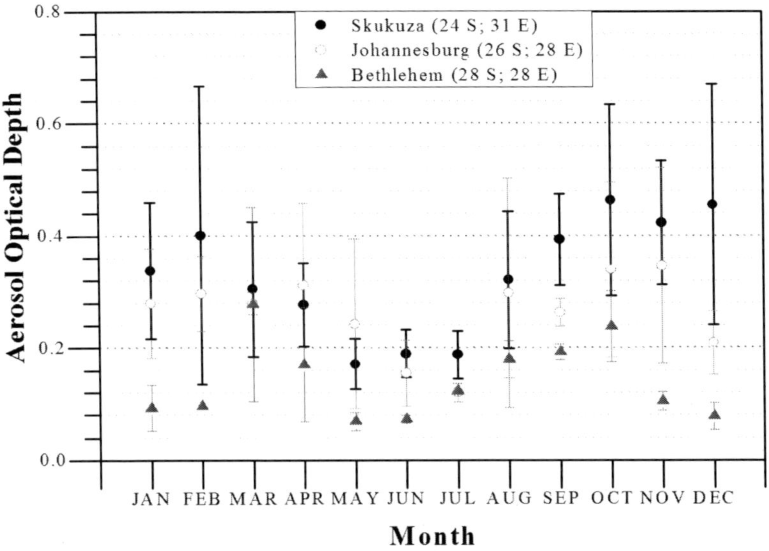

Fig. 3. A six-year's monthly average AOD_{500} over selected South Africa AERONET sites.

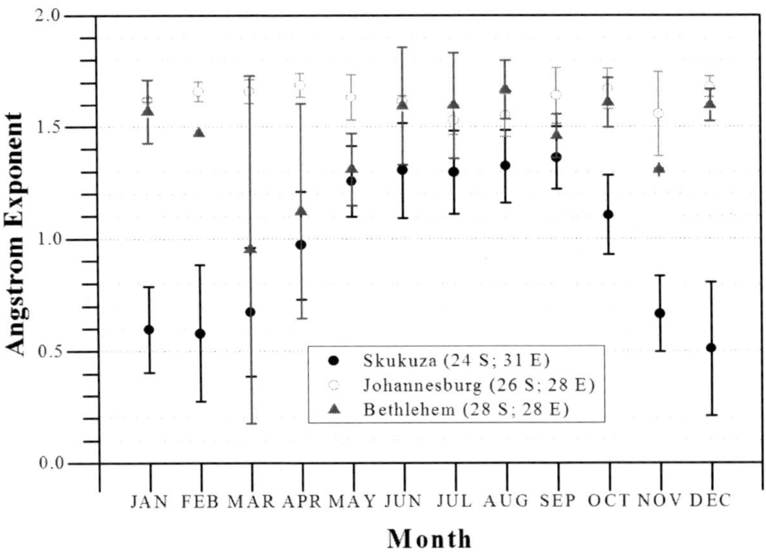

Fig. 4. Same as Fig. 3 but for Angstrom exponent for 440 nm to 675 nm wavelength regions.

AOD values, this might due to biomass burning [10, 12]. The individual monthly variations (measured interms of standard deviation) show a higher variability for Johannesburg and Skukuza than Bethlehem.

The interesting result from the determination of the Angstrom exponent in the selected wavelength regions (Fig. 4, Bethlehem and Johannesburg regions), indicates that the relative influence of fine and accumulation-mode aerosols in the VIS band is much more sensitive to scattering and/or absorption of solar and terrestrial radiation than coarse-mode aerosol particles. Furthermore, the interpretation of the information contained in monthly mean values of $\alpha_{440-675}$ based on a reported studies [11, 13] shows a higher α value in the longer wavelengths which illustrates anthropogenic combustion processes mainly resourced from the fossil fuels or urban/industrial aerosols. This explains the dominance of fine and accumulation mode particles over those regions. At the Skukuza site, the Angstrom exponent value is less than 1 which indicates the significant presence of coarse-mode particles. Short and long-range transport hygroscopic aerosols, combustion processes and their interaction with atmospheric moisture results in dominant coarse-mode particles.

3.3. *LIDAR measurements*

The extinction profile was derived from the LIDAR and compared/validated using ground based and satellite borne instruments. Figure 5 presents the height profile of the extinction coefficient derived from the LIDAR data taken during the nights of 23 February 2008. The profiles are overlapped by the Stratosphere Aerosol Gas Experiment (SAGE-II) extinction data at 525 nm collected over southern Africa regions. Here, we have used the corresponding monthly-mean extinction profiles (February). The extinction profiles derived from LIDAR and SAGE-II are in close agreement with respect to trend and magnitude. The LIDAR profile has been terminated above 4 km due to thick cloud passage. One is able to observe the boundary layer peak at ~2.5 km which is considered an important parameter for model and atmosphere mixing (including pollutants). The presence of a cloud results in a sharp enhancement in the extinction and backscatter co-efficient to a high value making the detection quite unambiguous.

The above mentioned height profile of aerosol extinction coefficients obtained using the LIDAR and SAGE-II satellite data are integrated appropriately to get the aerosol optical depth. Generally, we considered

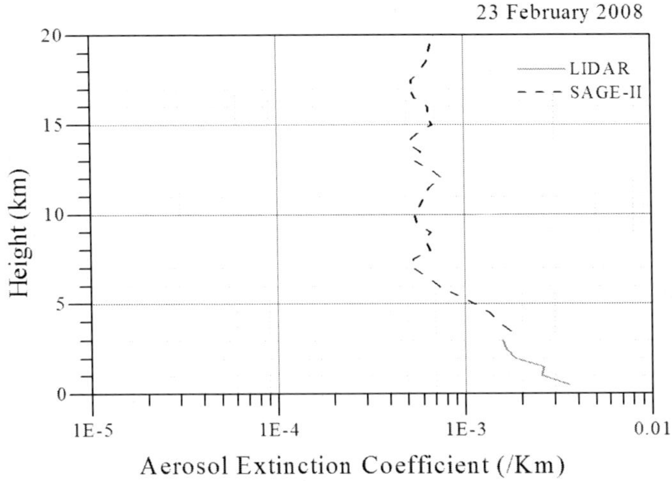

Fig. 5. Height profile of aerosol extinction derived from LIDAR signal returns and SAGE-II satellite data.

the LIDAR profile for the lower height region with respect to the SNR and at higher altitudes from the SAGE-II data. We found the value for February months is around ~0.224 and for April about 0.3227 which is in good agreement with AOD measured by the photometer over Johannesburg. The values are within the variations of the AOD reported (February: 0.2966 ± 0.06668 and April: 0.31234 ± 0.14707).

4. Future Perspectives and Concluding Remarks

The monthly variations of aerosol concentration over South Africa are moderate during the winter period and enhanced during the September month. The annual average values of AOD and Angstrom exponent climatology indicate that the urban/industrial areas of South Africa are dominantly loaded by fine and accumulation-mode aerosols produced from urban/industrial or biomass-burning activities. The measured aerosol optical depth by combined LIDAR and SAGE-II satellite are in good accordance with the value measured by sun-photometer.

Long-term observations from satellite remote sensors, complemented by LIDAR and *in-situ* measurements, significantly improve our understanding of the climatology of stratospheric and troposphere aerosols over Southern Africa regions.

Acknowledgments

The authors are also grateful to the different South African funding agencies: Council for Scientific and Industrial Research-National Laser Centre (CSIR-NLC), Department of Science and Technology (DST), National Research Foundation (NRF) (Grant no: 65086) and the African Laser Centre (ALC). We also express our thanks to SAGE-II, AERONET and radiosonde data centres for providing access to the required data.

References

1. Y. J. Kaufman, D. Tanrè, L. Remer, E. Vermote, A. Chu and B. N. Holben, *J. Geophys. Res.* **102** (1997) 17051.
2. M. Masmoudi, M. Chaabane, D. Tanré, P. Gouloup, L. Blarel and F. Elleuch, *Atmos. Res.* **66** (2003) 1.
3. V. Sivakumar, M. Tesfaye, W. Alemu, D. Moema, A. Sharma, C. Bollig and G. Mengistu, *South African J. Science* (2010), in press.
4. W. Chu, M. McCormick, J. Lenoble, *J. Geophys. Res.* **94** (1989) 8339.
5. L. Thomason, L. Poole and C. Randall, *Atmos. Chem. Phys.* **7** (2007) 1423.
6. B. N. Holben, T. F. Eck, I. Slutsker, D. Tanré, J. P. Buis, A. Setzer, E. Vermote, J. A. Reagan and Y. A. Kaufman, *Rem. Sens. Environ.* **66** (1998) 1.
7. B. N. Holben, D. Tanre, A. Smirnov, T. F. Eck, I. Slutsker, N. Abuhassen, W. W. Newcomb, J. Schafer, B. Chatenet, F. Lavenue, Y. J. Kaufman, J. Vande Castle, A. Setzer, B. Markham, D. Clark, R. Frouin, R. Halthore, A. Karnieli, N. T. O'Neill, C. Pietras, R. T. Pinker, K. Voss and G. Zibordi, *J. Geophys. Res.* **106** (2001) 12067.
8. A. Sharma, V. Sivakumar, C. Bollig, C. Van der Westhuizen and D. Moema, *South Africa J. Science* (2010), in press.
9. V. Sivakumar, A. Sharma, D. Moema, C. Bollig, C. Van der Westhuizen and H. van Wyk, *Proc. 24th Int. Laser Radar Conf.*, Boulder, Vol. 99 (2008).
10. P. Formenti, H. Winkler, P. Fourie, S. Piketh, B. Makgopa, G. Helas and M. O. Andreae, *Atmos. Res.* **62** (2002) 11.
11. T. F. Eck, B. N. Holben, D. E. Ward, O. Dubovic, J. S. Reid, A. Smirnov, M. M. Mukelabai, N. C. Hsu, N. T. O' Neil and I. Slutsker, *J. Geophys. Res.* **106** (2001) 3425.
12. S. Generoso, F.-M. Breon, Y. Balkanski, O. Boucher and M. Schulz, *Atmos. Chem. Phys.* **3** (2003) 1211.
13. J. S. Reid, T. F. Eck, S. A. Christopher, P. V. Hobbs and B. N. Holben, *J. Geophys. Res.* **104** (1999) 27473.

Advances in Geosciences
Vol. 16: Atmospheric Science (2008)
Eds. Jai Ho Oh *et al.*

MEAN OZONE AND WATER VAPOR HEIGHT PROFILES FOR SOUTHERN HEMISPHERE REGION USING RADIOSONDE/OZONESONDE AND HALOE SATELITE DATA*

VENKATARAMAN SIVAKUMAR
National Laser Centre, Council for Scientific and Industrial Research, P.O. Box 395, Pretoria, South Africa
Department of Geography, Geoinformatics and Meteorology, University of Pretoria Lynnwood Road 0002, South Africa
svenkataraman@csir.co.za

DESALEGNE TEFERA, GIZAW MENGISTU
Department of Physics, Addis Ababa University, P.O. Box 1176, Addis Ababa, Ethiopia

ONDEGO JOEL BOTAI
Department of Geography, Geoinformatics and Meteorology, University of Pretoria Lynnwood Road 0002, South Africa
joel.botai@up.ac.za

The aim of this work is to construct a model (mean) profile for ozone and water vapor in Southern hemisphere latitude using 14 years (1993–2006) of HALogen Occultation Experiment (HALOE) satellite data and about 10 years (1998–2007) of the Southern Hemisphere ADditional OZonesondes (SHADOZ) balloon measurement data from Nairobi (1.3°S; 36.8°E), Malindi (3.0°S; 40.2°E) and Irene (25.9°S; 28.2°E). A comparison of HALOE mean profile has made between 0° to 10° Southern Hemisphere latitude with Nairobi and Malindi SHADOZ ozonesonde data, 20° to 30° with Irene SHADOZ measurement data, respectively. A good agreement in terms of ozone and water vapour measurements has been found between SHADOZ ozonesonde and HALOE. The relative percentage of difference lies within 5% for the height region from 4.5 km to 30 km altitude whereas, the comparison of mean water vapour show high uncertainty.

1. Introduction

Water vapor (H_2O) and Ozone (O_3) are the most important trace gases in the earth's atmosphere. They play an important role in atmospheric

*This work is partially supported by Africa Laser Centre, South Africa.

dynamics, in numerous homogeneous and heterogeneous atmospheric chemical reactions as well as in the absorption of long and short wave radiation directly or indirectly. Research work has shown that the concentration of H_2O in the middle atmosphere increases [1]. The large abundance of H_2O in the atmosphere has a significant consequence on the earth's climate and due to its large energy transfer associated with phase transition, the short-term dynamics of the atmosphere is also affected [2]. Water vapor also has another crucial importance as positive feedback to atmospheric temperature and vice versa. An 1°C increase in atmospheric temperature warming will cause a 6% increase in H_2O (g) concentration which in turn would lead to further warming, thus initiating positive feedback [3]. Water vapor involves all its three phases in multitude of chemical reactions in the atmosphere. It involves the formation of the Polar Stratospheric Clouds (PSC), which are the reservoirs of halogenated molecules involved in the spring ozone depletion. Acid rain, in the form of (H_2CO_3, HNO_3, H_2SO_4, etc.), is formed by the reaction of CO_2, NO_2, SO_2 in their aqueous phases.

On the other hand, the Ozone layer in stratosphere protects us by absorbing the harmful solar Ultra-Violet (UV) radiation. The downward transport of ozone from the stratosphere, where it is produced, naturally contributes to the ozone abundance in the stratosphere, but it is also formed in the troposphere by sunlight driven chemical reaction cycles, involving oxides of nitrogen ($NO_x = NO + NO_2$), Carbon monoxide (CO), methane (CH_4) and other hydrocarbon compounds. Ozone acts as a green house gas with highest efficiency in troposphere and lowermost stratosphere by absorbing solar and terrestrial infrared radiation [4, 5]. Tropospheric ozone, particularly in the tropics, is not fully understood due to many complex processes taking place. Since both O_3 and H_2O are green house gases, they have great influences on global warming. Increases in their concentrations in the atmosphere causes further global warming effects. In particular, tropical tropospheric air experiences a slow subsidence towards the surface. More rapid upward motions within convective clouds provide sufficient mass to compensate for the sinking motion. This in turn has a consequence on the troposphere-stratosphere boundary.

Because of the crucial roles that water vapor H_2O and O_3 play in atmospheric processes, an accurate knowledge and understanding of the temporal and spatial distribution of these trace gases are important for both climate and weather prediction. An accurate profile of theses trace gases is also very important in atmospheric modeling application. The work of this paper is based on the data obtained from HALOE satellite measurement and the SHADOZ ozonesonde measurements to obtain a height profile of H_2O (g) and O_3 in Southern region of Africa. Such profiles can be used as a reference for comparisons with other measurements.

2. Data

2.1. *HALOE data selection*

HALogen Occultation Experiment (HALOE) was launched in September 1991 the upper Atmosphere Research Satellite (UARS) spacecraft and monitoring atmospheric trace gases since then. The data used here are those of version 19 of the retrieval algorithm on the HALOE website (http://haloedata.larc.nasa.gov/). The selected data is a series of measurements from 1991 to 2005 in consecutive orbits during or as close as possible to periods within latitudinal range from 0°S to 40°S. The data file consists of the ozone and water vapor mixing ratios in altitude levels and the pressure, temperature including the quality (random error) at each altitude. HALOE satellite is a solar occultation experiment designed to monitor vertical distribution of HCl, HF, CH_4 and NO by gas filter correlation radiometry and H_2O, O_3, NO_2 and temperature versus pressure using CO_2 absorption by broadband filter radiometry [6]. HALOE uses the Solar Occultation technique to make measurements of the vertical profile of atmosphere parameters. Here, we have considered the data which corresponds to altitudes greater than 10 km since it is less accurate at height regions lower than 10 km.

2.2. *SHADOZ ozonesonde data*

The SHADOZ network was initiated by NASA in 1998 to develop a coordinated ozonesonde network at tropics. Ozonesonde is used for measuring height profile of ozone from sea level to about 30 km altitude, adding ozone sensor into it. The recorded ozone measurements are taken in units of parts per million by volume, ppmv. The details about the data and quality of ozonesonde measurements can be found in the literature [7, 8]. We used about 10 years of ozonesonde data gathered from 1998 to 2007 of Irene station, from 1999 to 2006 of Malindi station and from 1998 to 2007 of Nairobi stations. The measurement data for a height up to 30 km altitude are collected from SHADOZ data which are archived at http://croc.gstc.nasa.gov/shadoz/site2.html/. The SHADOZ data measurement contains pressure, temperature, relative humidity and ozone. The mean value of ozone concentration is calculated from the data obtained and the mean value of water vapor mixing ratio in ppmv is found from the data using the relation below;

$$\chi_{H_2O} = 61121^* RH^* \exp(17.502\, T(z)/[240.97 + T(z)])/P(z)$$

where χ_{H_2O} is water vapor mixing ratio, RH is relative humidity, $T(z)$ is temperature in degree centigrade.

3. Results

The height profile of water vapor and ozone are obtained for the regions of southern latitude hemisphere. The mean values of 14 years of HALOE satellite data and 10 years of SHADOZ ozonesonde in-situ measurement data are further used for making comparison. The HALOE data measurement ranges up to 75 km while the SHADOZ data measurement ranges up to 30 km altitude. For the sake of uniformity the SHADOZ ozonesonde data, height resolution was stepped down to 300 m.

3.1. *HALOE mean ozone profile*

The height profiles of mean ozone concentration for the southern hemisphere tropical latitudes in the range 10°S are plotted in Fig. 1. It is noticed that HALOE measurement underestimate the ozone values in the lower troposphere region in particular below 5 km as expected from the satellite instruments that measure from top to bottom and uncertainties increase downward in the troposphere. The mean ozone profile from HALOE data in the four-latitude region shows that the maximum

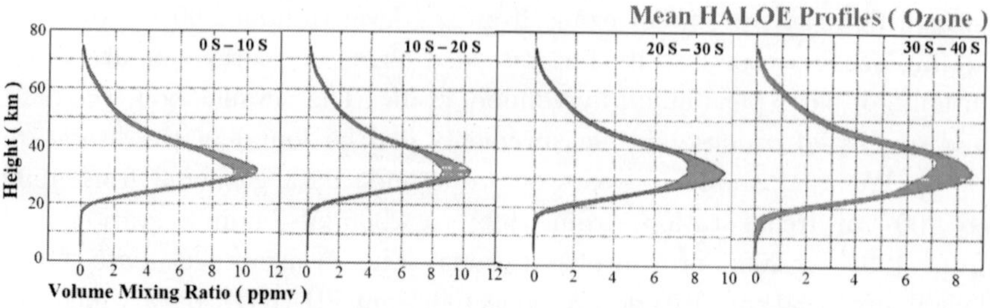

Fig. 1. The height profile of mean ozone obtained from HALOE datasets.

value occurred at around 30 km altitude. The relative variability of mean ozone profile in all four cases is high in the tropospheric region and low in the stratospheric region. The relative variability greater than or equal to 20% is observed below an altitude of 21 km (not shown in figure). The maximum values of variability observed in the latitude range of 0 to 10 degrees, indicates that the variation of ozone with time is very strong in tropical latitude between altitudes of 7 km to 12 km (noted from by zooming the plot). In addition, the satellite radiometer measurement is in general coarse due to the land surface emissivity, high scattering of aerosols and clouds cause a difficulty to obtain a profile of constituents of the lower layer of the atmosphere (troposphere). Over all variability of ozone profile in the tropical tropospheric latitude confirm that the instability of ozone in troposphere is due to different factors. Some of the different factors that caused instability of ozone are the seasonal and spatial variability of ozone in troposphere such as chemical production and destruction, convective and advective transport processes.

3.2. *HALOE mean water vapor*

The mean water vapor profiles calculated from HALOE satellite data are displayed in Fig. 2. It is noted here that the height profiles are only plotted for height regions above 10 km, due to the high error in the measurements below 10 km and lesser number of observations. From Fig. 2, we observe that the mean water vapor profile increases with altitude in the stratospheric region above the tropopause (17 km) of the HALOE data. This is reasonable when considering the chemical production of water vapor by oxidation of methane (CH_4). The profile also indicates a decrease in water vapor within the tropopause approximately from 12 km to 17 km. This can be

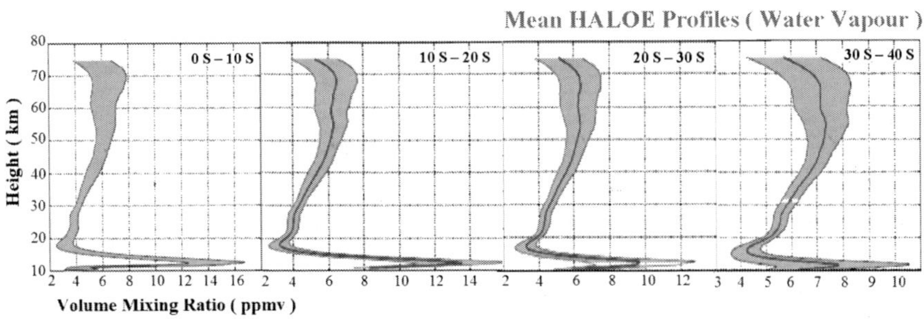

Fig. 2. The height profile of mean water vapour obtained from HALOE datasets.

viewed from the drying temperature of tropopause. The HALOE satellite measurement underestimate the value of water vapor below the tropopause as expected from the satellite observation. The relative variability of water vapor in the respective latitude ranges is found to be very high over all tropopause regions and low above the tropopause in the stratosphere. The maximum relative variability observed between 20 and 30-degree latitudes is ~32% and 30 and 40-degree latitudes ~44%. This shows the variation of water vapor with latitude, i.e., high concentration in low latitude and low in high latitude.

3.3. *SHADOZ mean ozone*

The mean ozone obtained from SHADOZ ozonesonde *in-situ* measurement; from Nairobi located at 1.27°S and 36.8°E, Malindi at 2.99°S and 40.2°E and Irene at 25.9°S and 28.22°E, stations are displayed in Fig. 3. It is found that the variability (standard deviation compared to mean) of ozone concentration below 19 km and in the troposphere region is highly variable. Though the figure does not display much variability below 15 km due to scaling, we have zoomed the region manually and identified the variability. The variability of ozone below 19 km, increases, and reaches a maximum of 45% (Irene at 16 km) and then slightly lower values. The variability is less above the tropopause at about 20 km and it varies from 4–20%. The larger relative variability found below 19 km, might be due to the complex chemical and dynamical processes occurred. Several processes contribute to the variability of tropical tropospheric ozone including the horizontal and vertical transport, convective lifting of ozone poor air from the surface and

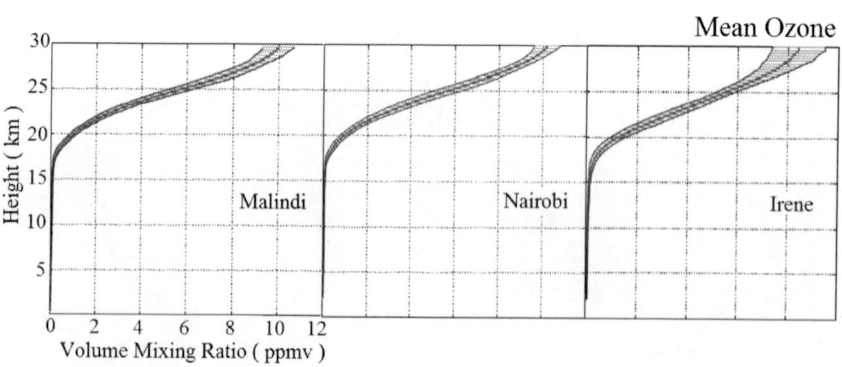

Fig. 3. Height profile of mean ozone obtained from SHADOZ datasets.

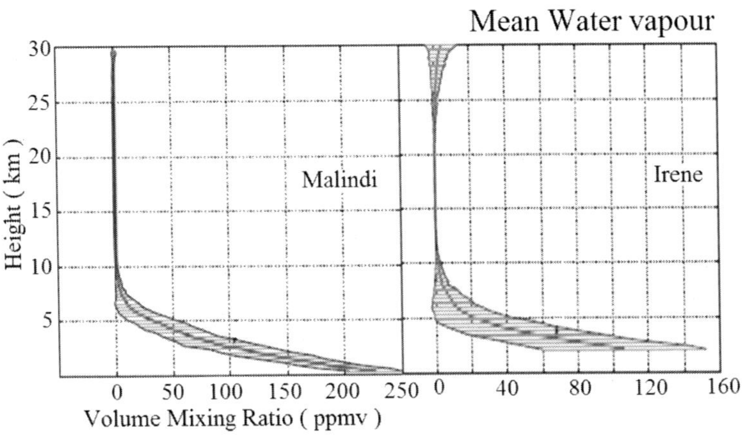

Fig. 4. Height profile of mean water vapour obtained from SHADOZ datasets.

the photochemical production by precursors. The low ozone variability in the region above 20 km height confirms the stability of ozone.

3.4. *SHADOZ mean water vapor*

The mean water vapour profile calculated from the SHADOZ ozonesonde measurement for stations at Malindi and Irene is displayed in Fig. 4. Figure 4 shows the variability of water vapor increases with altitude and is very high above 2 km. The relative variability of water vapour is greater than 20% in Malindi measurement and 43% in Irene measurement above 2 km. Such difference indicates that the variation of water vapor concentration varies with latitude region. The variation of relative humidity with temperature also contributes to the variability of water vapor in the stratospheric region. The questionable accuracy with altitude can affect the variability by large.

4. Concluding Remarks

The mean ozone and water vapor profile is calculated from 14 years of HALOE measurement from 0 to 40 degree South and about 10 years of SHADOZ sonde Nairobi, Malindi and Irene and a comparison was made with respect to the latitude range using the relative variability as a diagnostic criterion. The variability of mean ozone found in HALOE satellite is comparable to that of the SHADOZ Sonde measurement. The variability of mean water vapor in HALOE satellite is found higher in troposphere below 20 km and lower in stratosphere. The mean ozone profile difference found between SHADOZ sonde and HALOE satellite measurement data is less than 7%. The difference in water vapor measurement between SHADOZ and HALOE found maximum value in the lower troposphere below 12 km and upper mid-stratosphere. This confirms that measurement of water vapor is much complicated. In conclusion, the mean ozone profile obtained from both HALOE data and SHADOZ ozonesonde balloon measurement is helpful as a model profile.

References

1. A. Zahn, *et al.*, Modeling the budget of middle atmospheric water vapor isotope, *Atmos. Chem. Phys.* **6** (2006) 2073–2090.
2. M. S. Lohmann, *et al.*, Water vapor profile using LEO — LEO Intersatellite Links, Danish Metereological Institute, Atmospheric Ionosphere Research.
3. R. L. Jones and J. F. B. Mitchell, Climate change — Is water vapor is understood? *Nature* **353** (1991) 210.
4. I. Balin, Measurement and Analysis of Aerosols, Cirrus–Contrails, water vapor and temperature in the upper troposphere with the Jung–Fraujoch LiDAR System, Ph.D. thesis in atmospheric sciences — LiDAR applications at EPFL/LPAS-Lausamme-Switzerland.
5. J. Buchdahl, Atmospheric, Climate and Environment Information program: Air, http://www.doc.mmu.ac.uk/airc (1999).
6. J. M. Russel, *et al.*, The halogen occultation experiment, *J. Geophys. Res.* **98** (1993) 10777–10797.
7. F. Borchi, *et al.*, Evaluation of SHADOZ sondes, HALOE and SAGE II ozone profiles at the tropics from SAOZ UV-vis remote measurements onboard long duration balloons, *Atmos. Chem. Phys.* **5** (2005) 1381–1397.
8. V. Sivakumar, *et al.*, Stratospheric ozone climatology and variability over a southern subtropical site: Reunion Island (210 S; 550 E), *Annales Geophysicae* **25** (2007) 2321–2334.

Advances in Geosciences
Vol. 16: Atmospheric Science (2008)
Eds. Jai Ho Oh *et al.*

GROUND BASED LIDAR OBSERVATIONS OF ANOMALIES IN MIDDLE ATMOSPHERIC TEMPERATURE PROFILES OVER A TROPICAL STATION GADANKI (13.5°N, 79.2°E)

C. NAGESWARA RAJU
Department of Physics, S.V.Degree College, Kadapa, A.P, India
raju_svdc@yahoo.com

M. KRISHNAIAH
Department of Physics, Sri Venkateswara University, Tirupati, A.P, India

Y. BHAVANI KUMAR
National Atmospheric Research Laboratory, P.B. No. 123, Tirupati, A.P, India

A Rayleigh backscattering lidar has been in regular operation at the National Atmospheric Research Laboratory, Gadanki, a rural site in the Tropical part of India, since 1998. Using this system, profiles of middle atmospheric temperatures were obtained in the height range covering 30 to 80 km. The derived height profiles of temperatures manifest presence of significant cooling at upper stratospheric heights and strong gravity wave activity at mesospheric heights. The occurrence of stratospheric cooling over the tropical site Gadanki is compared with that of Advanced Microwave Sounding Unit (AMSU) data. It is reported that these are the unusual features in temperature profiles that represent sudden stratospheric warming (SSW) occurrences at polar latitudes. The lidar observation of sturdy wave activity at mesospheric heights strongly correlates with the frequent occurrence of temperature inversion at these heights. Upward propagating gravity waves undergo dissipation at these heights and cause stratified layers of temperature inversions.

1. Introduction

Over the past thirty years, a great deal of development have been made in the field of remote sensing of Earth's atmosphere. Several measurement techniques have been extensively used to better understand the global atmosphere. Among them, the "lidar technique" has been found to be a potential remote sensing method for probing the atmosphere. A typical backscatter lidar transmitter employs pulsed laser light for a vertical sounding of the atmosphere to make range-resolved remote measurements. The laser radiation interacts with various constituents of the atmosphere in a number of ways. Different types of optical processes in the atmosphere are related to different types of light scatterings (measures, 1984). In the case of backscattering lidar, laser radiation gets scattered elastically from molecules and particles in the atmosphere. The receiver part of the lidar collects the backscattered laser radiation, converts light into an electrical signal, and processes it to provide a range-resolved spatial distribution of the atmospheric composition.

Over the past two decades, it has been observed that the application of the Rayleigh–Scatter lidar to middle atmosphere investigations is the only reliable technique capable of measuring middle atmospheric temperatures on a long-term basis (Siva Kumar *et al.*, 2003; Bhavani Kumar *et al.*, 2006). Recently an operational lidar system has been installed at the National Atmospheric Research Laboratory (NARL) in Gadanki (13.5°N, 79.2°E), which is a Tropical site in India, under a Joint Scientific Collaboration Programme between NARL, India, and the Communications Research Laboratory (CRL), Japan, for regular investigations of the tropical atmosphere. The system became operational in March 1998. The purpose of this paper is to present the observations obtained by using this system during 1998–2001. The technical details of the system are also given for the benefit of the users of the system in the future.

2. System Description

The lidar system was installed at NARL in March 1998. The complete system configuration is shown in Fig. 1. The lidar transmitter subsystem is an Nd:YAG pulsed laser with a second harmonic generator (Powerlite 8020: Continuum). The output energy of frequency doubled 532 nm laser pulse is 550 mJ in repetition of 20 Hz. The laser pulses are transmitted to the Zenith by a steering mirror after the beam divergence is reduced to lower than 0.1 mrad through a beam expander (×10).

The lidar employs two independent receivers for collecting the laser-backscattered returns from the atmosphere. One of these is Rayleigh-scatter receiver that collects molecular backscatter in a height range of 30–80 km, whereas the other one is a Rayleigh–Mie scatter receiver that operates in the height range of 4 to 40 km to receive the backscatter from air molecules and

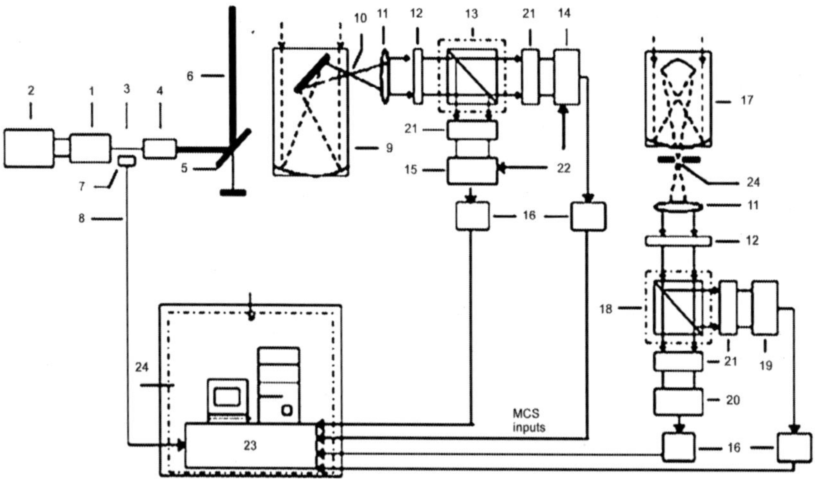

1. Nd: YAG Laser source
2. Laser control system
3. Laser beam (9 mm)
4. Beam expander (10X)
5. Streering mirror
6. Expanded laser beam (90 mm)
7. Photo diode sampler
8. Trigger signal
9. 750 mm Telescope
10. PIN hole
11. Collimating Lens
12. IF Filter (1nm)
13. Non polarizing beam splitter
14. 'U' PMT - Low sensitivity
15. 'R' PMT - High sensitivity
16. Pulse Discriminators
17. 360 mm Telescope
18. Polarizing Beam splitter
19. 'P' PMT (Co-polarized)
20. 'S' PMT (Cross polarized)
21. ND filters
22. Gating pulse to PMT's
23. MCS card - PC based
24. DELL computer system

Fig. 1. Block diagram of Indo-Japanese lidar system.

aerosol/cloud particles. The Rayleigh-scatter receiver employs Newtonian configured telescope as a front end. It's primary mirror diameter is 750 mm and F-ratio of 3:1. The Field of View (FOV) is limited to 1.0 mrad by an aperture put at the focus point of the telescope. On the one hand, the large FOV allows much of the background sky noise and on the other hand, it permits large tolerance for axis adjustment. The light through this aperture is collimated and passes a narrow band filter of center wavelength of 532 nm with full width at half maximum (FWHM) bandwidth of 1.07 nm. To extend the dynamic range of the detectors, the lidar system uses differential non-polarising beam splitter (9:1) to direct the incoming radiation to two similar photo multiplier tubes of dissimilar gains. The 10% channel (low sensitive — designated as U Channel) is used to collect the signals from the lower altitudes (25 to 50 km), where as 90% channel (high sensitive — designated as R Channel) collects returns from higher altitudes (35 to 90 km). The detector is a low noise PMT (Hamamatsu R3234-01). As the telescope is situated about 2 mts away from the steering mirror, the laser returns from the altitudes below 5 km is rejected by the aperture. Moreover, to protect the PMTs against intense light signal, electrical gating up to an altitude of 20 kms is employed. Because of adoption of low signal induced noise type PMT and biaxial configuration, there is no need for any mechanical chopper. This makes the system simple and easier to operate.

The data is accumulated the data for 250 sec for each measurement corresponding to 5,000 laser firings, which constitute the basic lidar signal. Figures 2(a) and 2(b) are, respectively, the sample basic signal profiles from the P and U Channels of Mie and Rayleigh-scatter receivers covering a height range of 150 km. Usually, the background noise is estimated for each measurement and subsequently subtracted from the signal for correction. Several of such noise-corrected signal profiles are integrated further for lidar signal analysis. A detailed system and data processing is given by Bhavani Kumar *et al.* (1999).

3. Results and Discussion

The basic technique used for temperature retrieval from the density profiles is based on the theory described by Hauchecorne and Chanin (1980)

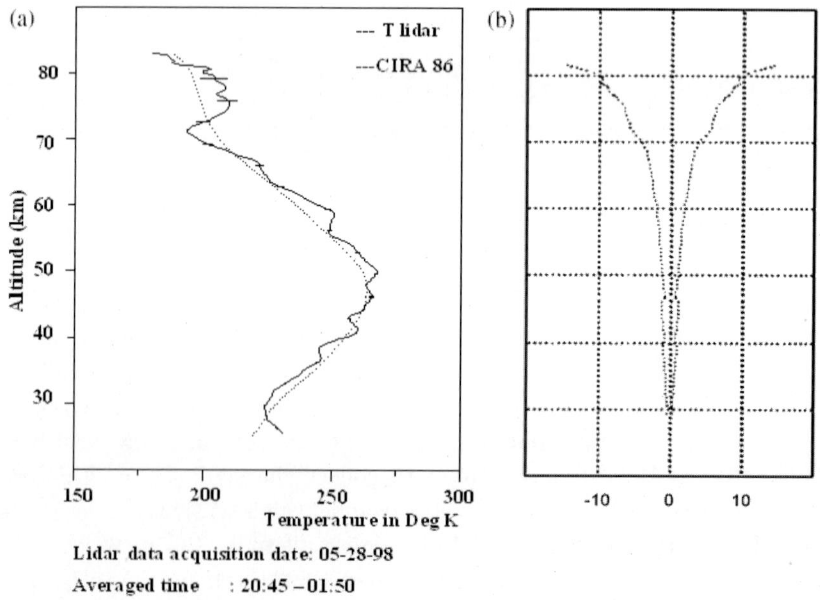

Fig. 2. A composite temperature profile (solid line) derived using Rayleigh-scatter lidar data shown along with model atmosphere data (dotted line) and standard error in temperature.

and Wilson *et al.* (1991). At altitudes above 30 km, the basic scattering arises from the molecular part of the atmosphere. Laser backscatter from atmospheric molecules undergoes Rayleigh scattering and is directly proportional to the molecular number density. The derived relative density profiles are simply corrected for range-squared dependence and normalized at a reference altitude. Absolute temperatures are derived from the density measurements by assuming that the atmosphere is in hydrostatic equilibrium and by integrating the Ideal Gas Law from the greatest altitude downward. The downward integration is triggered from the top using a model pressure value. For this purpose, we have used an atmospheric model, COSPAR International Reference Atmosphere-1986 (CIRA-86) has been used. Figures 2(a) and 2(b) are, respectively, the profiles of temperature and standard error derived from Rayleigh-scatter lidar data integrated over a period of five hours. It is a composite temperature profile constructed using data from the low sensitivity channel (U) for altitudes below 40 km, from the high sensitivity channel (R) above 47 km, and from both channels between 40 and 47 km. A weighing function based on the convex convergence technique (Bhavani Kumar *et al.*, 2000) is used for smooth transition between 40 and 47 kms. Using the weight factors, the temperature and standard error in temperature for the transition region are expressed as:

$$T_{(40-47)}(z) = \frac{[T_R(z)^* \delta T_U(z)] + [T_U(z)^* \delta T_R(z)]}{[\delta T_R(z) + \delta T_U(z)]}, \tag{1}$$

$$T_{(40-47)}(z) = \frac{[T_R(z)^* \delta T_U(z)] + [T_U(z)^* \delta T_R(z)]}{[T_R(z) + T_U(z)]}, \tag{2}$$

where T_R, δT_R, T_U and δT_U are derived temperature and standard error in temperature from R and U channel data, respectively. The data used in the present study are taken from the observations conducted during the period from March 1998 to December 2001. It is estimated that the standard error in hourly mean temperatures is about 0.2 K and 5 K at 30 and 70 km, respectively.

Sudden stratospheric warming is a dramatic meteorological phenomenon of the Northern Hemispheric middle atmosphere. High latitude stratospheric warming is associated with cooling in the low latitude stratospheric region. This is a typical case for the physical and dynamical aspects of the coupling between the low and high latitude middle atmosphere. The study reveals that temperature in the upper stratosphere and lower mesosphere undergoes strong perturbations during the SSW occurrence period.

The unusual feature was observed in the profiles of temperature during the winter period of 1998 and 1999. A substantial cooling in the height profiles of temperatures has been noticed on 23 December 1998 and 28 February 1999 as shown in Fig. 3. This is about 15 to 20 K lower than the corresponding temperature of the CIRA-86, which represents a monthly zonal mean for December and February at 15°N. It was confirmed that during this period, a strong midwinter warming occurred in the Northern hemisphere (NH) stratosphere based on zonal mean temperatures for 1998 and 1999 from NOAA (National Oceanic and Atmospheric Administration) AMSU (Advanced Microwave Sounding Unit) satellite data (http://www.cpc.noaa.gov).

Figure 4 shows the AMSU and CPC zonal mean 90–65 N temperatures for 1998 and 1999 at 1.0 hPa stratospheric level (about 50 km). The zonal mean temperatures were especially high during the stratospheric warming episodes in December 1998 and February 1999. In the Northern hemisphere (65 to 90 N) stratospheric mid-winter warming has been observed between December and March. Sudden stratospheric warming (SSW) is initiated by the propagation of planetary wave disturbances from the troposphere into the stratosphere and mesosphere and their interaction with the mean stratospheric flow [Matsuno, 1971]. It is also reported [Hauchecorne and Chanin, 1982] from mid-latitude observations that a long wave perturbation such as planetary wave could influence the temperatures of middle atmosphere. Figure 5 shows major stratospheric cooling events from April 1999 to May 2001.

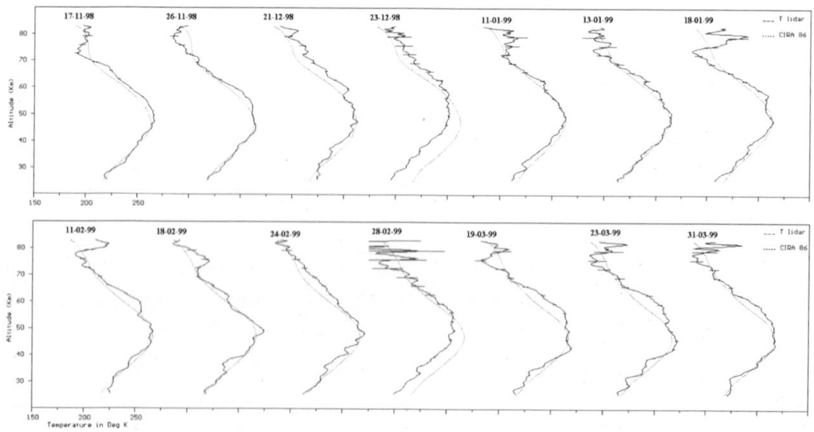

Fig. 3. Sample height profiles of temperature showing the depth of cooling at upper stratospheric heights on 23-12-98 (Upper panel) and on 28-02-1999 (Lower panel).

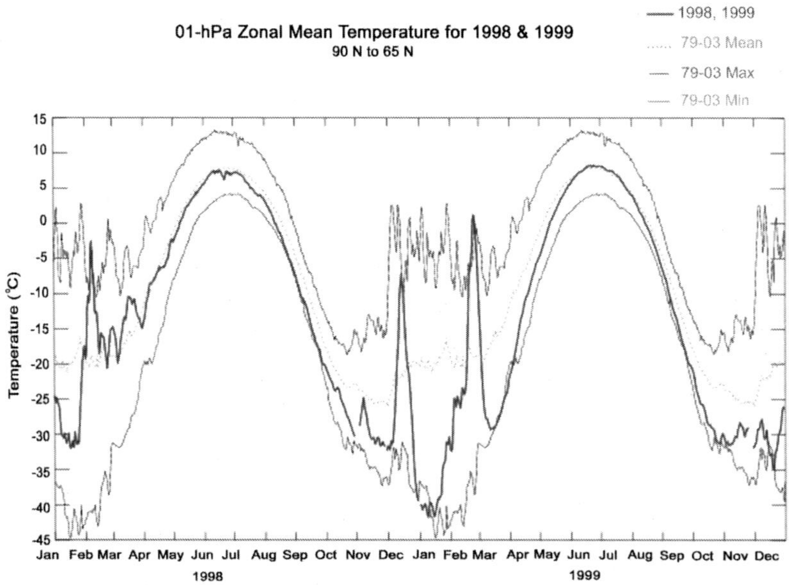

Fig. 4. AMSU zonal mean 65–90 N temperatures for 1998 and 1999.

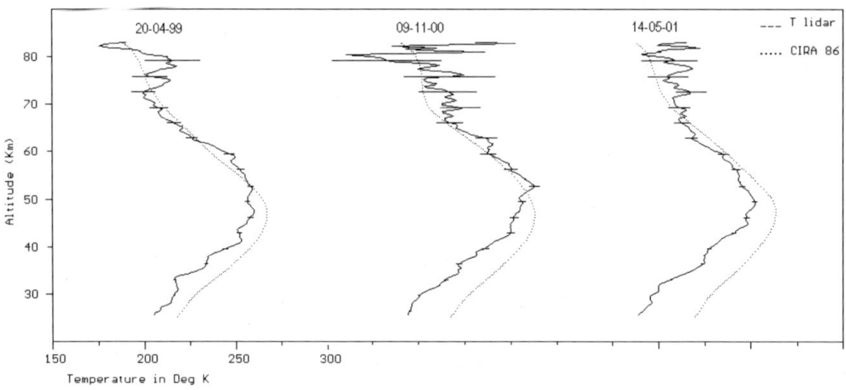

Fig. 5. Stratospheric cooling events from April 1999 to May 2001.

During SSW event, a sudden enhancement is reported in winds [Appu and Mohan Kumar, 2006] a few days prior to the peak intensity of the polar warming the wind weakens along with the weakening of the warming events. The sudden enhancement of the mesospheric winds can be considered as a precursor to the peak phase of SSW. The results obtained from the present study clearly indicate the existence of a strong coupling between Tropical

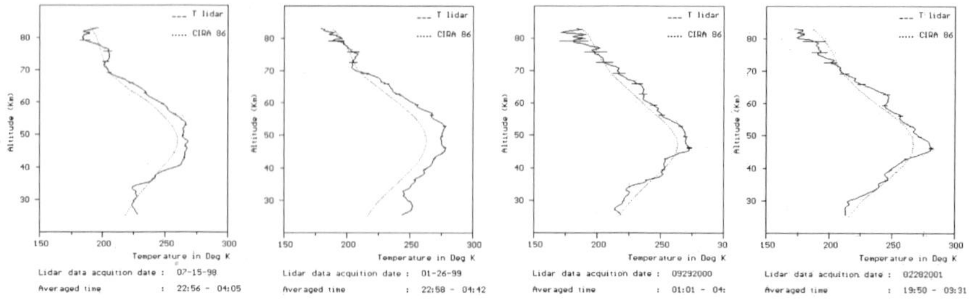

Fig. 6. Stratopause warming between July 1998 to February 2001.

and Polar Regions during SSW period. The global characteristics of the wind and temperature perturbations are responsible for such phenomena. After the cooling event, there is consistent warming observed in the stratopause. This can be clearly seen in the temperature profiles shown in Fig. 6

The profiles of atmospheric temperature at the upper stratosphere and lower mesosphere region on 18 March 1998 and 01 February 1999, presented at intervals of every one-hour on each night, are depicted in Fig. 7 (left panel) from 30 km to 80 km altitude region. The right panel gives the mean altitude profile with rms deviation observed, shown as horizontal bars, in the height profiles over a nocturnal period. The sequence of height profiles indicated in Fig. 7 show a gradual broadening of stratopause.

Gravity waves are primarily responsible for short-scale fluctuations in the thermal profiles. Atmospheric internal gravity waves play an important role in the dynamics of the middle atmosphere and have been observed in the profiles of neutral atmospheric temperatures (Chanin and Hauchecorne, 1981; Shibata *et al.*, 1986; Wilson *et al.*, 1991). The Indo-Japanese lidar observation of gravity wave activity in the nocturnal temperatures is shown in Fig. 8. Sequences of successive basic temperature profiles are plotted over a period of 6 hrs to indicate wave like structures in the temperature profiles of the night of 8–9 May 1998. Significant wave perturbations observed were considered for the sample wave analysis.

The temperature profiles obtained by Rayleigh-scatter lidar frequently exhibit a strong temperature inversion of 20–40 K at mesospheric heights. Figure 9 shows the presence of mesospheric temperature inversion in the observed temperatures profiles during the study period. This was attributed to the gravity wave breaking at these altitudes. The observations of Hauchecorne *et al.* (1987) showed that the occurrence of the inversion has semi-annual variation with maxima in summer and winter in the

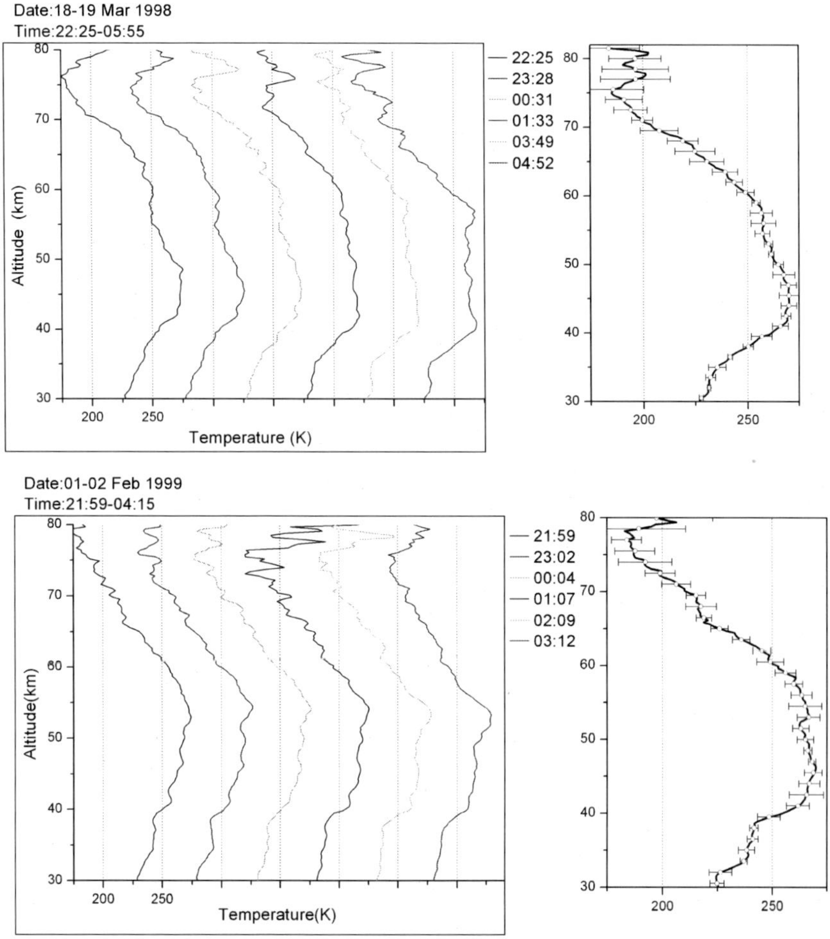

Fig. 7. Sequence of one-hour integrated height profiles of atmospheric temperatures observed 1998 over Gadanki.

height range of 70–83 and 55–72 km, respectively. The observed inversion was explained by the heating of the turbulent layers generated by the continuous breaking of the upward propagating internal gravity waves. In our observations, this inversion always persists above 70 km; sometimes two minima were observed at 4–5 km apart. The mid-latitude observations of inversion were reported below 70 km. Selected MTI observations from March 1998 to December 2001 over Gadanki site are shown in Fig. 9.

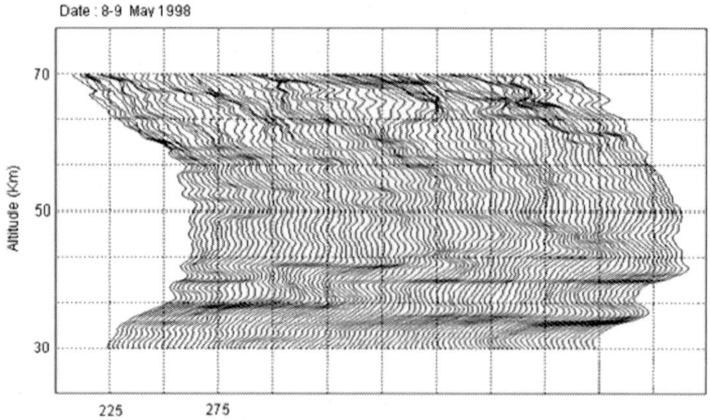

Fig. 8. Gravity Wave signatures in the Rayleigh-scatter lidar temperatures profiles observed in the night of 8–9 May 1998. Sequences of basic temperature profiles of time duration each 250 S are plotted.

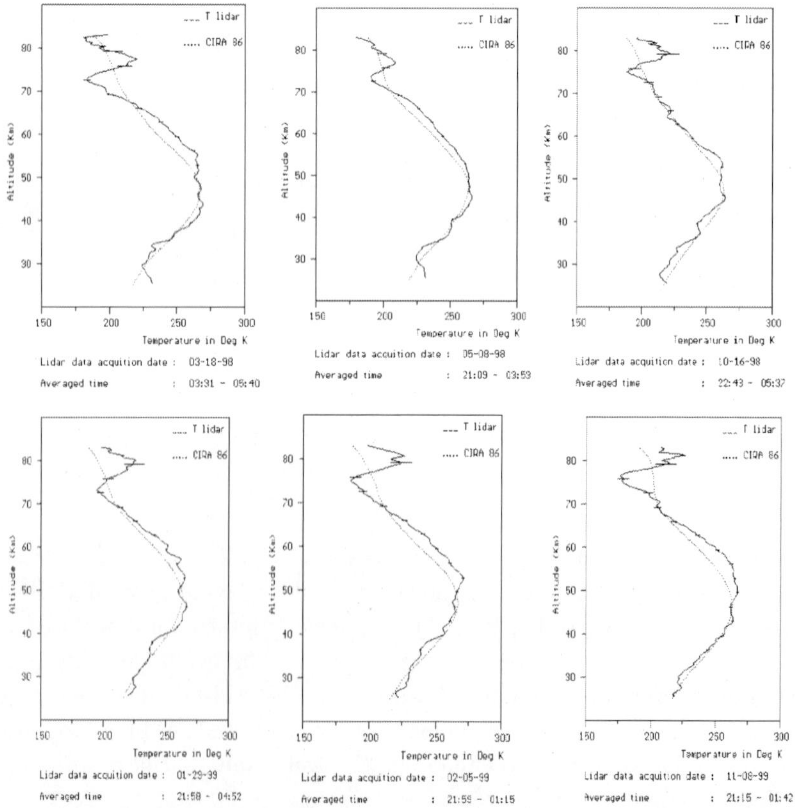

Fig. 9. Presence of MTI during observational period from 1998 to 2001.

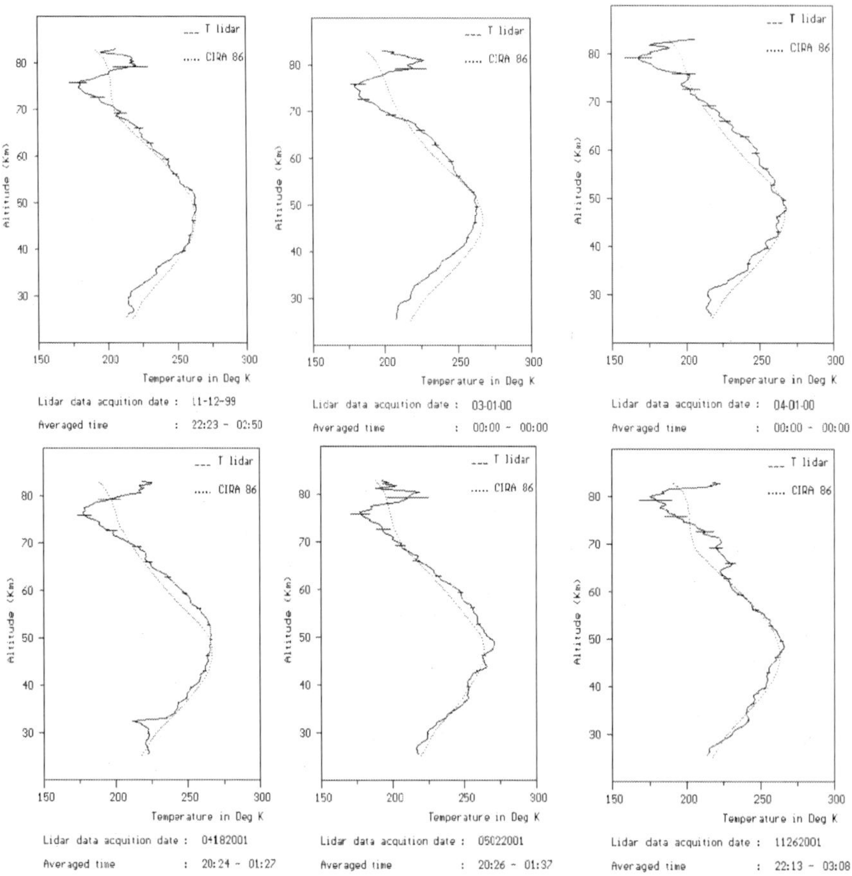

Fig. 9. (*Continued*)

4. Summary

The Rayleigh-scatter lidar observations of temperatures in the upper stratosphere and the lower mesosphere have been found to show prominent wave activity. Gravity wave activity in the nocturnal temperatures was found. The Results obtained from the present study clearly indicate the existence of a strong coupling between Tropical and Polar Regions during SSW period. The sequence of height profiles have shown a gradual broadening of stratopause. Features such as temperature inversions at mesospheric heights, noticeable cooling in stratosphere temperatures during stratwarm (SSW) period and stratospheric warming were observed over the tropical site Gadanki.

Acknowledgments

This work was supported by the University Grants Commission, SERO, Hyderabad in the form of a Minor Research Project sanctioned to the author C. Nageswara Raju, who would like to thank the Deputy Secretary (UGC-SERO) Hyderabad. The author also thanks the Director and the staff of National Atmospheric Research Laboratory, Gadanki for their kind encouragement to this work.

References

1. K. S. Appu and K. M. Kumar, Interaction between low and high latitude middle atmospheric circulation during stratwarm events, *Proc. MST-11*, NARL, Gadanki (2006), p. 365.
2. Y. B. Kumar, K. Raghunath, V. S. Kumar, P. B. Rao, A. R. Jain, K. Mizutani, T. Aoki, M. Yasui and T. Itabe, Indo-Japanese Lidar system: Part-I — System description and data processing, *Proc. Radar Symp.*, India (1999), pp. 560–570.
3. Y. B. Kumar, V. S. Kumar, P. B. Rao, M. Krishnaiah, K. Mizutani, T. Aoki, M. Yasui and Itabe, Middle atmospheric temperature using ground based instrument at low latitude, *Ind. J. Radio. Space Phys.* **29** (2000) 249–257.
4. Y. B. Kumar, C. N. Raju and M. Krishnaiah, Indo-Japanese Lidar observations of the tropical middle atmosphere during 1998 and 1999, *Adv. Atmos. Sci.* **23** (2006) 711–725.
5. M.-L. Chanin and A. Hauchecorne, Lidar observation of gravity and tidal waves in the stratosphere and mesosphere, *J. Geophys. Res.* **86** (1981) 9715–9721.
6. A. Hauchecorne and M.-L. Chanin, Density and temperature profiles obtained by Lidar between 35 and 70 km, *Geophys. Res. Lett.* **7** (1980) 565–568.
7. A. Hauchecorne and M.-L. Chanin, A mid-latitude ground-based Lidar study of stratospheric warmings and planetary wave propagation, *J. Atmos. Terr. Phys.* **44** (1982) 577–583.
8. A. Hauchecorne, M.-L. Chanin and R. Wilson, Mesospheric temperature inversion and gravity wave breaking, *Geophys. Res. Lett.* **14** (1987) 933–936.
9. T. Matsuno, A dynamical model of the stratospheric sudden warming, *J. Atmos. Sci.* **28** (1971) 1479–1494.
10. R. M. Measures, *Laser Remote Sensing, Fundamentals and Applications* (J. Wiley and Sons, NY, 1984).
11. T. Shibata, T. Fukuda and M. Maeda, Density fluctuations in the middle atmosphere over Fukuoka observed by an XeF Rayleigh Lidar, *Geophys. Res. Lett.* **13** (1986) 1121–1124.

12. V. Sivakumar, Y. Bhavanikumar, K. Raghunath, P. B. Rao, K. Mizutani, T. Aoki, M. Yasui and T. Itabe, Lidar measurements of mesospheric temperature inversion at a low latitude, *Annales Geophysicae* **19** (2001) 1039–1044.
13. R. Wilson, M.-L. Chanin and A. Hauchecorne, Gravity waves in the middle atmosphere observed by Rayleigh Lidar 1. Case studies, *J. Geophys. Res.* **96** (1991) 5153–5167.

Advances in Geosciences
Vol. 16: Atmospheric Science (2008)
Eds. Jai Ho Oh *et al.*

STATISTICAL DISTRIBUTION MODELS FOR URBAN AIR QUALITY MANAGEMENT

A. DEEPA
Department of Civil Engineering,
Environmental and Water Resources Engineering Division,
Indian Institute of Technology Madras,
Chennai-600 036, India
deepsharsh@gmailcom

S. M. SHIVA NAGENDRA
Department of Civil Engineering,
Environmental and Water Resources Engineering Division,
Indian Institute of Technology Madras,
Chennai-600 036, India
snagendra@iitm.ac.in, shivanagendra@yahoo.com

The air quality data collected for four years (February 2005 to December 2008) from the continuous ambient air quality monitoring station set up by Tamilnadu Pollution Control Board (TNPCB) at Royapuram in Chennai city (India) has been analyzed. During peak hour traffic flows, it is found that carbon monoxide (CO), oxides of nitrogen (NOx) and suspended particulate matter (SPM) concentrations are violating the national ambient air quality standards (NAAQS) at the study region. The Emission inventory at the study region indicates that two wheelers are the dominating source of air pollution.

The statistical distribution models have been fitted to the Royapuram air quality data using the SYSTAT software. Two parameter statistical distributions such as normal, lognormal, Weibull, chi square, gamma and exponential have been fitted using the method of moments and maximum likelihood methods presented in Jakeman *et al.* (1986). The goodness of fit test has been used to identify the suitable statistical distribution model. The result shows that the two parameter lognormal distribution model is best fitted to the Royapuram air quality data.

1. Introduction

Urban air pollution is a major environmental problem both in developed and developing countries of the world. It is the result of a complex interaction between natural and anthropogenic environmental conditions.

In India, air pollution problem is one of the major health concern in urban areas where vehicles contribute significantly to the high level ambient pollutant concentrations. Since, vehicular emissions are the ground level sources and thus have the maximum impact on the public using the urban corridors. The pollutants such as carbon monoxide (CO), oxides of nitrogen (NOx), hydrocarbons (HC), and particulate matter (PM) concentrations in many urban areas are sufficiently high to cause increased mortality, morbidity, deficits in pulmonary function and cardiovascular and neurobehavioral effects. In addition to these health effects, air pollution is also seriously damaging the material resources of the cities such as buildings, various works of art and vegetation. Therefore, episodes of poor air quality in cities indicated the need of local air quality management system (LAQMS) to protect humans and material from the adverse effects of air pollution.

In the past, statistical distribution models have been widely used for air quality management (Jakeman *et al.*, 1986). In this paper, we discussed the identification of suitable statistical distribution models for fitting to CO, NOx and suspended particulate matter (SPM) concentrations data collected at a residential area (Royapuram) in Chennai city.

2. Statistical Distribution Models Applied to Air Pollutant Data

Larsen (1971) found that log-normal distribution is accurate in fitting the air quality data for the pollutants such as CO, nitric oxide (NO), nitrogen dioxide (NO_2), NOx, ozone (O_3), sulfur dioxide (SO_2) and HCs monitored at eight cities during the years 1962–1968. Ott and Mage (1976) were reported that log normal distribution is accurate in fitting to the ambient air quality data. Bencala and Seinfeld (1976) concluded that pollutant concentration follows log normal distribution when the wind speed is log normally distributed. Mage and Ott (1984) were analyzed using the effectiveness of parameter estimation methods namely method of moments, fractiles, maximum likelihood and empirical rollback approach. The study concluded that the maximum likelihood method is best suited for fitting log normal distribution. Simpson (1984) examined the cumulative frequency distribution of air quality using four different types of data set: (i) continuous monitoring for a year (ii) continuous monitoring for one week (iii) random sampling at any time, (iv) random sampling on weekdays between 9 am and 5 pm. The analysis showed that continuous monitoring for one week data sets has yielded good fitting. Taylor *et al.* (1986) presented the procedures for identification of distributional model using the goodness of fit test method. Later, the methodology was evaluated on the SPM, NO, NO_2, NOx, O_3, SO_2 and CO data recorded at 19 sites in Melbourne, Australia. The results concluded that lognormal distribution was appropriate for SPM, NO, NOx, and SO_2 data sets, gamma distribution was best for both CO, NO_2 and Weibull distribution was appropriate for CO and O_3 data sets. In another study, Cernuschi *et al.* (1998) developed a model using simple rollback relationship for assessment of reduction in CO concentration. The weibull and two parameter lognormal distributions were fitted to the 8 hr average CO concentration using method of moments. The goodness of fit test analysis (Kolmogorov–Smirnoff test) concluded that weibull distribution is accurate in fitting the CO data. Rumburg *et al.* (2001) analyzed the probability distribution of PM data collected in Spokane, Washington. This study concluded that the $PM_{2.5}$ and PM_8 concentration data were best fitted by a three-parameter lognormal distribution and a generalized extreme value distribution.

3. Statistical Distribution Models Applications in India

Sharma *et al.* (1999) used extreme value theory for making the prediction of number of violation of the national ambient air quality standards (NAAQS) for the 1- and 8-hr average CO concentration in Delhi, India. The model parameters were estimated by method of moments, maximum likelihood, least square and Gumbel's fitting method and found that the Gumbel's fitting method was better than other methods. Nagendra and Khare (2002) presented a review of Line Source Modeling (LSMs) approaches namely deterministic, numerical, statistical and artificial neural network (ANN) and discussed the common practical problems and limitations associated with each techniques. Gokhale and Khare (2007) developed the statistical distribution model fitting to CO concentrations for the heterogeneous traffic pattern at the urban hotspots in Delhi, India. The results showed that the log logistic distribution model (LLD) was the best fit for the CO concentration data at both the intersection and the roadway. In another study, Gokhale and Khare (2007) combined two different modeling approaches — deterministic and stochastic for predicting the "average" as well as "extreme" air pollutant concentration distribution at two study regions in the Delhi city, India. Jain and Khare (2008) carried out vulnerability analysis to evaluate the stresses due to air pollution at the study area in Delhi city. The results showed that at urban roadways/intersection the vulnerability index were medium to high and it was low at the residential areas.

4. Air Quality Trend in Royapuram

Tamil Nadu Pollution Control Board (TNPCB) runs several air quality monitoring stations in Chennai city to assess the city air quality. Figure 1 shows the current air quality monitoring network in the city. These stations also record the meteorological parameters such as wind, speed and direction, relative humidity and temperature. Among the monitoring stations, only few are continuous monitoring stations and rest are manual stations. In manual station the monitoring of pollutants is carried out for 24 hours (4-hourly sampling for gaseous pollutants and 8-hourly sampling for particulate matter) with a frequency of twice a week, to have 104 observations in a year at selected locations.

Figure 1 also describes the sampling location at the Royapuram study area. This region is at latitude of 13°6' 26" and longitude of 80° 17' 43". The first railway station of south India constructed in the year 1850 is located in this area. Chennai's Port Trust is also very close to the study area. The study area hosts many schools and colleges, hospitals and religious worship places. Therefore, the region is having intense human activity and significant vehicular emissions. Table 1 presents the instruments used for measuring the pollutants concentrations such as CO, NOx and SPM at the study site. These instruments are calibrated once in three months.

The air quality data collected for four years (February 2005 to December 2008) from the continuous ambient air quality monitoring station run by TNPCB at Royapuram has been analyzed. The collected 30 minutes raw data has been processed to obtain 1-hr and 8-hr average of CO, 24-hr averages of SPM and NOx concentrations. Figures 2 to 4 show the trend of 24-hr average NO_x, SPM and 1-hr average CO concentrations at the study region, respectively. From the figures it is observed that the minimum and maximum 24-hr average NO_x and SPM concentrations are 8 ppb and, 69.01 ppb (124.22 μg/m^3) and 10 μg/m^3 and 257.16 μg/m^3, respectively. Similarly, the minimum and maximum 1-hr average CO concentrations are found to be 0.1 ppm and 7 ppm

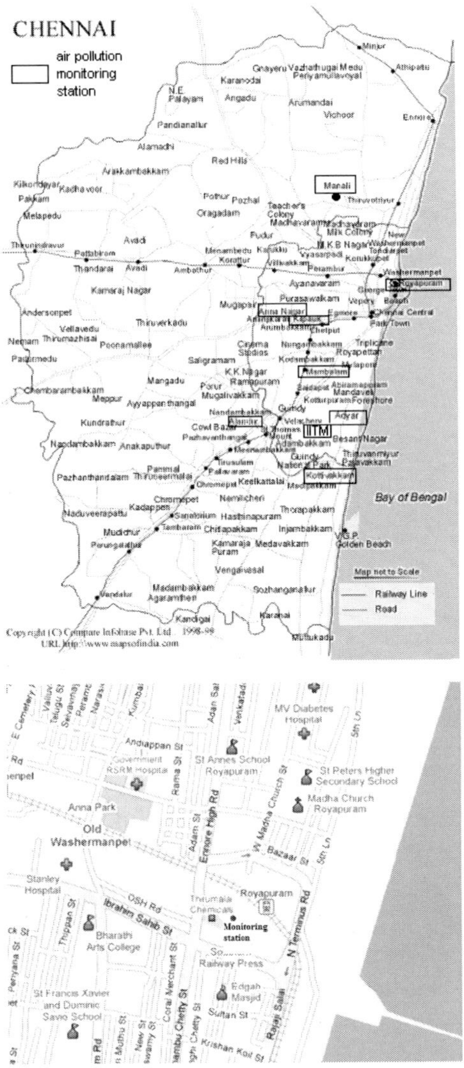

Fig. 1. Details of continuous ambient air quality monitoring stations in Chennai city and the Royapuram study area.

($8.015\,\mathrm{mg/m^3}$), respectively. The analysis of 24 hr average NOx and SPM concentrations indicated that the annual average of these pollutants are well within the prescribed limit of NAAQS standards. However, 1-hr average CO and 24-hr average of NOx and SPM concentrations are exceeding the prescribed NAAQS.

Table 1. Instruments used for measuring the pollutants concentrations at Royapuram monitoring station.

Air pollutant	Time average	Monitoring technique/ method	Principle	Range
Nitrogen oxides (NO-NO_2-NO_x)	1 min	NO_x analyser	Gas phase chemiluminiscence	0–2 PPM
Suspended particulate matter (SPM)	8 hours	High volume sampler	Sampling by high volume sampler average flow rate not less than $1.1\,m^3$/min	5–10,000 mg/m^3 0.3–100 mm particle size
Carbon monoxide (CO)	1 min	CO analyser	Non dispersive infra red (NIDR) spectroscopy	0.1–10 PPM

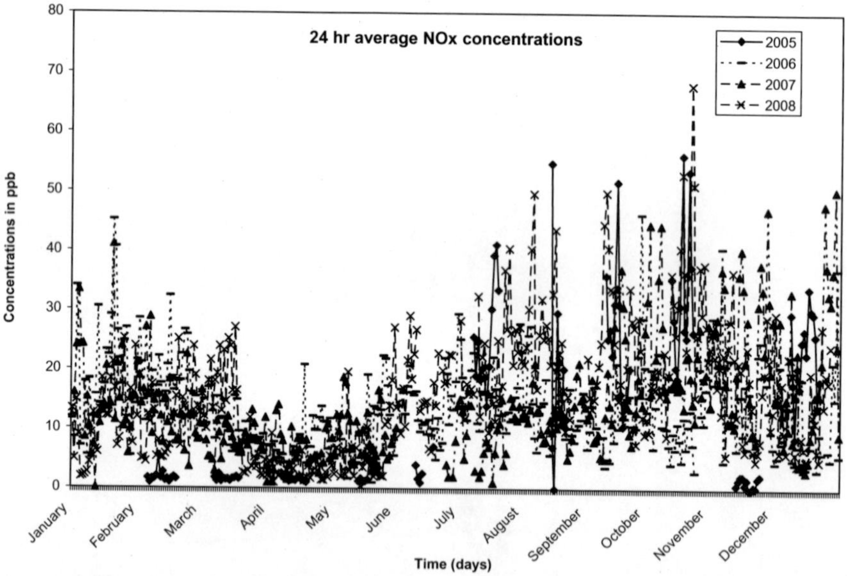

Fig. 2. 24-hr average NO_x concentrations at the study region during the year 2005 to 2008.

4.1. *Seasonal and diurnal analysis of air quality data*

According to the Regional Meteorological Centre, the weather of Chennai is classified into three seasons namely winter (January and February), summer (March–May) and monsoon (June–December) periods. In the

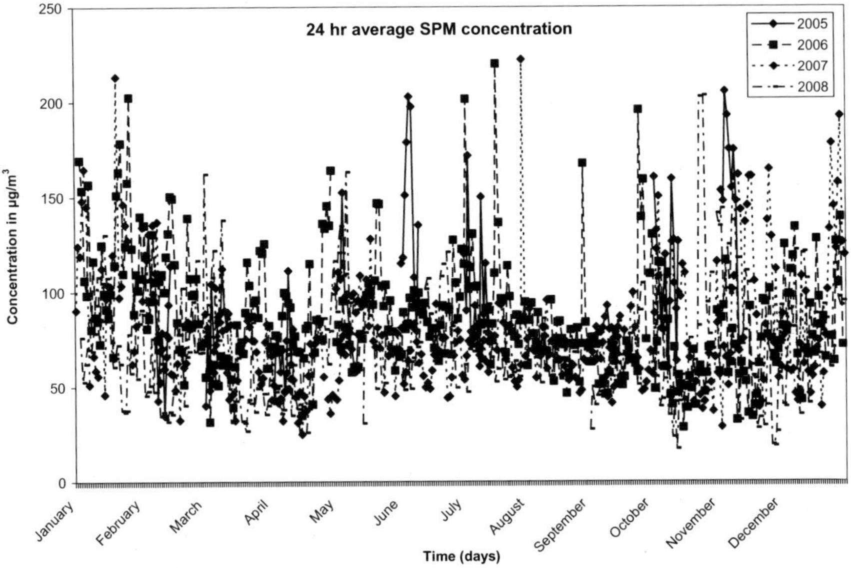

Fig. 3. 24-hr average SPM concentrations at the study region during the year 2005 to 2008.

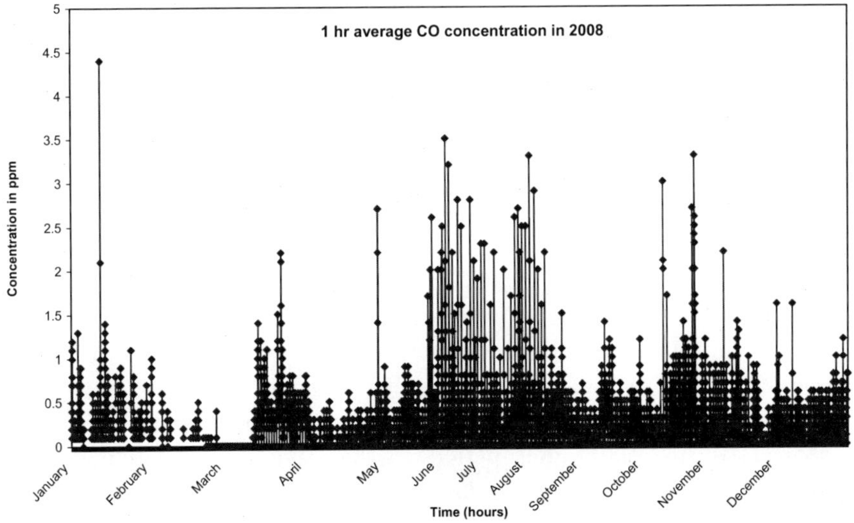

Fig. 4. 1-hr average of CO concentration at the study region during the year 2008.

present work, air quality data has been classified under three seasons as per the specification.

CO concentrations in diurnal cycle are ranged between 0.1 ppm and 4.4 ppm (5.038 mg/m^3), 0.1 ppm and 3.5 ppm (4.1 mg/m^3), 0.1 ppm and 1.97 ppm (2.25 mg/m^3), respectively, for winter, summer and monsoon seasons. In the weekly cycles it is observed that CO concentrations are generally higher during weekdays (Monday to Friday) and minimal during weekends (Saturday and Sunday). Further, in diurnal cycle, CO concentrations are maximum corresponding to peak hour traffic flow and minimum during non-traffic hours (afternoons and night time). During peak hour traffic flow CO (between 8 to 11 am) concentrations are found to be violating the NAAQS at the study region. From the weekly and diurnal cycles of 8-hr average CO concentration during different seasons (winter, summer and monsoon seasons) it is observed that most of the time 8-hr average CO concentrations at the study region are well below the specified NAAQS for the residential area (2 mg/m^3).

NOx concentrations in diurnal cycle are ranged between 2.3 ppb and 69.67 ppb (125.46 μg/m^3), 1.5 ppb and 57 ppb (102.6 μg/m^3), 4.5 ppb and 58 ppb (104.5 μg/m^3), respectively, for winter, summer and monsoon seasons. Similarly in the weekly cycles it is observed that NOx concentrations are generally higher during weekdays (Monday to Friday) and minimal during weekends (Saturday and Sunday). Further, in diurnal cycle, NOx concentrations are maximum corresponding to peak hour traffic flow (between 8 to 11 am) and minimum during non-traffic hours (afternoons 1 to 3 pm and night time 11 pm to 6 am). During peak hour traffic flow NOx concentrations are found to be violating the NAAQS (44.5 ppb specified for residential site) at the study region.

The weekly and diurnal cycles of 1-hr average SPM concentration during winter, summer and monsoon seasons are ranged between 11 μg/m^3 and 257 μg/m^3, 10 μg/m^3 and 220.21 μg/m^3, 10 μg/m^3 and 224 μg/m^3, respectively. During the winter SPM concentrations are found to be violating the NAAQS (200μg/m^3 specified for residential site). The seasonal analysis indicated that during the winter season CO, NOx and SPM concentrations are high and low during summer seasons. Further, the atmospheric stability (it is a function of wind speed and temperature gradient) during the winter period at the study region fall under the category of unstable or highly unstable conditions, i.e., more than 80% of the estimated stability class falls under the category of unstable class and about 10% falls under slightly stable class at Royapuram study region.

Hence, pollutant concentrations are higher only during prevailing inversion conditions (winter period) and sufficient source emissions. While better dispersion conditions (high wind speeds) in monsoon and summer seasons dilute the pollutants concentrations in the ambient environment.

5. Emission Inventory and Traffic Flow Analysis

Emission inventory was carried out for quantification of the air pollution load at 2 km × 2 km grid area at Royapuram. The preliminary survey indicated that vehicular emissions are the main sources of air pollution at the study region. Therefore, detailed emission estimation has been made only to the vehicular sources. The air quality monitoring station is located adjacent to the traffic junction, which is connected by three main roads namely Mannarsamy Koil Street, Ibrahim Road and Thambu Chetty Street. A traffic census was conducted at the study area between morning 6 am to night 8 pm (at 15 minutes interval) for three days — Monday, Saturday and Sunday. The maximum traffic flow was observed between 8 am to 10 am in the morning and 6 pm to 8 pm in the evening on Monday. Similarly, on Saturday the morning and evening peak flow occurred between 9 am to 11 am and 6 pm to 8 pm, respectively. On Sunday the morning and evening peak flow occurred between 10 am to 12 am and 7 pm to 9 pm, respectively. The traffic flow characteristics at the Royapuram during weekday and weekends are presented in Fig. 5. From the analysis it is observed that two wheeler are dominating the traffic, followed by three wheelers, cars, buses and trucks, respectively.

5.1. *Emission load analysis*

For vehicular sources, emission factors for various types of vehicles were obtained from the Automotive Research Association India (ARAI), Pune. The basic equation (Eq. (1)) used for quantification of vehicular source is given by:

$$\begin{aligned}\text{Emission load} &= (\text{length of the road} \times \text{no of vehicles})\\ &\quad \times \sum(\text{ventage of vehicles} \times \text{emission factor}) \qquad (1)\end{aligned}$$

Table 2 presents the emission load of CO, NOx and SPM at the study region for Saturday, Sunday and Monday, respectively. From analysis, it is found that CO is significantly contributed from 2 wheelers followed by cars, auto and buses.

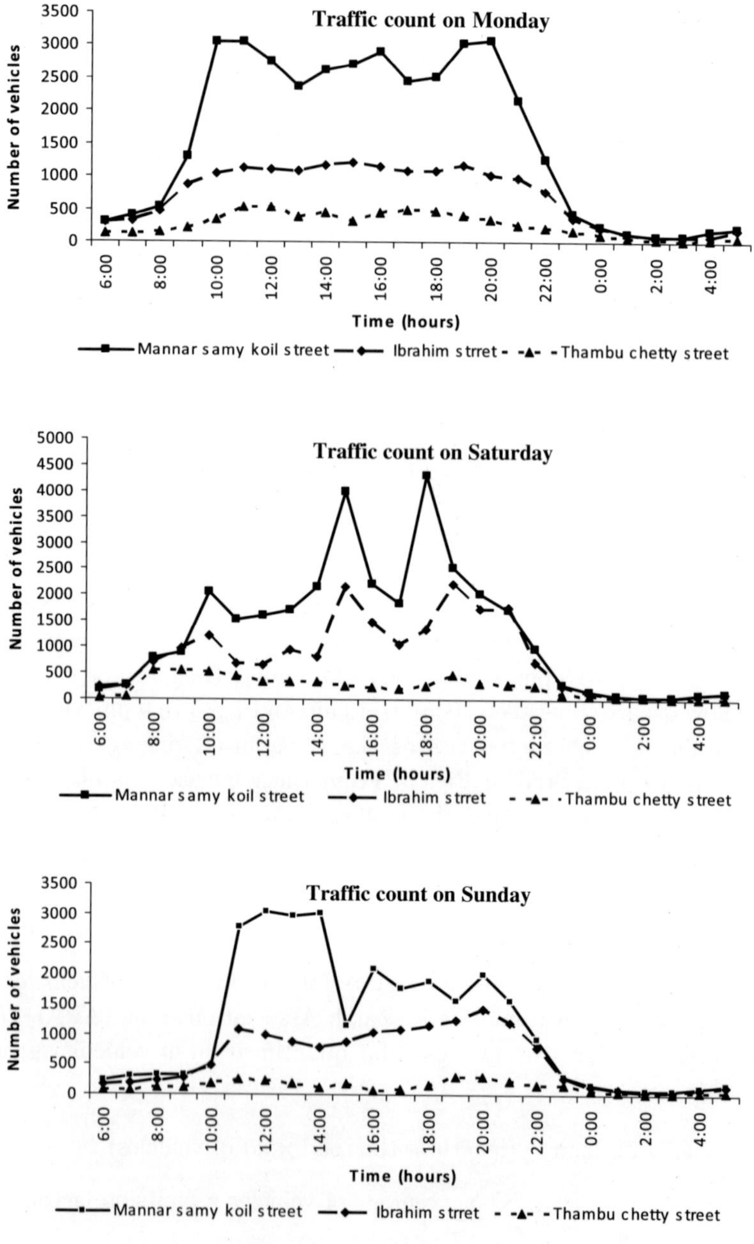

Fig. 5. Hourly traffic flow on weekday and weekends at Royapuram.

Table 2. Emission load of CO, NOx and SPM at the study region for weekday and weekends (Kg/day).

	Monday			Saturday			Sunday		
Vehicle type	CO	NOx	SPM	CO	NOx	SPM	CO	NOx	SPM
2w-SM*	912.3	39.4	18.5	526.9	39.8	9.4	530.5	42.9	6.7
2w-MC*	752.7	136.6	29.5	432.1	137.9	58.4	708.8	115.7	55.7
3w-A*	413.7	65.5	38.2	173.2	58.8	27.7	225.8	38.4	21.6
4w-C*	313.4	21.4	0.5	162.1	16.8	18.9	258.2	15.5	15.6
4w-B*	257.9	380.1	23.5	127.9	386.1	1.3	204.2	310.7	1.7
4w-TL*	62.6	58.1	17.3	35.5	38.9	4.9	30	25.3	5.3
Total load	2711.4	700.9	127.4	1458	678.2	123.6	1957	549.4	106.5

*SM: Scooter & mopeds; MC-Motor Cycles; A- Auto; C-cars; B-Buses; TL: Trucks and Lories.

6. Statistical Distribution Models

6.1. *Development of distribution model*

Suitable statistical distribution model fitting to the air quality data collected from the two air quality control region have been identified using the SYSTAT software following the procedure described in Jakeman *et al.* [1].

Parameter estimation: Parameters for each statistical distribution model have been estimated using the method of moments and maximum likelihood method. The parameters for 1 hr and 8 hr average CO concentrations and 24-hr average NOx and SPM concentrations have been estimated for the four years data (from February 2005 to December 2008) collected at the study region. Three data sets of 8 hr average CO concentrations and 24-hr average NOx and SPM concentrations having sample sizes of $n = 100$, $n = 150$ and $n = 200$ and for 1 hr average CO concentrations having the sample sizes of $n = 100$, $n = 200$ and $n = 300$ have been taken for fitting statistical distribution models. The model parameter such as scale and location parameters have been estimated for normal, lognormal and exponential, weibull, Gamma and chi square distributions.

6.2. *Identification of the appropriate distributional model*

Mage and Ott (1984) was introduced goodness of fit test to determine which distributional model is applicable to the air quality data. The goodness of fit test methods are developed based upon the observed data, to test

Table 3. Percentage frequency explained by the various distribution models on Royapuram air quality data.

Pollutant	Data set	Normal	Gamma	Lognormal	Weibull	Chi-square	Exponential
CO (1hr)	100	86.67	12.9	96	16.57	47.8	20.95
CO (1hr)	200	79.12	17.72	89.69	21.81	51.37	28.27
CO (1hr)	300	93.74	8.15	98.48	11.85	46.16	25.84
CO (8hr)	100	98.8	8.304	90.79	7.99	35.14	35.37
CO (8hr)	150	98.94	9.28	94.57	6.58	—	29.47
CO (8hr)	200	98.73	—	—	—	—	24.3
NOx	100	94.25	4.09	96.8	6.13	17.83	23.67
NOx	150	95.97	5.72	96.42	3.98	19.8	20.41
NOx	200	92.67	4.1	97.37	5.79	17.93	20.97
SPM	100	95.43	6.5	97.77	9.99	19.64	32.25
SPM	150	89.84	8.92	97.45	12.89	19.55	32.95
SPM	200	87.62	8.04	98.14	12.4	20.4	29.34

the hypothesis, that the sampled population of air quality data is normal, lognormal, exponential, gamma, beta, weibull and chi square or any other distribution. In this work, the statistical models have been identified based on Kolmogorov goodness of fit test. Kolmogorov statistic is valid only for the observations are taken from continuous data sets. In this method, the null hypothesis (H_0), i.e., air pollution data are log normally distributed is tested against the alternative hypothesis (H_1) that the data follow some other distributional form. Table 3 presents the summary of percentage frequency (Kolmogorov–Smirnov test statistic is expressed in percentage) explained by each type of distribution model for the air quality data collected from Royapuram. Based on the goodness of fit test the two parameter lognormal distribution is fitted well for the 1 hr average CO, 24 hr average NOx and SPM data sets at the study region.

7. Conclusions

The analysis of four year (February 2005–December 2008) air quality data (CO, NOx and SPM concentrations) collected at Royapuram in Chennai city indicated that 24 hr average NOx and SPM concentrations are exceeding the NAAQS limit ($80\,\mu g/m^3$ and $200\,\mu g/m^3$). The 1 hr average CO concentrations are also violating the NAAQS specified for the residential area ($4\ mg/m^3$). However, the 8-hr average CO concentrations are well below the NAAQS ($2\,mg/m^3$) during the study period. Diurnal and weekly analysis showed maximum CO and NOx concentrations during peak hour traffic flow. The weekly cycles indicated that pollutants concentrations are higher during week days than weekends. The emission inventory analysis indicated that CO is significantly contributed by the two wheelers followed by cars, auto and buses. Further, it is found that CO loading during Monday is 1.25 times higher than that of weekend loading.

Statistical distribution models have been fitted to the Royapuram air quality data sets of 1 and 8 hr average CO concentrations and 24-hr average NOx and SPM concentrations. Based on the goodness of fit test it is found that the two parameter lognormal distribution is fitted well for the 1hr average CO, 24 hr average NOx and SPM concentrations data.

Acknowledgments

We wish to thank the Tamil Nadu Pollution Control Board, Chennai for providing data.

References

1. A. J. Jakeman, J. A. Taylor and R. W. Simpson, *Atmos. Environ.* **20** (1986) 2435–2447.
2. R. I. Larsen, *AP-89, U.S. EPA* (1971).
3. W. R. Ott and D. T. Mage, *Comp. Oper. Res.* **3** (1976) 209–216.
4. K. E. Bencala and J. H. Seinfeld, *Atmos. Environ.* **10** (1976) 941–950.
5. D. T. Mage and W. R. Ott, *Atmos. Environ.* **18** (1984) 163–171.
6. R. W. Simpson, *Atmos. Environ.* **18** (1984) 353–360.
7. J. A. Taylor, A. J. Jakeman and R. W. Simpson, *Atmos. Environ.* **20** (1986) 1781–1789.
8. S. Cernuschi, M. Giugliano, G. Lonati and F. Marzolo, *Sci. Tot. Environ.* **220** (1998) 147–156.
9. B. Rumburg, R. Alldredge and C. Claiborn, *Atmos. Environ.* **35** (2001) 2907–2920.
10. P. Sharma, M. Khare and S. P. Chakrabarti, *Tran. Res.* **D4** (1999) 201–276.
11. S. M. S. Nagendra and M. Khare, *Atmos. Environ.* **36** (2002) 2083–2098.
12. S. Gokhale and M. Khare, *Environ. Mod. Soft* **22** (2007) 526–535.
13. S. Gokhale and M. Khare, *Atmos. Environ.* **41** (2007) 7887–7894.
14. S. Jain, and M. Khare, *Environ. Mon. Ass.* **136** (2008) 257–265.

Advances in Geosciences
Vol. 16: Atmospheric Science (2008)
Eds. Jai Ho Oh *et al.*

FIRST OZONE SOUNDING RESULTS OVER THE UAE*

TARIQ MAJEED[†,‡], MAZHAR IQBAL[‡], ONUOD AL-MARZOUQI[‡]
[†] *University of Michigan, 2455 Hayward St., Ann Arbor, Michigan, 48109-2143 USA*
[‡] *American University of Sharjah, P.O. Box 26666, Sharjah, UAE*
[†] *tmajeed@aus.edu*

DAVID W. TARASICK and JONATHAN DAVIES
Environment Canada, North York, Ontario, Canada

SAMUEL J. OLTMANS[§] and ANNE M. THOMPSON[¶]
[§] *United State Department of Commerce, NOAA, Boulder, CO 803, USA*
[¶] *Pennsylvania State University, University Park, PA 16802, USA*

ABDULLAH MANDOOS MOHAMMAD Al-ABRI,
S. KHALID ZAIDI and PHIL ROGERS
National Center for Meteorology and Seismology, Abu Dhabi, UAE

The first ever campaign of observing vertical profiles of ozone over the UAE began on January 24, 2006, with a couple of balloon flights carrying ECC ozonesondes to understand the characteristics of the chemical and dynamical structures at this unique tropical location (latitude 24.45N; longitude 54.22E). The campaign was continued sporadically over the past several years. A total of six ozone profiles are available to date. Preliminary results of January profiles indicate that the stratospheric ozone is quite stable while thick polluted layers confined to the lower tropospheric heights show enhancements in ozone contents. Excessive values of ozone concentration in the range 45–50 ppbv at the ground levels are observed. These values are much larger than those observed at other subtropical sounding stations using the same technique. The difference in values probably reflects either the elevated levels of air pollution due to petrochemicals and automobile emissions in the UAE or the effects of photochemical production and destruction of ozone by horizontal transport and convective uplift from the surface. These processes are affected by higher levels of solar radiation and large variations in the relative humidity in the tropical/subtropical troposphere.

*This work is supported by the Ministry of Presidential Affairs, Abu Dhabi, UAE and internal University grant.

1. Introduction

An accurate knowledge of the vertical distribution of ozone is required to evaluate current understanding of the processes responsible for changes in stratospheric and tropospheric ozone. The vertical profile of ozone loss determines how global stratospheric temperature is affected by ozone depletion and how surface temperatures respond to this change in the entire profile. More importantly for a subtropical location like the UAE, it also determines solar UV levels at the ground. The primary sources of information on profile trends are ozonesonde measurements. The vertical distribution of ozone over the UAE is examined in detail by measuring *in-situ* density profiles spanning a period of sporadic observation started in the winter of 2006.

The precise information on the vertical distribution of ozone is obtained with ozonesonde flights in the free troposphere and stratosphere below about 30 km. Information on photochemical and transport processes of ozone is obtained by interpreting height profiles of ozone and associated meteorological profiles using the GEM-AQ (Global Environmental Multiscales — Air Quality) model [1]. This model provides the production and loss rates of ozone from complex photochemical reactions, and as well as complete meteorological fields (GEM is the operational Canadian weather forecast model operated by Environment Canada). Transport processes have been shown to play an important role in explaining the measurements of ozone in the tropical western, central and eastern Pacific [2] as well as the tropics covered by the SHADOZ network [3]. Significant inter-hemispheric transport also occurs: ozone from biomass burning in South America and Africa reaches the upper troposphere over Japan, the North Pacific and North America, as well as being distributed widely in the southern hemisphere [4]. Similarly, surface ozone from Asia is distributed throughout the northern hemisphere, and the free tropospheric ozone from Asia is transported to the southern Pacific. Overall, transport from polluted source regions generally accounts for more than 40% of ozone abundance even in remote locations [4]. Recent analysis using TES data has demonstrated the importance of transport in the summertime build-up of tropospheric ozone over the Middle East [5].

This paper reports first results of an observational campaign of ozone altitude profiles by using balloon-borne Electro-Chemical Cell (ECC) ozonesondes in the vicinity of Abu Dhabi. This is an important subtropical location representative of an area for which *in-situ* measurements

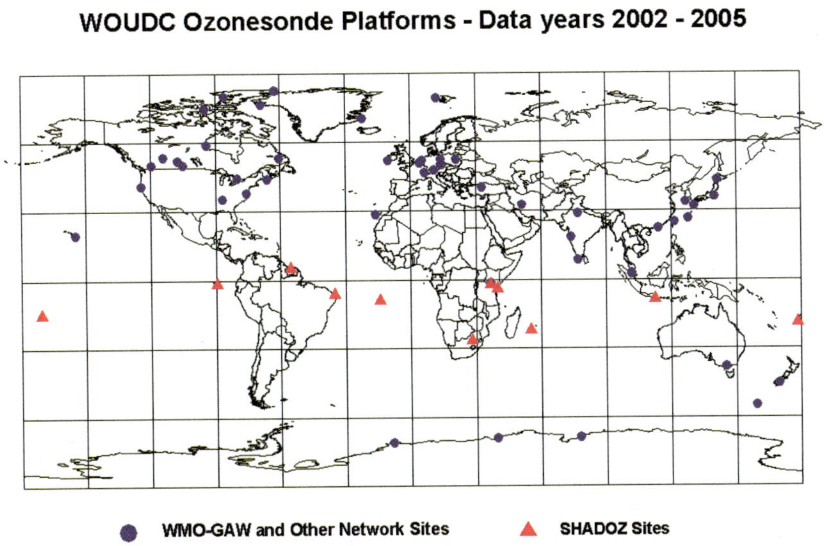

Fig. 1. *In-situ* observation stations worldwide from the World Ozone and Ultraviolet radiation Data Center (WOUDC). Sites in India produce data of poor quality. The single site in the Middle East (Isfahan) operates intermittently.

of ozone are highly warranted. Figure 1 shows all the ozone sounding sites in the world. Although it indicates ozone soundings from Isfahan, Iran and from several sites in India, these are intermittent, and there are essentially no regular ozonesonde launches in a large region between Europe and Southeast Asia. Thus our observational campaign of ozone profiles provides us an excellent opportunity for establishing a long-term ozone monitoring station, representative of the Gulf region, which we believe to be recognized by the World Ozone and Ultraviolet radiation Data Center (WOUDC) and by the Southern Hemisphere ADditional OZonesondes (SHADOZ) as the first ozone station (for in-situ measurements) representing the gulf region. Furthermore, acquired ozone data will be used to characterize the air quality chemistry and climate of the sub-tropical atmosphere of the UAE, to validate NASA satellite instruments such as TES (Tropospheric Emission Spectrometer), OMI (Ozone Monitoring Instrument) and MLS (Microwave Laboratory Sounder), and to compare with models such as GEM (Global Environmental Model) — AQ (Air Quality) in order to better understand the processes controlling tropospheric and stratospheric ozone.

2. Processes Affecting Ozone in the Troposphere

Ozone at and near ground level, although it absorbs some UV radiation and is important to the oxidation and removal of other pollutants, is generally harmful to living organisms. Ground level ozone is produced by sunlight acting on motor vehicle exhaust gases (primarily CO and NO_x) and Volatile Organic Compounds (VOCs), and is a key component of urban pollution. Ozone is a principal factor in air quality, as it has adverse effects on human respiratory health [6]. Our preliminary results indicate that the UAE is highly affected by VOCs emissions from the oilfields of Saudi Arabia. These emissions are transported in summer by the persistent cyclonic circulation over the Gulf region and react with other pollutants to form ozone during the periods of high UV index. This ozone build-up over the Middle East was also identified in previous work [7] with the GEOS-CHEM model.

The physical processes responsible for determining the ozone concentration in the subtropical atmosphere including transport, are quite different than those taking place in high and mid-latitude locations. The ozone concentration in the subtropical atmosphere is affected by photochemical production and destruction, by quasi-horizontal transport from mid-latitudes and by convective uplift from the surface. All of these processes are affected by the higher temperatures, higher levels of solar radiation, and the higher altitude of the tropical tropopause relative to higher latitudes [8, 9, 1]. Simultaneous Balloon-borne ECC observations of water vapors and ozone in the tropical western, central and eastern Pacific as well as over the equatorial eastern Brazil indicate that the air which crosses the tropical tropopause in the late northern summer may be dehydrated during the late northern fall [2]. This may yield an increase in an average height of the tropical tropopause which cools down the lower stratosphere. Furthermore, the data indicate that ozone is shown to be quite sensitive to relative humidity. Thompson *et al.* [10] reported identical observations in soundings of water vapor and ozone at Lusaka, Zambia (15.5°S, 28°E), showing high values of relative humidity and very low ozone amounts in the upper troposphere above a dry middle troposphere. The general upward motion in the transition region primarily maintains a high relative humidity within this layer effecting the distribution of ozone in the troposphere [11]. Thus, the observation of lower values of ozone in the upper troposphere suggests that the air which has been lifted to the upper troposphere by deep convection [12] may be responsible for transporting ozone from the troposphere to stratosphere. Clearly, the deep convection in the tropical/subtropical region is responsible for transporting lower tropospheric air to the tropopause, dehydrating the air to very low ozone mixing ratios, while maintaining high relative humidity and low ozone concentrations.

Ozonesonde data from the northern subtropics show a spring maximum, with higher values in the middle troposphere than above and below [13]. However, in the tropics (12.5 N to 12.5 S) satellite-derived tropospheric ozone column (TOC) measurements show a maximum in October or November and a minimum in February–April [14, 15]. The maximum is much larger over the Atlantic region, and is often described as having a zonal wave number 1 pattern, larger in the southern tropics, with an amplitude of about 20–30 Dobson Units (DU) there. Although largest in October, the wave-1 pattern persists year-round and extends to the southern subtropics [16]. It is present in both lower tropospheric (1000–400 hPa) and upper tropospheric (400–100 hPa) ozone [17] and is also seen in MOZAIC data for the upper troposphere [18]. First identified by Logan [13], it has been the subject of intense study [18, 19] and is believed to be caused by emissions from tropical biomass burning augmented by upper tropospheric ozone production from lightning NO_x [20–22]. Near the surface individual sites show departures from the general pattern due to the competing influence of pollution sources (including biomass burning) and seasonal variations in prevailing winds; this can lead to large variations in summer concentrations. For example, sites in southern Japan show a summer minimum in lower tropospheric ozone caused by flow from the Pacific Ocean during the summer monsoon; in winter the prevailing flow is from the Asian continent [13]. Until recently there was an important lack of data in the tropics, but this has been greatly improved by the establishment of 14 new stations in the tropics under the Southern Hemisphere ADditional OZonesondes (SHADOZ) program [3, 10, 18]. Unfortunately ozone profile data in subtropical regions is still lacking.

3. Results and Discussion

The initial observational campaign of the vertical profiles of ozone over the UAE started in January, 2006, with limited resources through the UAE Ministry of Presidential Affairs. We launched two balloons carrying ECC ozonesondes from the Abu Dhabi international airport. The campaign of observing the vertical profiles of ozone along with meteorological parameters such as relative humidity, temperature, the magnitude winds and directions has been continued sporadically over the past several years. All daytime measurements of the vertical profiles of ozone were made near 12 UT while nighttime measurements were made near 00 UT. A total of six vertical profiles of ozone are available to date.

Figure 3 shows the vertical profile of daytime ozone concentration observed at the Abu Dhabi international airport on January 24, 2006, in comparison with the measured ozone profile at Fiji and estimated ozone profile using Brewer–Umkahr algorithm. Note that the peak ozone concentration in the stratosphere at Abu Dhabi is about 15% larger than that observed at Fiji.

However, the peak ozone concentration derived from the data observed by the Brewer spectrometer, located at the National Center for Meteorology and Seismology (NCMS) in Abu Dhabi, is underestimated by about 20%. The situation near the ground is even more complicated than the peak in the stratosphere. The direct comparison of the surface ozone with that derived from Brewer–Umkehr at Abu Dhabi has not been made because Umkehr inversion algorithm does not estimate ozone below an altitude of 2 km. While the Abu Dhabi ozone concentration is in good agreement with the surface ozone concentration observed at near-by emirate of Dubai, it is about a factor of five larger than that observed at Fiji. This large difference in ozone concentration appears to be due to the relative humidity observed at the two locations. At Abu Dhabi, the relative humidity at the surface is about 55% which rapidly reduces to 0% at about 1.5 km with a maximum of nearly 75% at 0.5 km. At Fiji, however, the surface humidity is about 62% which reduces to almost 0% at about 4 km with a peak in the relative humidity of ~90% at 1.5 km (see Fig. 4). Clearly, the atmosphere near the ground at Abu Dhabi is completely dehydrated more rapidly than at Fiji. More detailed comparison will be discussed in Fig. 4 later.

In Fig. 2 we compare the day-night ozone mixing ratio (ppbv) profiles below 4 km altitude at Abu Dhabi with an ozone profile estimated from the Brewer–Umkehr algorithm. Considerable difference in ozone mixing ratios is observed between the day and night-time profiles below 1 km altitude. The surface mixing ratio of ozone at night is about a factor of almost two smaller than that at day as a result of increased relative humidity and absent solar sources for ozone. It is also important to note that Brewer–Umkehr yielded lower estimates of ozone ranging from 50% to 35% between 4 and 2 km altitudes. The lower values of ozone are possibly due to Umkehr inversion algorithm employed to retrieve ozone profile from the total column of ozone measured by the Brewer instrument. Errors may have been introduced while

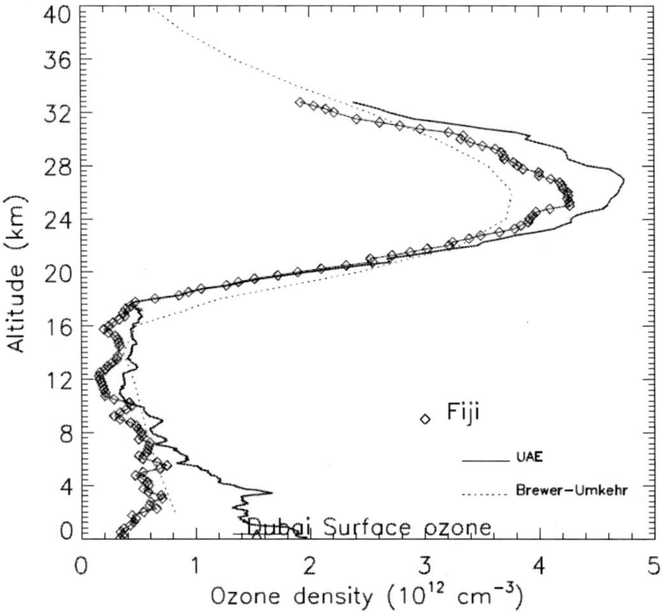

Fig. 2. The measured vertical density profile of ozone at the Abu Dhabi International airport is compared with that observed at Fiji and an estimated profile using Brewer–Umkahr algorithm. The surface ozone datum obtained by Dubai municipality is also shown in comparison with the ECC data.

inverting the slant column of ozone to the vertical column, and finally to the vertical profile of ozone. As shown in Fig. 2, the Brewer–Umkehr profile of ozone is quite comparable with the observed ECC ozone profile between altitudes of 5 and 16 km.

In Fig. 4, we show a detailed comparison of the observed vertical profiles of ozone, relative humidity, and temperature at Abu Dhabi with those observed at other subtropical stations. The surface temperatures at these locations are similar in the range 20–24°C. The peak in the relative humidity below 5 km at Abu Dhabi is observed to be about 20% and 13% smaller than those observed at Hong Kong and Hilo, respectively. The amount of ozone in this region of the atmosphere is quite smaller at Hong Kong and Hilo (<20 ppbv) compared to that at Abu Dhabi ($\sim$50 ppbv). For the atmosphere above 5 km, the distribution of ozone is found to be anti-correlated with the distribution of relative humidity except near the tropopause. The height of the tropopause is found to be in the range 15–18 km at all four stations. The variation in the height of the tropopause affects the distribution of ozone in the transition region between the lower

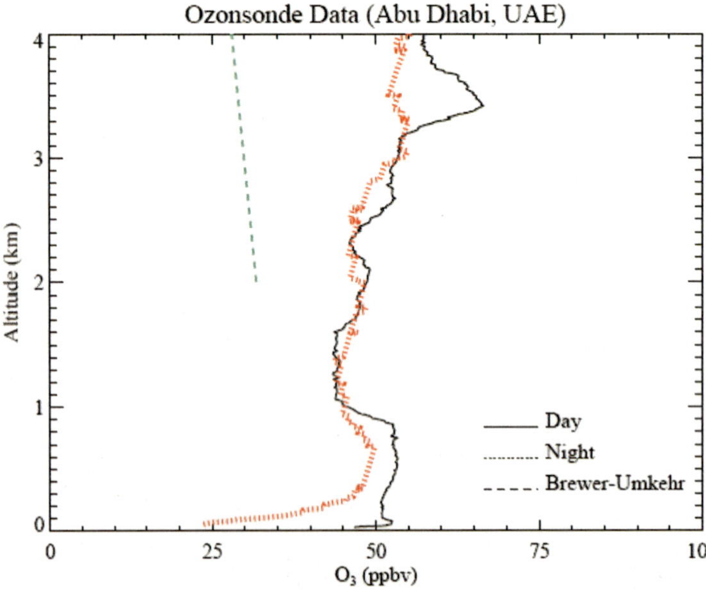

Fig. 3. Day and night times profiles of Ozone mixing ratios are shown in comparison with those retrieved from Brewer–Umkehr instrument.

stratosphere and upper troposphere [3]. According to Vomel *et al.* [2], a large increase in the relative humidity near the transition region is due to the deep convection process which lifts lower tropospheric air up to the trpopause causing transport of ozone from the troposphere to stratosphere. As a result, lower values of mixing ratios are observed in the transition region as shown at Fiji and Hilo. However, such a behavior is not seen at Abu Dhabi and Hong Kong. The behavior of relative humidity is quite similar at these two stations above about 10 km. The mixing ratios of ozone are observed to increase gradually towards the stratosphere.

Figure 5 shows the observations of daytime ozone mixing ratios, temperature, and relative humidity at Abu Dhabi on October 17, 2008 (left panel) and November 29, 2008 (right panel), respectively. These observations represent how the ozone distribution in response to the relative humidity behaves in the autumn in the UAE. Note that the autumn relative humidity profiles are quite different than that observed on January 24, 2006. The values of the autumn relative humidity above 5 km altitude are ranging from 30% to almost zero. However, near the surface (altitude less than 5 km) the autumn atmosphere is quite humid compared to that of winter. The

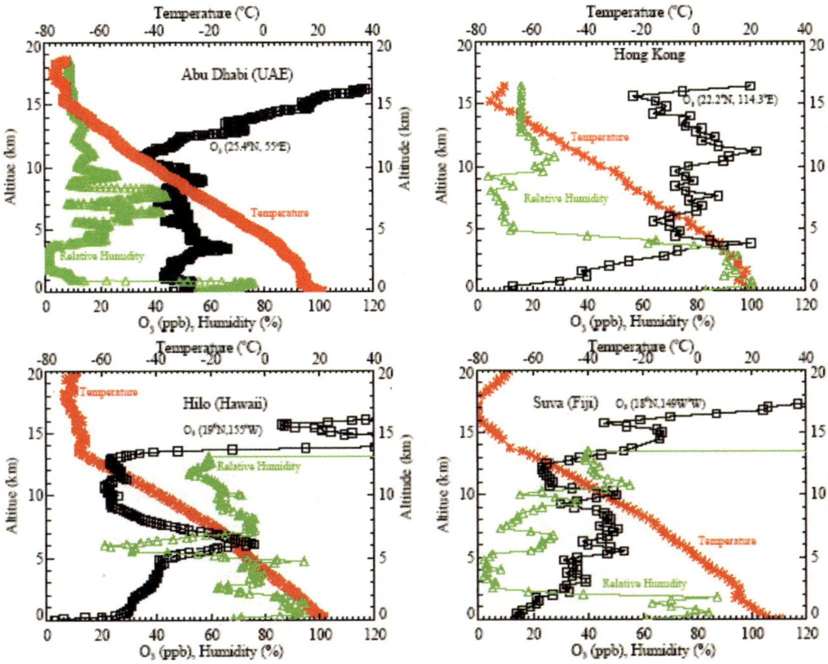

Fig. 4. Ozone mixing ratios, temperature, and relative humidity at Abu Dhabi are shown in comparison with those observed at Hong Kong, Hilo, and Fiji on January 24, 2006.

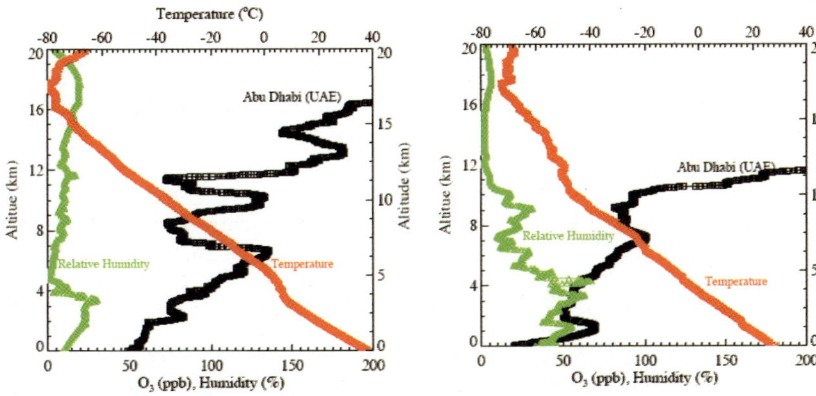

Fig. 5. Profiles of ozone mixing ratios are shown in comparison with those retrieved from Brewer–Umkehr instrument.

surface temperature of 39.9°C was observed on October 17, 2008, compared to that of 28.5° on November 29, 2008. The corresponding relative humidity on October 17 was observed to be around 6% while on November 29 it was ~30%. The behavior of ozone near the surface is observed to be according to the variation in the relative humidity. The surface zone mixing ratio for low value of the relative humidity is about 50 ppbv while for high value of the relative humidity the surface mixing ratio reduces to about 20 ppbv. 10%. Note that the ozone mixing ratio increases rapidly in response to relative humidity above 12 km altitude.

4. Conclusions

The first sounding results of the vertical distribution of ozone over the United Arab Emirates (UAE) are reported. We find that the height profiles of ozone along with the relative humidity and temperatures measured on January 24, 2006 are quite different than those measured on October 17, 2008 and on November 29, 2008. A comparison of the vertical profiles of the above parameters with other subtropical stations revealed that the vertical distribution of zone is strongly anti-correlated with the relative humidity. However, with the limited ozone data, it is rudimentary to suggest that horizontal transport and convective processes are responsible for such a behavior of ozone distribution over the UAE. We intend to present more ozone data and their analysis at the AOGS meeting in Hyderabad, India, next year as regular flights are scheduled for measuring the vertical profiles of ozone at Abu Dhabi from January 1, 2010.

Acknowledgment

This work is supported by the Ministry of Presidential Affairs, Abu Dhabi, UAE, Emirates Foundation and internal university grant.

References

1. J. W. Kaminski, *et al.*, *Atmos. Chem. Phys.* **8** (2008) 3255–3281.
2. H. Vomel., *et al.* Balloon-borne observations of water vapor and ozone in the tropical upper troposphere and lower stratosphere, *J. Geophys. Res.* **107** (2000), doi: 10.1029/2001JD000707.
3. A. M. Thompson, J. C. Witte, S. C. J. Smit, S. J. Oltmans, B. J. Johnson, V. W. J. H. Kirchhoff and F. J. Schmidlin, *J. Geophys. Res.* **112** (2007) D03304, doi:10.1029/2005JD007042.

4. K. Sudo and H. Akimoto, *J. Geophys. Res.* **112** (2007) D12302, doi:10.1029/2006JD007992.
5. J. J. Liu, D. B. A. Jones, J. R. Worden, D. Noone, M. Parrington and J. Kar, *J. Geophys. Res.* **114** (2009) D05304, doi:10.1029/2008JD010993.
6. M. Jerrett, R. T. Burnett, C. A. Pope III, K. Ito, G. Thurston, D. Krewski, Y. Shi, E. Calle and M. Thun, *N. Engl. J. Med.* **360** (2009) 1085–1095.
7. Q. Li, A. M. Fiore, B. N. Duncan, H. Liu, P. Ginoux and V. Thouret, *J. Geophys. Res.* **107** (2002) 4351, doi:10.1029/2001JD001480.
8. J. de Grandpre, *et al.*, *J. Geophys. Res.* **26** (2000) 475–526.
9. F. Borchi and J.-P. Pommerea, *Atmos. Chem. Phys.* **7** (2007) 2671–2690.
10. A. M. Thompson, J. C. Witte, R. D. McPeters, S. J. Oltmans, F. J. Schmidlin, J. A. Logan, M. Fujiwara, V. W. J. H. Kirchhoff, F. Posny, G. J. R. Coetzee, B. Hoegger, S. Kawakami, T. Ogawa, B. J. Johnson, H. Vömel and G. Labow, *J. Geophys. Res.* **108** (2003) 8238, doi:10.1029/2001JD000967.
11. J. A. Logan, *J. Geophys. Res.* **104** (1999) 16,115–16,150.
12. J. R. Ziemke, S. Chandra and P. K. Bhartia, *J. Geophys. Res.* **106** (2001) 9853–9867, doi:10.1029/2000JD900768.
13. M. J. Newchurch, D. Sun, J. H. Kim and X. Liu, *Atmos. Chem. Phys.* **3** (2003) 683–695.
14. J. Fishman, A. E. Wozniak and J. K. Creilson, *Atmos. Chem. Phys.* **3** (2003) 893–907.
15. S. E. Bortz, M. J. Prather, J.-P. Cammas, V. Thouret and H. Smit, *J. Geophys. Res.* **111** (2006) D05305, doi:10.1029/2005JD006512.
16. A. M. Thompson, J. C. Witte, R. D. Hudson, H. Guo, J. R. Herman and M. Fujiwara, *Science* **291** (2001) 2128–2132.
17. I. Jonquières, A. Marenco, A. Maalej and F. Rohrer, *J. Geophys. Res.* **103** (1998) 19,059–19,074.
18. D. P. Edwards, J.-F. Lamarque, J.-L. Attie, L. K. Emmons, A. Richter, J.-P. Cammas, J. Gille, G. L. Francis, M. N. Deeter, J. Warner, D. C. Ziskin, L. V. Lyjak, J. R. Drummond and J. P. Burrows, *J. Geophys. Res.* **108** (2003) 4237, doi:10.1029/2002JD002927.
19. R. V. Martin, D. J. Jacob, J. A. Logan, I. Bey, R. M. Yantosca, A. C. Staudt, Q. Li, A. M. Fiore, B. N. Duncan, H. Liu, P. Ginoux and V. Thouret, Interpretation of Toms observations of tropical tropospheric ozone with a global model and in-situ observations, *J. Geophys. Res.* **107** (2002) 4351, doi:10.1029/2001JD001480.
20. D. W. Tarasick, J. Davies, K. Anlauf, M. Watt, W. Steinbrecht and H. J. Claude, *J. Geophys. Res.* **107** (2002) 4308, doi:10.1029/ 2001JD001167.

Advances in Geosciences
Vol. 16: Atmospheric Science (2008)
Eds. Jai Ho Oh *et al.*

MAGNETIC MEASUREMENTS OF ATMOSPHERIC DUST DEPOSITION IN SOILS*

ALEŠ KAPIČKA†, EDUARD PETROVSKÝ and HANA GRISON
Institute of Geophysics, Acad. Sci. Czech Republic,
Boční II/1401, 141 31 Praha 4, Czech Republic
†*kapicka@ig.cas.cz*

VILÉM PODRÁZSKÝ and PAVEL KŘÍŽEK
Department of Silviculture, Faculty of Forestry and Wood Sciences,
Czech University of Life Sciences, Kamýcká 129,
165 21 Praha 6, Czech Republic

Atmospheric dust of anthropogenic origin contains significant portion of minerals characterized by ferrimagnetic properties [1,2]. These minerals, mostly iron oxides, can serve as tracers of industrial pollutants in soil layers. Moreover, recent results, e.g., [3,4] show significant correlation between concentration-dependent magnetic parameters (e.g., low-field magnetic susceptibility) and concentration of heavy metals (e.g., Pb, Zn, Cd). In our paper we have investigated magnetic properties of depth soil profiles from Krušné hory Mountains (Czech Republic), which belong to a highly contaminated, so-called Black Triangle in central Europe. Emissions are determined by considerable concentration of big sources of pollution (power plants burning fossil fuel, metallurgical and chemical industry). Increased values of magnetic susceptibility ($25 - 200 \times 10^{-5}$ SI) were clearly identified in the top-soil layers. Thermomagnetic analyses and SEM observation indicate that the accumulated anthropogenic ferrimagnetics dominate these layers. Magnetic enhancement is limited to depths of 4–7 cm below the soil surface, usually in F-H or top of Ah soil horizons; deeper soil horizons contain mainly magnetically weak materials and are characterized by much lower values of susceptibility (up to 30×10^{-5} SI). Significant magnetic parameters (e.g., Curie temperature Tc) and SEM results of contaminated topsoils are comparable with magnetic parameters of atmospheric dust, collected (using high-volume samplers) at the same localities.

1. Introduction

Results of regular monitoring provide information on temporal development of air pollution and spatial distribution of the pollutants concentration

*This work is supported by grant 205/07/0941 of the Grant Agency of the Czech Republic.

over the territory of Czech Republic. Distribution of the pollutants is very complex and depends upon several factors. Emissions from local heating, traffic and partly medium sized sources are of local extent, while those from major sources can be transported over long distances, often even outside the country territory.

Dust particles are of a significant threat to human health and, thus, represent major problem in air pollution. Industrial dust particles are emitted, among others, by combustion (of fossil fuel in stationary sources and fuel I traffic). About two-thirds of PM10 and half of finer, more harmful PM2.5 fraction are produced by electric plants. Traffic is another significant source of these particles and, moreover, causes redistribution of already deposited particles back to the atmosphere.

Generally, two different approaches are used to measure the amount of dust deposited on a surface; determination of the quantity of dust deposited in terms of weight, or determination of the soiling of a surface, by a change in its properties. The former approach uses high-volume samplers. This method determines average dust concentrations and comprises the collection of dust by drawing a constant flow rate of ambient air through a collector. Data are usually collected over a 24-hour period and results are expressed in mass of dust per unit volume of air per 24 hrs. A selective inlet may be fitted to a high-volume sampler to restrict the particle size being sampled (for example, to ensure only PM10 particles are sampled). Measurement of magnetic properties of contaminated soil surface exemplifies the second approach for monitoring of air pollution.

Currently ambient air pollution caused by suspended particles represents a major problem not only in the Czech Republic but almost throughout the whole of Europe. With regard to health risks, especially fine particles represent the most difficult challenge. Analyses of measured data show that the PM10 limit values are markedly exceeded in a number of sites throughout the country [5].

Contamination due to deposition of particulate matter shows significant variability within the Czech Republic. National parks, such as Krkonoše (NE Bohemia) and Šumava (SW Bohemia) belong to the cleanest areas. There are no major sources of atmospheric pollution in the surrounding area and relatively low concentrations of atmospheric dust are due to long-range transport. The average annual concentration of PM10 is in both the regions $< 15\,\mu g/m^3$, which is well below the allowance limit. On the other hand, the Krušné hory region, which is part of the "black triangle" area [6], belongs to the most polluted parts of the Czech Republic. This area is on the junction between the Czech Republic, Germany and Poland and contains numerous sources of atmospheric pollution, such as major power plants burning high-sulfur brown coal, chemical industry, incinerators and big heat plants, intense automobile and rail traffic, deposits of power-plant ashes, waste dumps, deposits of overburdens, etc. Average annual concentration of PM10 (35–46 $\mu g/m^3$) in the area of Krušné hory foredeep belongs to the highest within the Czech Republic. Situation on the top of Krušné hory is somewhat better, with average annual concentration of PM10 at about 20 $\mu g/m^3$ [5].

In this study, we demonstrate the application of magnetic method to assess contamination of soils due to atmospheric deposition in the Krušné hory region. Soil magnetometry has several advantages. It is fast and, contrary to stationary monitoring stations, allows data acquisition at large number of sites, enabling thus better delineation of areas with different amount of deposited dust. Areas with higher imissions can be thus targeted for sampling for more detailed and standardized chemical analyses.

2. Magnetic Study of Soil Contamination due to Atmospheric Deposition of Pollutants

The major part of atmospheric fallout is of industrial origin and contains significant portion of ferrimagnetic particles [2]. Deposited dust accumulates mostly in topsoils and sediments. However, in order to assess the contamination due to industrial activities, discrimination between anthropogenic and natural (weathering of geological basement, pedogenic processes, etc.) contributions are necessary. This can be achieved by analyzing magnetic properties of contaminated samples collected in the areas in concern.

Magnetic susceptibility (k) measured in low magnetic field is one of the most important parameters used in environmental magnetism. In case of soil samples, this integral parameter represents combine contributions of diamagnetic (iron-free silicates and carbonates), paramagnetic (silicates containing Fe and Mn), antiferromagnetic (e.g., hematite) and ferrimagnetic (e.g., maghemite and magnetite) minerals. Magnetic susceptibility is composition and concentration dependent parameter and primarily can be considered to reflect concentration of (strongly magnetic) ferrimagnetic substances.

Interpretation of magnetic data depends on the samples examined. It is relatively easy and straightforward in the case of passive biomonitors such as tree leaves, needles, or peat bogs, where atmospherically particles are deposited on (magnetically weak) diamagnetic substances (e.g., [7]). However, availability of these collectors is rather limited to certain localities. The other disadvantage is low concentration of the deposited material and, thus, need for highly-sensitive instruments, which are not available for the *in-situ* monitoring. Contrary to that, soil surface represents the most obvious trap to atmospheric fallout. Consequently, topsoil measurements are often used in assessing the soil contamination due to atmospheric deposition of dust particles (e.g., [8]). Figure 1 shows typical vertical distribution of magnetic susceptibility in soil in the industrial region and relatively clean area of the Czech Republic.

Relative contribution of ferrimagnets of natural origin in topsoils is mainly due to two factors: weathering of geological basement and neoformation of ferrimagnetic minerals during pedogenic processes. Mineral composition of underlying rocks is the primary factor influencing mineral populations in the developed soils. However, vertical distribution of magnetic susceptibility can easily identify sites with significant geological contribution to magnetic enhancement of topsoils [9]. The other contribution results from oxidation-reduction processes in soils, transforming magnetically weak low-crystalline Fe-oxides and

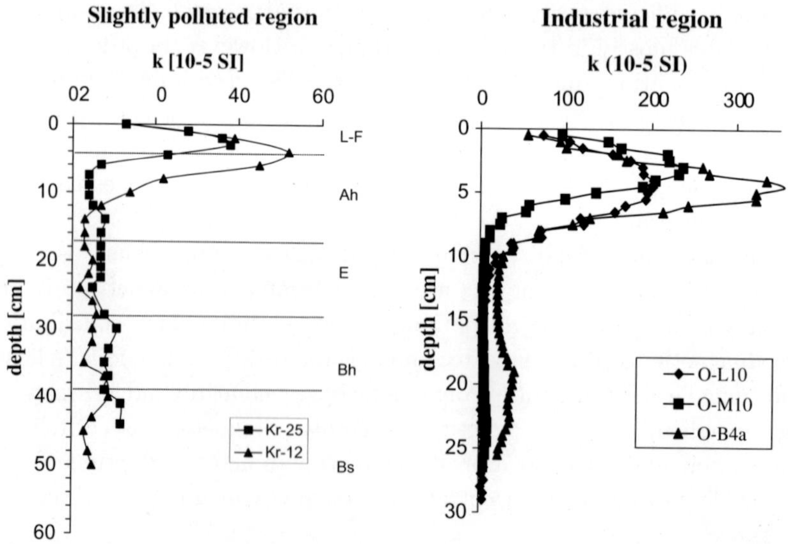

Fig. 1. Typical depth soil profiles of magnetic susceptibility from relatively clean area and industrial region in Czech Republic.

Fe-hydroxides to magnetically strong maghemite/magnetite. However, these particles of pedogenic origin are prevailingly in a form of very fine, superparamagnetic (SP) or single-domain (SD) grains and their relative significance can be estimated using frequency-dependent magnetic susceptibility k_{FD} [10].

Atmospheric dust of industrial origin contains Fe-oxides, magnetite (Fe_3O_4), maghemite (γ-Fe_2O_3) and hematite (α-Fe_2O_3). These are most often produced during combustion of fossil fuel (in, e.g., power plants) by decomposition of pyrite or Fe-rich clay minerals present in coal, followed by high-temperature oxidation of iron [1]. Also emissions from industrial units such as steel and cement works, as well as traffic represent significant sources of anthropogenic ferrimagnets (e.g., [11–13]). Magnetic properties of industrially derived ferrimagnets are different from those of natural origin. In terms of morphology, they are typically of spherical shape, with Fe-oxides sintered on Al-Si phase. Their typical size varies between about 2 and 50 μm and from magnetic point of view multi-domain (MD) structures prevail [14].

3. Results and Discussion

Monitoring of PM10 is performed using stationary high-volume samplers located at the Krušné hory piedmont at the altitude of 300 m a.s.l. and close to the summit (780 m a.s.l.). On both sites, sampling is during 24 hours and the air flow through the sampler is about 720 m^3/day. Long-term monitoring clearly showed that the highest concentrations of atmospheric dust in this region are in the winter period (December–February) [5]. Therefore, for demonstration of PM10 concentration variations with different altitude, typical values obtained in January 2009 were used. Figure 2 shows examples of filters with PM10 sampled on 13, 15 and 17 January 2009. PM10 concentration in both the localities are listed in Table 1. High concentrations in the piedmont are attributed to major sources of air pollution in the close neighborhood (power plants, smelters). Comparison with the summit concentrations suggests limited transport of PM10 to higher altitudes.

Vertical distribution of magnetic susceptibility was measured to depths of 30–40 cm on forest soils along the whole altitude profile from piedmont to the summit using MAGPROX SM400 kappameter (ZHInstruments, Brno, Czech Republic, [15]). Profiles, measured in the altitudes of 443 m, 815 m and 851 m a.s.l. are in Fig. 3. In all the cases the profiles show

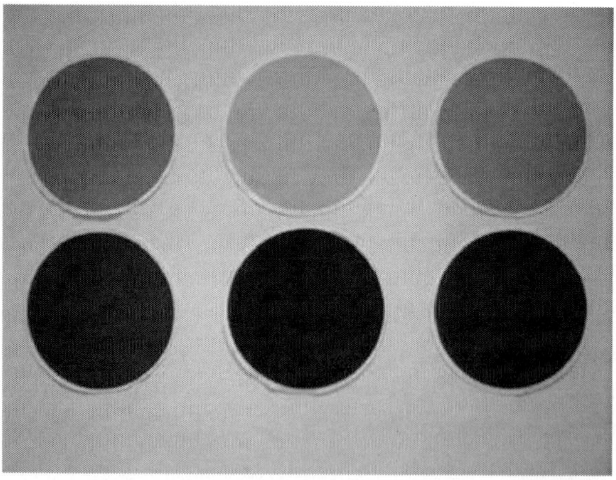

Fig. 2. Filters from high volume samplers in Krušné hory Mts (January 13, 15 and 17, 2009). Upper row — summit of the mountains (780 m a.s.l.), bottom row — piedmont of the mountains (300 m a.s.l.).

Table 1. PM10 concentrations in sampler filters from two localities in Krušné hory Mts.

Locality	Date	Mass [mg]	Concentration. [μg·m $^{-3}$]
Summit	Jan 13, 2009	20.63	29
Piedmont		92.61	138
Summit	Jan 15, 2009	5.99	8
Piedmont		152.01	225
Summit	Jan 17, 2009	12.14	17
Piedmont		62.16	92

clearly magnetically enhanced superficial (L-F) pedozones. At the same time, magnetic susceptibility in deeper mineral horizons is much lower and practically constant. This pattern suggests dominance of atmospherically deposited anthropogenic ferrimagnetics and negligible effect of natural minerals of lithogenic origin (e.g., [16]). The maximum value of magnetic susceptibility along the depth profiles seems to depend upon the altitude, the higher the altitude, the lower magnetic enhancement in the topsoil layer. This finding reflects qualitatively the differences in PM10 concentrations in the piedmont and summit sites.

Increased concentration of atmospherically deposited ferrimagnetics of anthropogenic origin was found in the topsoil layers. However, we still have to prove the presence of ferrimagnetics in the collected PM10 samples and

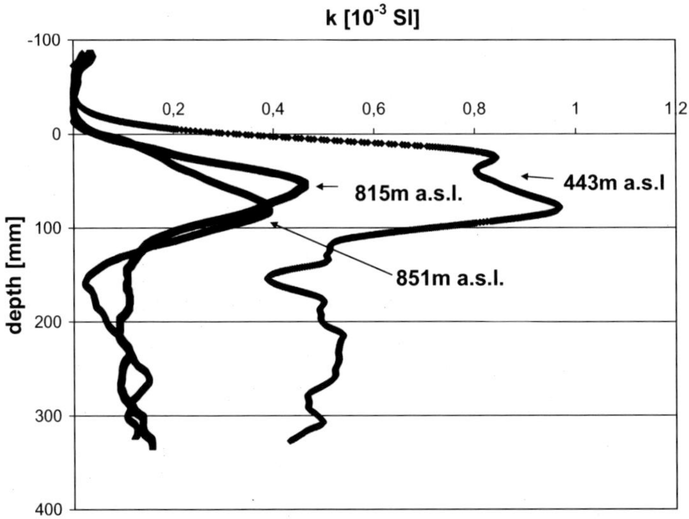

Fig. 3. Depth soil profiles of magnetic susceptibility at sites with different altitude (443 m, 815 m and 851 m a.s.l) in Krusne hory Mts.

to show that this substance has magnetic properties similar to those of the topsoil layer.

Therefore, thermomagnetic analyses (temperature dependence of magnetic susceptibility) were carried out using KLY-4S Kappabridge equipped with CS-3 furnace (AGICO, Brno, Czech Republic). Both samples of quartz microfiber filters with PM10 and magnetic extracts of soil samples were measured at temperatures from 20 to 700 degC in order to detect temperatures of magnetic phase transitions of major ferrimagnetic minerals. Samples of the *L-F* soil horizons were dominated by magnetite with Curie temperature of about 580 degC. Magnetic composition of bottom soil layers is always more complex, showing whole sequence of transformations of magnetically weak minerals at elevated temperatures. In the case of PM10 samples, despite very minute concentration, magnetite could be identified as well. This finding is of crucial importance for justification of magnetic monitoring of atmospherically deposited dust using soil magnetometry in the region of Krušné hory Mts.

Anthropogenic ferrimagnetic particles in atmospheric dust have typical morphology. Those resulting from combustion processes are typically spherules with the size from 5 to about 50 μm (e.g., [14]). We examined using SEM the PM10 samples as well as magnetic extract of the topsoil

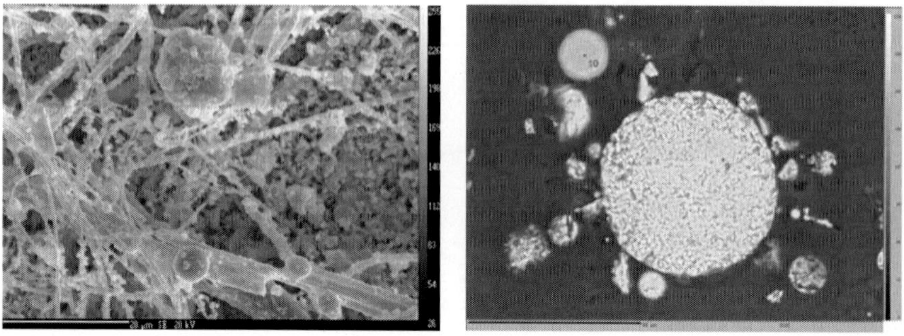

Fig. 4. SEM of ferrimagnetic spherules on quarz microfiber filter (PM10) from high volume sampler (left) and in topsoil magnetic extract (right).

samples. Typical results are depicted in Fig. 4. Despite the fact that the PM10 samples contain only smaller particles, below 10 μm, spherical particles were easily identified. Similar particles, with more variable grain size, were found also in magnetic extracts of topsoil samples. Deeper mineral soil horizons were free of these Fe-rich spherules, which are presumably of industrial origin.

4. Conclusion

Our results prove that measurements of topsoil magnetic susceptibility can help in assessing the spatial distribution of soil contamination due to atmospheric deposition of pollutants over the area of Krušné hory Mts. Topsoil horizons are magnetically enhanced due to presence of magnetic mineral phase which shows similar characteristics as that found in the collected PM10 samples, and which is presumably of industrial origin. Magnetic susceptibility of the enhanced topsoil layer shows qualitatively similar altitude dependence as concentration of atmospheric PM10. If certain rules are obeyed (estimating the significance of ferrimagnetic particles of lithogenic and/or pedogenic origin), magnetic method can be used for relatively fast and cheap assessment of soil contamination due to atmospheric deposition of pollutants. Contrary to routinely used stationary PM10 samplers, soil magnetometry allows acquisition of large datasets, covering the area in concern with much larger density. Hence, it allows delineation of significantly contaminated areas for more targeted sampling for detailed chemical analyses, which are more expensive and time consuming. However, soil magnetometry can only be considered as proxy and approximate method, which is site specific and has to be always calibrated using available environmental data.

Acknowledgments

This study was supported by Grant Agency of the Czech Republic through grant No. 205/07/0941.

References

1. P. J. Flanders, *J. Appl. Phys.* **75** (1994) 5931.
2. A. Kapička, N. Jordanova, E. Petrovský and S. Ustjak, *Phys. Chem. Earth* (*A*) **25** (2000) 431.
3. X. S. Wang, Y. Qin, *Environ. Geol.* **49** (2005) 10.
4. C. Spiteri, V. Kalinski, W. Rösler, V. Hoffmann, E. Appel and MAGPROX team, *Environ. Geol.* **49** (2005) 1.
5. MZP CR, *Statistical Environmental Yearbook of the Czech Republic 2006* (MZP CR, Praha, 2007), in Czech.
6. S. Hykysova and J. Brejcha, *WIT Trans. Ecology and the Environment* **123** (2009) 387.
7. C. X. Zhang, B. C. Huang, J. D. A. Piper and R. S. Luo, *Sci. Tot. Environ.* **393** (2008) 177.
8. E. Petrovský, A. Kapička, N. Jordanova, M. Knab and V. Hoffmann, *Environ. Geol.* **39** (2000) 312.
9. A. Kapička, N. Jordanova, E. Petrovský and V. Podrázský, *Water, Air and Soil Poll.* **148** (2003) 31.
10. B. A. Maher and R. M. Taylor, *Nature* **336** (1988) 368.
11. Y. Zheng and S. H. Zhang, *Chinese Sci. Bull.* **53** (2008) 408.
12. V. Hoffmann, M. Knab and E. Appel, *J. Geochem. Explor.* **66** (1999) 313.
13. W. Kim, S. J. Doh, Y. H. Park and S. T. Yun, *Atmos. Environ.* **41** (2007) 7627.
14. Z. Strzyszcz, T. Magiera and F. Heller, *Studia Geophysica et Geodaetica* **40** (1996) 276.
15. E. Petrovský, Z. Hůlka and A. Kapička, *Environ. Technol.* **25** (2004) 1021.
16. A. Kapička, E. Petrovský, N. Jordanova and V. Podrázský, *Phys. Chem. Earth* (*A*) **26** (2001) 917.

Advances in Geosciences
Vol. 16: Atmospheric Science (2008)
Eds. Jai Ho Oh *et al.*

IMPACT OF VEGETATION ON THE INDIAN SUMMER MONSOON: MODEL SENSITIVITY EXPERIMENTS

TRILOCHAN PATTANAIK
Pukyong National Univerisy, Dept. Env Atmos. Sci,
Busan, South Korea
trilochan21@gmail.com

H. S. CHAUDHARI
Indian Institute of Tropical Meteorology,
Pune, India
hemantkumar@gmail.com

JAI HO OH
Pukyong National Univerisy, Dept. Env Atmos. Sci,
Busan, South Korea

ASHISH DHAKATE
Indian Institute of Tropical Meteorology,
Pune, India

G. P. SINGH
Banaras Hindu Univeristy,
Varanashi, India
singgpin@yahoo.co.in

This study examines response of vegetation on the simulation of July rainfall in 1998 and 2002 using meso-scale model MM5. The model was integrated with two nested domains at 90 and 30 km resolution. The outer domain covers from 30°S–50°N, 30°E-120°E and inner domain from 5°N-40°N, 65°E-105°E over the Indian region. Two experiments are performed, each consisting of the model integration from middle of June to end of July for both 1998 excess and 2002 deficient monsoon years. In the first run (Expt1), the model was integrated with United States Geological Survey (USGS) vegetation. In the second run (Expt2), the model was integrated with Indian Space Research Organization (ISRO) satellite derived vegetation. Results indicate that Expt1 (USGS) has greater tendency of overestimation of rainfall. It is seen that RMSE (Root Mean Square Error) in the month of July for all India rainfall is lower for Expt2 (ISRO). Model bias of July is also closer to unity for ISRO derived vegetation (Expt2). Tibetan anticyclone is better simulated by ISRO vegetation (Expt2).

The success of ISRO derived vegetation can be attributed to greater spatial coverage over Indian region. Still it is difficult to conclude the superiority of vegetation fraction between USGS and ISRO. Intensive study is required for exploring the detailed impact of vegetation on Indian region.

1. Introduction

The Indian subcontinent witnessed a severe monsoon drought in 2002, which largely resulted from a major rainfall deficiency in the month of July. The Indian Summer Monsoon Rainfall (ISMR) from June to September for India as a whole during 2002 was 81% of its long period average. Much of the rainfall decrease occurred in the core rainy month of July 2002, when the rainfall distribution over the country was nearly 50% below the long-term normal and was the lowest in historical records during the last 130 years (Chaudhari *et al.*, 2008). Excess ISMR in 1998, featured abundant rains (1,200–1,600 mm) in the typically heavy rainfall area of western India, which is 200–600 mm above the long-term mean.

Simulation of Indian summer monsoon has been the focus of many studies, due to its anomalous characteristics in the tropical circulation. It has been well established that the AGCMs (Atmospheric General Circulation Models) are qualitatively able to capture the mean Asian summer monsoon rainfall distribution. In recent years, attempts have been started to simulate the climate features on a smaller scale in order to explore the possibility of simulating the regional climate characteristics which could not be simulated by the rather coarse AGCM resolution. In general, the initial and boundary conditions for the integration of the regional climate models (RCMs) are taken from AGCM output at regular time intervals. The success of RCM depends on the accuracy of initial state of the atmosphere and time-varying lateral boundaries of the model domain.

The change of vegetation cover at the interface between the surface and the atmosphere, influences climate through changes in surface albedo, land roughness and soil hydrological and thermal features (Xuejie *et al.*, 2008).

This paper investigates the influence of vegetation on the simulation of July rainfall of two contrasting ISRM of 1998 and 2002. The initial and boundary conditions for the integration of MM5 is taken from National Centre for Medium Range Weather Forecasting (NCMRWF, India) T80 global model.

Table 1. Configuration of MM5 used for this study.

Model Elements		Non-hydrostatic model
Domain of integration		30°S–40°N, 30°E–120°E (outer domain, 90 km) 5°N–40°N, 65°E–105°E (inner domain, 30 km)
Physics	Cumulus parameterization	Grell
	PBL parameterization	MRF PBL
	Explicit moisture schemes	Simple-ice
	Radiation schemes	Cloud-radiation interaction
	Land surface processes	Five layer soil model

2. Brief Description of NCAR MM5

The NCAR (National Center for Atmospheric Research) MM5 used in this study is a non-hydrostatic model developed and distributed through the National Center for Atmospheric Research (NCAR). The present version of the NCAR MM5 has 23 sigma levels and the option to choose the region of the model domain and the horizontal resolution (Table 1). The NCAR MM5 also provides different options for the physical processes of radiation, explicit cloud scheme, convection parametrization, planetary boundary layer parametrization and soil physics.

3. Design of Numerical Experiments

Sensitivity experiments are performed with NCAR-MM5. The initial and boundary conditions for the integration of MM5 is derived from National Centre for Medium Range Weather Forecasting (NCMRWF) T80 global model. The model was integrated with the nested domains at 90 km and 30 km resolution. The outer domain covers from 30°S–40°N, 30°E–120°E (90 km resolution) and inner domain from 5°N–40°N, 65°E–105°E (30 km resolution) over the Indian region (see Fig. 1). The model integration is performed from middle of the June to end of July.

The main focus of this study is to simulate the July rainfall in 1998 and 2002. Two sets of model experiments are performed in this study: first run (Expt1) is with United States Geological Survey (USGS) vegetation and second run is performed with Indian Space Research Organization (ISRO) satellite derived vegetation (Expt2).

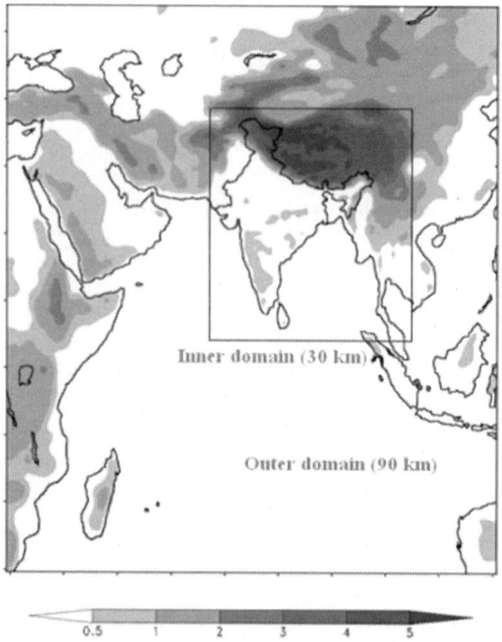

Fig. 1. Domains of MM5 simulations. Shading indicates topography in km.

4. Analysis of Simulated Results

4.1. *Precipitation distribution over India*

In this study an attempt has been made to highlight the effect of vegetation on model simulations. For the comparison purpose, Tropical Rainfall Measuring Mission (TRMM) data has been utilized. TRMM rainfall (observed rainfall) for July 1998 and July 2002 is presented in Figs. 2(a) and 2(b).

Model simulated rainfall patterns for the month of July 1998 are presented is Figs. 3(a) and 3(b). We have presented 30 km MM5-domain results. Difference between observation (TRMM rainfall) and Expt1 (USGS vegetation) is presented in Fig. 3(c). Difference between observation (TRMM rainfall) and Expt2 (ISRO vegetation) is also presented in Fig. 3(d). On similar lines, model simulated rainfall patterns and differences for the month of July 2002 are presented in Figs. 3(e)–3(h).

In the month of July for 1998, Expt1 (USGS) has performed better near foothills of Himalayas but it has overestimated over Bay of Bengal region as compared with Expt2 (ISRO). From the difference plots it is clear

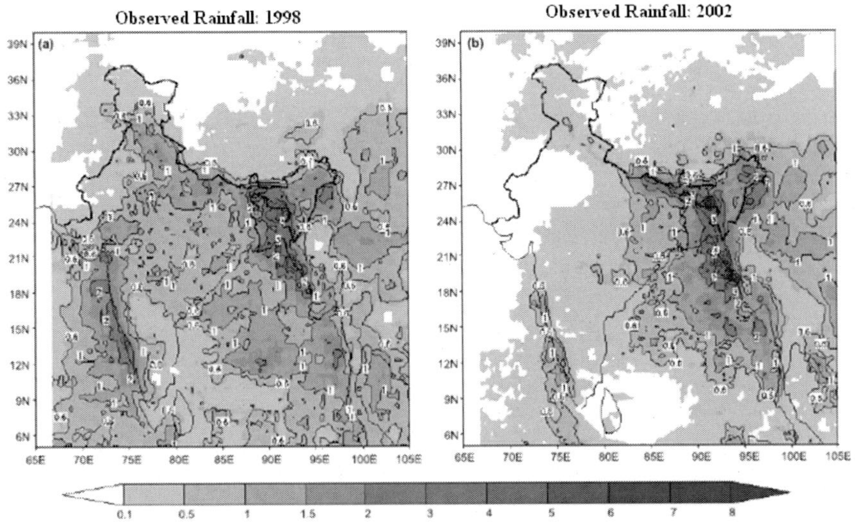

Fig. 2. July rainfall (cm/day) by TRMM for (a) 1998 and (b) 2002.

that the rainfall over western coast is in agreement with observed rainfall for both Expt1 and Expt2. Expt1 (ISRO) is better in July 1998 over the region. Over the Arabian Sea, model has overestimated rainfall in both the years.

Figure 4 shows the plot of model biases (Fig. 4(a) and RMSE (Root Mean Square Error; Fig. 4(b). Bias is defined as the ratio of the model simulated value to observed value. If the bias is less that 1, it indicates that the model is underestimating and if greater than 1, it implies that the model is over estimating the value. RMSE is the measure of the differences between the values predicted by a model and the values actually observed. It is clear that bias of July simulation of Expt2 (ISRO) is close to unity. It implies that Expt2 with ISRO vegetation is better simulated than Expt1 of USGS vegetation. From plot of model biases, it is also clear that Expt1 (USGS) has a greater tendency of overestimation as compared to Expt2 (ISRO). It is seen that RMSE of all India total rainfall for Expt2 (ISRO) is less as compared to Expt1 (ISRO).

5. Wind Patterns

In this section, we have discussed wind patterns at 200 hPa. For the brevity, we have not presented wind pattern at 850 hPa. Model simulated

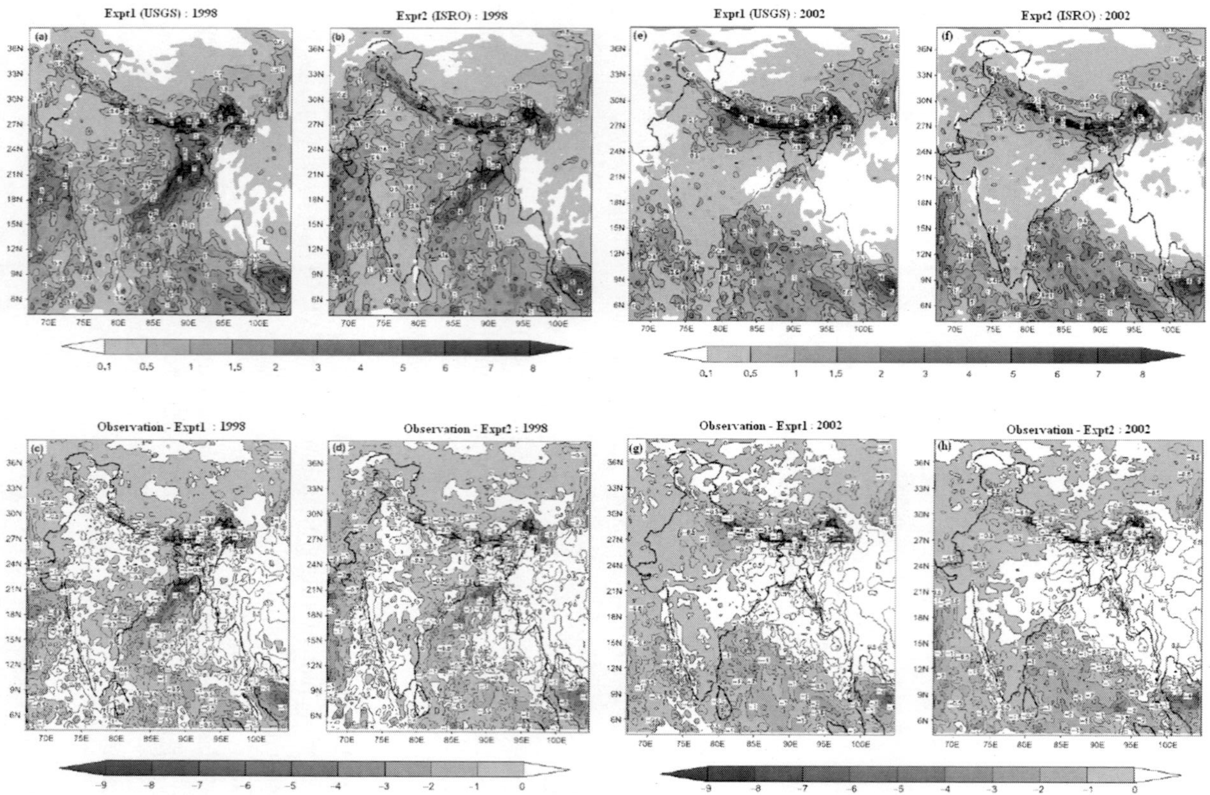

Fig. 3. Model simulated mean rainfall (cm/day) of July for (a) Expt1 (USGS) 1998, (b) Expt2 (ISRO) of 1998, (e) Expt1 (USGS) of 2002 and (f) Expt2 (ISRO) of 2002. The difference from observation is presented in (c) for Expt1 (USGS) of 1998, (d) Expt2 (ISRO) of 1998, (g) for Expt1 (USGS) of 2002 and (h) Expt2 (ISRO) of 2002.

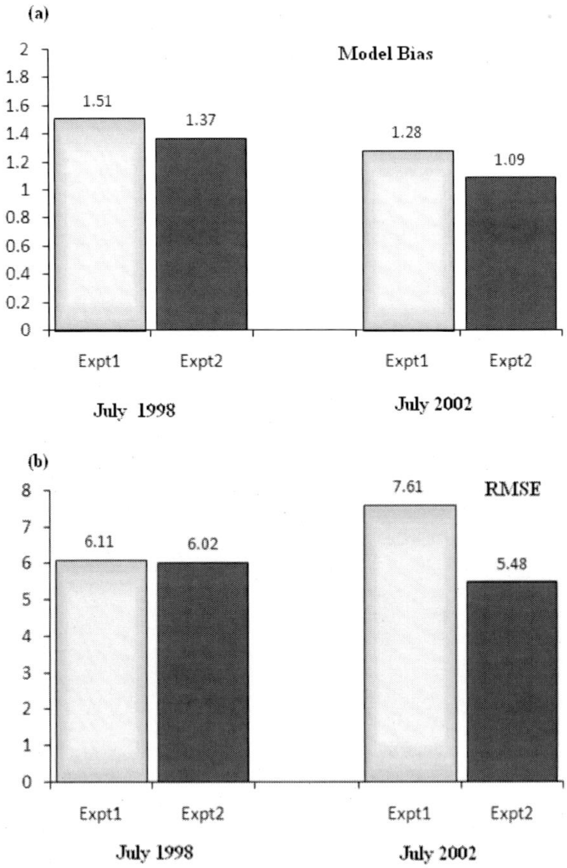

Fig. 4. (a) Model bias and (b) RMSE (mm) of MM5 simulations with respect to TRMM observation.

wind patterns (90 km MM5 domain) for the month of July for 1998 and 2002 at 200 hPa are presented in Figs. 5(c)–5(f). Corresponding observations (NCEP reanalysis wind patterns at 200 hPa) are also presented in Figs. 5(a) and 5(b). Comparison of Figs. 5(c)–5(f) with Figs. 5(a) and 5(b) shows that in July 1998 and July 2002, Expt2 of ISRO vegetation has simulated the Tibetan anticyclone closer to that of observation (NCEP wind patterns). Both NCEP and Expt2 (ISRO vegetation) have produced the anticyclone as a ridge over the Himalayan belt whereas Expt1 (USGS vegetation) depicts closed circulation with a defined center. In July 1998, Expt1 and NCEP (observation) show strong and widespread Tibetan anticyclone.

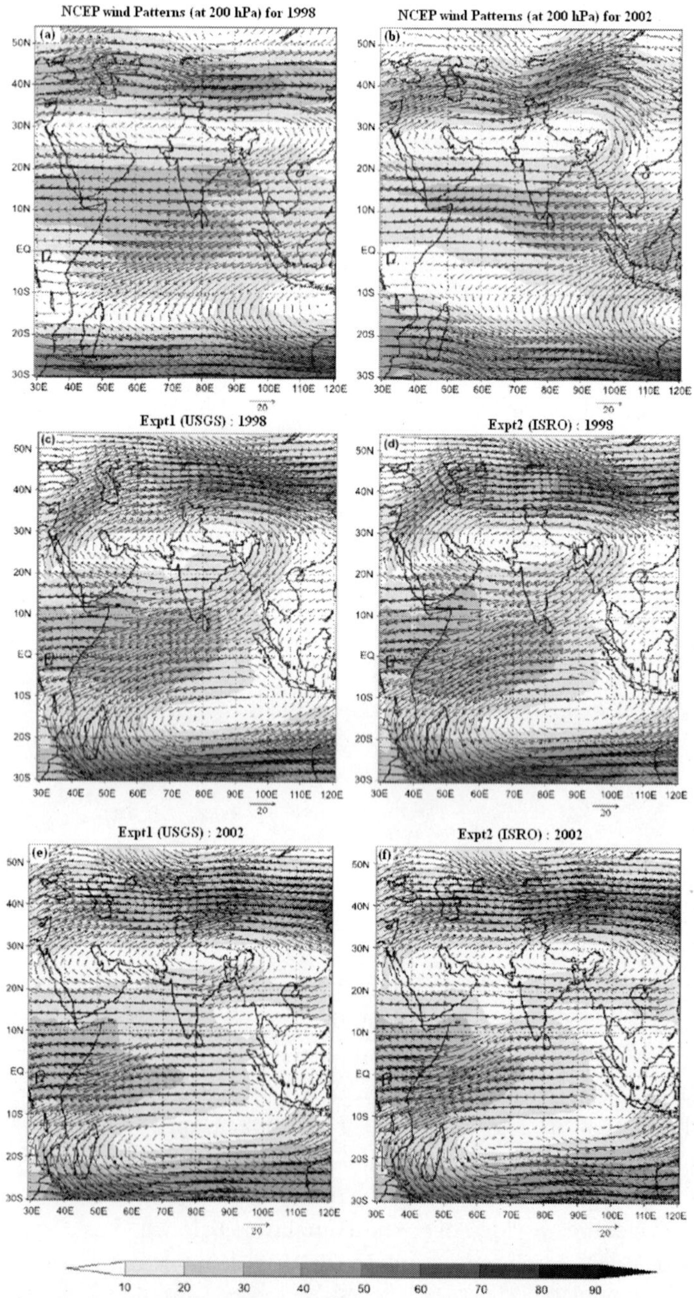

Fig. 5. Wind patterns at 200 hPa from NCEP for (a) 1998 and (b) 2002. MM5 simulated mean horizontal wind (m/s) for (c) Expt1 (USGS) of 1998, (d) Expt2 (ISRO) of 1998, (e) Expt1 (USGS) of 2002 and (f) Expt2 (ISRO) of 2002.

6. Conclusions

This study examines impact of vegetation on the simulation of Indian summer monsoon of 1998 and 2002 using regional model. Four sets of model experiments are performed, each consisting of the model integration from middle of June to end of July for both 1998 excess and 2002 deficient monsoon years. Two types of experiments are performed, each consisting of the model integration from middle of June to end of July for both 1998 excess and 2002 deficient monsoon years. In the first experiment (Expt1), the model was integrated with USGS vegetation and for the second experiment (Expt2), the model used ISRO satellite derived vegetation. Results indicate that Expt1 (USGS) has greater tendency of overestimation of rainfall. It is seen that RMSE in the month of July for all India rainfall is lower for Expt2 (ISRO). Model bias of July is close to 1 for ISRO derived vegetation (Expt2). Tibetan anticyclone is better simulated in ISRO vegetation (Expt2). The success of ISRO derived vegetation can be attributed to greater spatial coverage over Indian region. Wind fields at 200 hPa and 850 hPa (Figure not shown) are better simulated by ISRO derived vegetation (Expt2). Simulation of Tibetan anticyclone is more realistic in ISRO derived vegetation (Expt2). Intensive study is required for exploring the detailed impact of vegetation on Indian region including focused approach of understanding the differences in vegetation fraction/types and their spatial coverage.

Acknowledgments

The authors are thankful to Dr. P. C. Joshi and ISRO for providing vegetation data. One of the authors, Mr. T. Pattanaik is thankful to Dr. Someshar Das, NCMRWF and A. K. Bohra, NCMRWF for their suggestions and guidance during his tenure at NCMRWF.

References

1. H. S. Chaudhari , M. A. Shinde and J. H. Oh, *Quart. Int.* (2009), doi:10.1016/j.quaint.2008.05.009.
2. G. Xuejie, L. Yong, L. Wantao, Z. Zongci and F. Giorgi. *Adv. Atmos. Sci.* **20** (2008) 592.

Advances in Geosciences
Vol. 16: Atmospheric Science (2008)
Eds. Jai Ho Oh *et al.*

LIDAR OBSERVATIONS OF STRATOSPHERIC AEROSOL OVER GADANKI

C. NAGESWARA RAJU
Department of Physics, S. V. Degree College,
Kadapa, A. P, India
raju_svdc@yahoo.com

M. KRISHNAIAH
Department of Physics, Sri Venkateswara University,
Tirupati, A. P, India

Y. BHAVANI KUMAR
National Atmospheric Research Laboratory,
P.B No: 123, Tirupati, A. P, India

A Rayleigh–Mie backscattering lidar has been in operation at the National Atmospheric Research Laboratory, Gadanki (13.5° N, 79.2° E), a rural site in the tropical part of India, since 1998. The Lidar system is sensitive enough to profile atmospheric aerosol in the height region covering 10 to 30 km. Using this system, observations of upper troposphere clouds and aerosols at stratospheric heights have been made on regular basis to monitor the tropical upper troposphere and lower stratosphere structure. The Lidar system is sensitive to the degree of depolarization in the laser backscattering. Five years of lidar measurements covering 1998 to 2002 indicate that the aerosol distribution in the upper troposphere and lower stratosphere has shown benign aerosol profile with minimum backscatter at stratospheric heights. This paper presents results from the lidar observations and discusses physics behind such state of the atmosphere.

1. Introduction

Stratospheric aerosol can affect the global climate system in a variety of ways. These aerosols play a significant role in the Earth's radiative balance[1] and the attenuation of UV radiation. Also, they provide a surface for heterogeneous chemical reactions that are important for ozone loss in the middle atmosphere. Stratospheric aerosol is mainly composed of super cooled sulphuric acid droplets. They may also contain small amounts of other components such as ammonium sulphate. Physical properties of the stratospheric aerosol are either measured by *in-situ* or remote-sensing techniques. While *in-situ* techniques are best suited to measure aerosol size distribution and the aerosol's chemical composition, remote sensing techniques are better in monitoring the aerosol's optical properties. The link between both sets of measurements is then eventually provided by applying the Mie theory to calculate either for measured optical properties and assumed chemical composition the size distribution or vice versa. An established *in-situ* technique to measure the aerosol size distribution is by selective particle counting by the instruments deployed on stratospheric balloons.[2–3] Remote-sensing techniques either rely on active instruments like LIDARs or passive instruments that use direct sun, moon or star light or scattered skylight. In the present paper, the lidar observations during period from 1998 to 2002 are described. The characteristics are presented in terms of height distribution, their variability with atmospheric temperatures and their temporal and spatial variability characteristics.

2. System

The lidar system was installed at this site during 1998 under Indo-Japanese collaboration[4] for monitoring the tropical atmosphere. The complete system configuration is shown in Fig. 1. The lidar transmitter subsystem is a Nd:YAG pulsed laser with a second harmonic generator (Powerlite 8020: Continuum). The output energy of frequency doubled 532 nm laser pulse is 550 mJ in repetition of 20 Hz. The laser pulses are transmitted to the zenith by a steering mirror after the beam divergence is reduced to lower than 0.1 mrad through a beam expander (×10).

The National Atmospheric Research Laboratory (NARL) lidar employs two independent receivers for collecting the laser-backscattered returns from the atmosphere. One of these is a Rayleigh-scatter receiver that collects molecular backscatter in height range of 30–80 km, where as the other one is a Mie scatter receiver that operates in the height range of 4 to 40 km to receive the backscatter from air molecules and aerosol/cloud particles. The measured photon count profiles are verified for its validity before it is subjected to the analysis. This is performed by normalizing

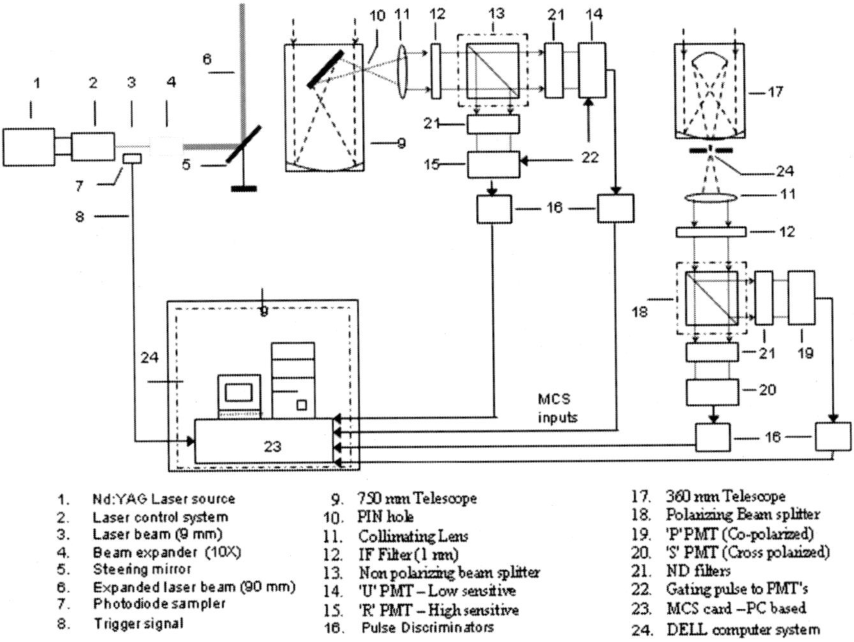

Fig. 1. Block diagram of Indo-Japanese lidar system.

the range corrected measured photon count at a particular altitude to a molecular density profile obtained from standard atmosphere model (CIRA-86) data to derive the expected photon count profile. The linear regression of the measured photon count with the expected photon count indicates the usability of the data.[5] This is verified for every height profile of the data.

3. Results and Discussion

The Indo-Japanese Lidar (IJL) has been measuring the stratospheric aerosol layer backscatter at 532 nm wavelength with dual polarization information since March 1998. The basic lidar signal represents the total backscatter that contains returns from both molecules and particle part of the atmosphere. The particulate part of the atmosphere represents either aerosol or clouds composed of ice crystals or hydrometeors. The strength of aerosol or cloud backscatter is usually represented in terms

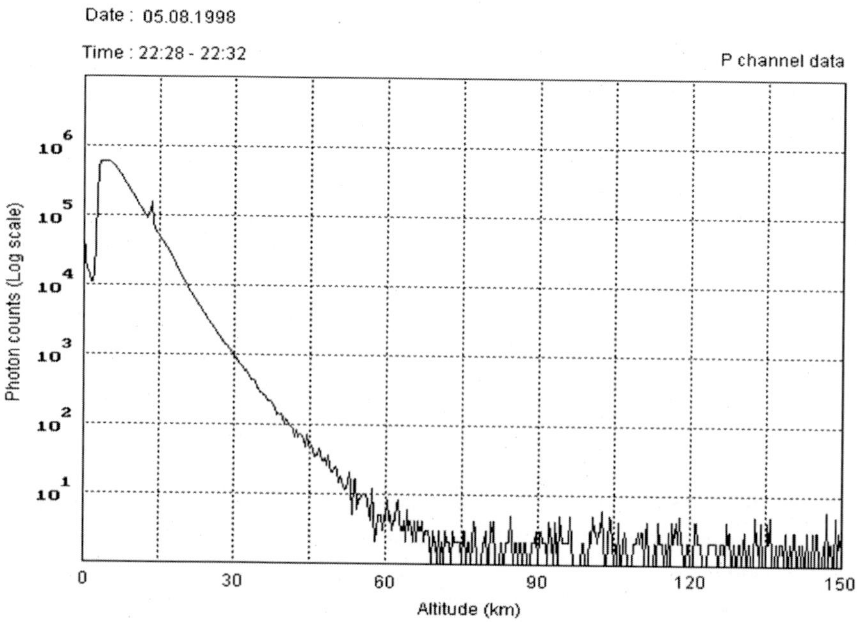

Fig. 2. Typical basic photon count profiles from Mie-scatter lidar. Each profile is 5000-laser shot integration over 250-sec duration.

of the backscatter ratio. The noise-corrected lidar signal is subjected to range normalization and inversion[6] to derive the backscatter ratio profile. Fernald's approach for inverting the lidar signal to derive aerosol backscattering profiles assumes that the aerosol size distribution and composition are not varying with height and extinction-to-backscatter ratio of the aerosol, $S_a = \alpha_a(z)/\beta_a(z)$, is constant over the range. The inversion method basically takes reference or calibration altitude at which the backscattering coefficient is mainly due to molecules and then doing downward integration to get the height profiles of $\beta_a(z)$. Here S_a is taken as 40[7] and the inversion is done downwards from the reference altitude, which is usually taken above 35 km altitude. The typical height profile of aerosol backscatter coefficient derived from lidar system *p*-channel (co-polarized signal) is shown in Fig. 3. The figure represents altitude profiles of total backscatter, molecular backscatter and derived aerosol backscatter coefficient for the night observation of 29, January 1999 integrated over one hour between 21:58 and 22:57 LT.

The backscatter ratio (R) is calculated from the sum of the Mie (particle or aerosol) and Rayleigh scattering coefficient (molecular) divided by the

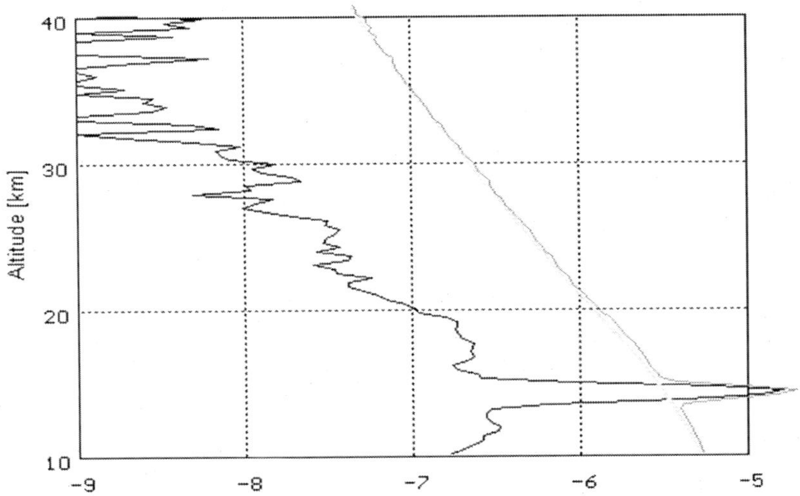

Fig. 3. Height profiles of total backscatter (right line), molecular backscatter (middle line), and aerosol backscatter coefficients (left line) derived from NARL lidar data for the night of 29 January 1999 for an integration period of one hour.

Rayleigh backscatter coefficient and is given by

$$R(z) = \frac{\beta_{\text{air}}(z) + \beta_{\text{aerosol}}(z)}{\beta_{\text{air}}(z)}, \tag{1}$$

where $\beta_{\text{air}}(z)$ and $\beta_{\text{aerosol}}(z)$ represent the volume backscatter cross-sections for particles and molecules at altitude z, respectively. Radiosonde pressure and temperature data obtained from the nearest meteorological station, Chennai, South-East of lidar site approximately 125 km away, has been used for deriving the air molecular density data. Each lidar data profile was calibrated by performing a linear regression of the measured signal to a theoretical molecular backscatter profile derived using the radiosonde data.

Another lidar derived quantity, volume depolarisation (D), is estimated from linear and orthogonal aerosol backscatter signals. The signal received from P channel represent co-polarized component that arise from molecules and aerosol, which are isotropic scatterers. The signal from S channel is mainly due to scattering from the anisotropy of the scatterers. The lidar derived volume depolarization $D(z)$, computed from the backscatter ratios of co- and cross-polarized signal components, is defined as

$$D(z) = K\left(\frac{S_{\perp}(z)}{S_{\|}(z)}\right), \tag{2}$$

where the subscripts refer to perpendicular and parallel polarized signal components. The calibration component, K, is equal to volume depolarization of 1.4% where no particles are present, such as the case of pure molecular atmosphere. It may be noted here that higher values of D, greater than air depolarization (0.014), indicate non-spherical particles presumably ice crystals.

3.1. *Altitude structure of background stratospheric aerosol observed above Gadanki site*

The lidar has been operating since March 1998 on a bi-weekly basis depending on fair weather conditions. The Mie lidar observations taken between 1998 and 2002 are presented, depicting month-to-month profiles in terms of backscatter ratio (R). Only those nights when there were no clouds (Low altitude clouds) during the whole observation time are presented. Height profiles up to 35 km are for a time resolution of 30 min and altitude resolution of 300 m. The lidar backscatter ratio profiles in Fig. 4(a) show the temporal variation of one hour integrated aerosol backscatter ratio profiles observed during the investigation period.

The right side given height profile is the mean altitude profile for the given night with horizontal lines indicating standard deviation (SD) over the mean value at selected altitudes. The statistical variability is obtained using the SD technique normally used in signal analysis.[8] The altitude structure of aerosol BSR is given from the upper troposphere (above 10 km) to the lower stratosphere (up to 35 km) for the observational period. In the upper troposphere region, some of the profiles show spiky backscatter in the measurements, which corresponds to the occurrence of optically thin clouds such as cirrus. There are no noticeable sharp enhancements in aerosol backscatter observed at the stratospheric heights. This confirms that during the time of our observations the tropical upper troposphere and stratosphere over Gadanki appears free from volcanic aerosol and what we have observed now is the background variable aerosol.

Five nights of mean integrated scattering ratio profiles estimated from the contribution of upper troposphere and lower stratospheric aerosol are given in Figs. 4(a) to 4(e). The maximum scattering ratio in these profiles varies between 1.1 and 1.2. This is slightly high when compared to the

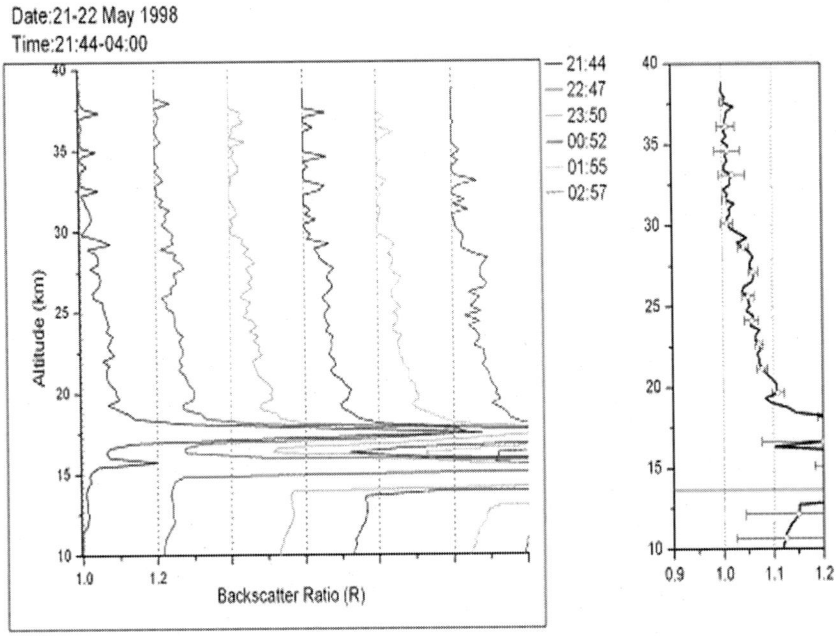

(a)

Fig. 4. Sequence of one hour integrated altitude profiles of aerosol backscatter ratio (R) observed by polarization Mie lidar over Gadanki site on the night of 21–22 May 1998. The right side profile shows the mean profile for the night with horizontal bars representing the variability in the layer as seen by lidar during the night observation. The aerosol peak observed around 16 km altitude.

observations at Ahmadabad (23°N, 72.5°E) (nearly 1.1) as reported by Jayaraman *et al.*[9] over Mount Abu, Gujarat, India. This may be attributed to the time of measurement and geographical location. The values are similarly comparable with those of the lidar measurements[10] at Chung-Li. Taiwan (25°N, 121°E). In addition to this, during the equinox period, an enhancement observed in the height range 17–19 km clearly shown in Figs. 4(a) to 4(e). This one hour integrated aerosol profiles show that the summer season is highly turbid as compared to the rest of the year. Moreover, the observations over Gadanki site reveal double peaked aerosol altitude structure on some nights.

The aerosol layer's persistent decrease has been caused by the sedimentation and transport in the atmosphere. It is interesting to point out that the aerosol extends into troposphere region where high altitude clouds such as tropical cirrus were formed. It is generally thought that the stratospheric aerosol can act as the condensation nuclei for high altitude

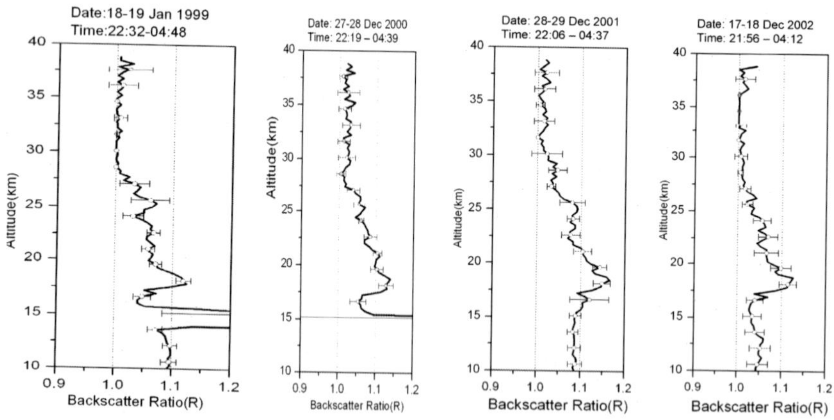

(b) (c) (d) (e)

Fig. 4. 4(b)–4(e). The mean profile with horizontal bars representing the variability in the layer by lidar during the night observation.

clouds. Moreover, the dynamical conditions such as winds in the upper troposphere are also another parameter that can cause transport and sedimentation agency for this aerosol in this altitude region. There is an interesting enhancement in BSR values on the night of 21–22 May 1998. The reason for such enhancement could be the lowering of tropical atmosphere temperatures, favoring hetero-molecular nucleation at these altitudes. At very low temperatures with large water vapor pressures, the aerosol grows by condensing water vapor on their surfaces, diluting the H_2SO_4–H_2O mixture.

3.2. *Altitude structure of stratospheric aerosol with respect to temperature*

The monthly mean aerosol extinction and temperature data constitute the basic data for the present investigation. The typical height profiles of aerosol backscatter ratio integrated over two hour period, as observed by IJL, on the nights of 8th May, 16th September, 21st December and 19th January 1999 are shown in Figs. 5(a)–5(d). Each lidar profile is shown with a corresponding radiosonde temperature profile that is obtained from the nearest meteorological station, Chennai, for identification of the local tropopause altitude. The general circulation of the atmosphere between the stratosphere and troposphere is called tropopause folding. Even though we have two apparently separate layers (Troposphere and Stratosphere), there is interaction and how and when interaction occurs is an important for understanding the chemistry of the Earth's atmosphere.

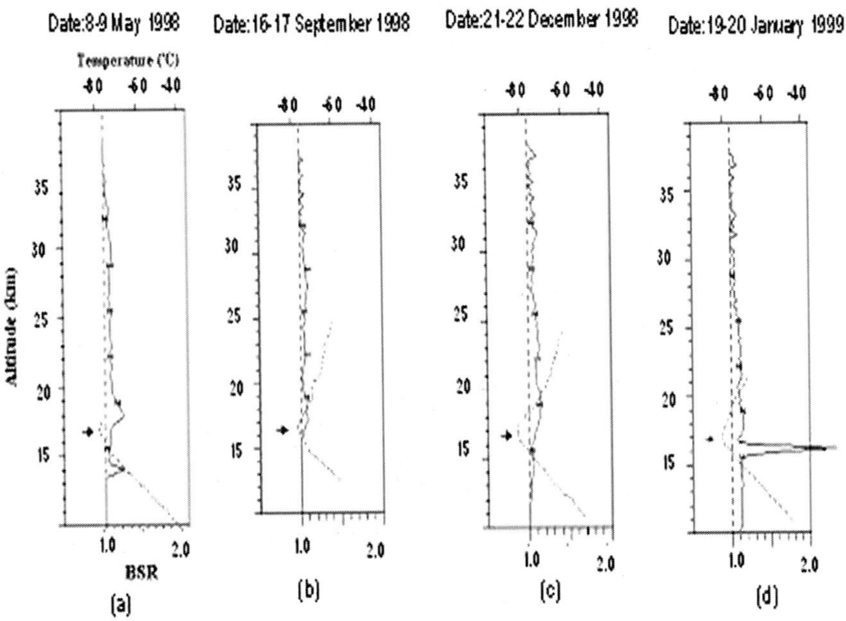

Fig. 5. Height profile of stratospheric aerosol obtained on the nights of 8 May 1998, 16 September, 21 December 1998 and 19 January 1999 shown along with altitude profile of temperature derived from Radiosonde data. The marked region indicates the local tropopause height, which shows the minimum BSR region.

The horizontal arrows indicate the altitude of cold-point tropopause obtained from these temperature measurements. A significant observation in the backscatter profiles is a noticeable minimum in BSR at the height marked as the local tropopause. A sharp gradient in the temperature accompanied by the tropical tropopause may be the probable reason for such minimum in the aerosol concentration. The horizontal bars show standard error dR in the measurements of R for a few sample heights for clarity. Typically, for clear sky conditions, the BSR standard deviation at low altitudes (10–11 km) is $\sim$0.02 and increases to 0.035 at high altitudes (34–35 km). This rise in the standard deviation is due to exponential decay in the signal counts with altitude.

3.3. *Temporal variation of integrated aerosol backscatter coefficient (IABC) and tropopause temperature*

A strong negative correlation between tropopause temperature and integrated aerosol backscatter coefficient is very clearly seen in Fig. 6. The correlation between the tropopause temperature and ambient

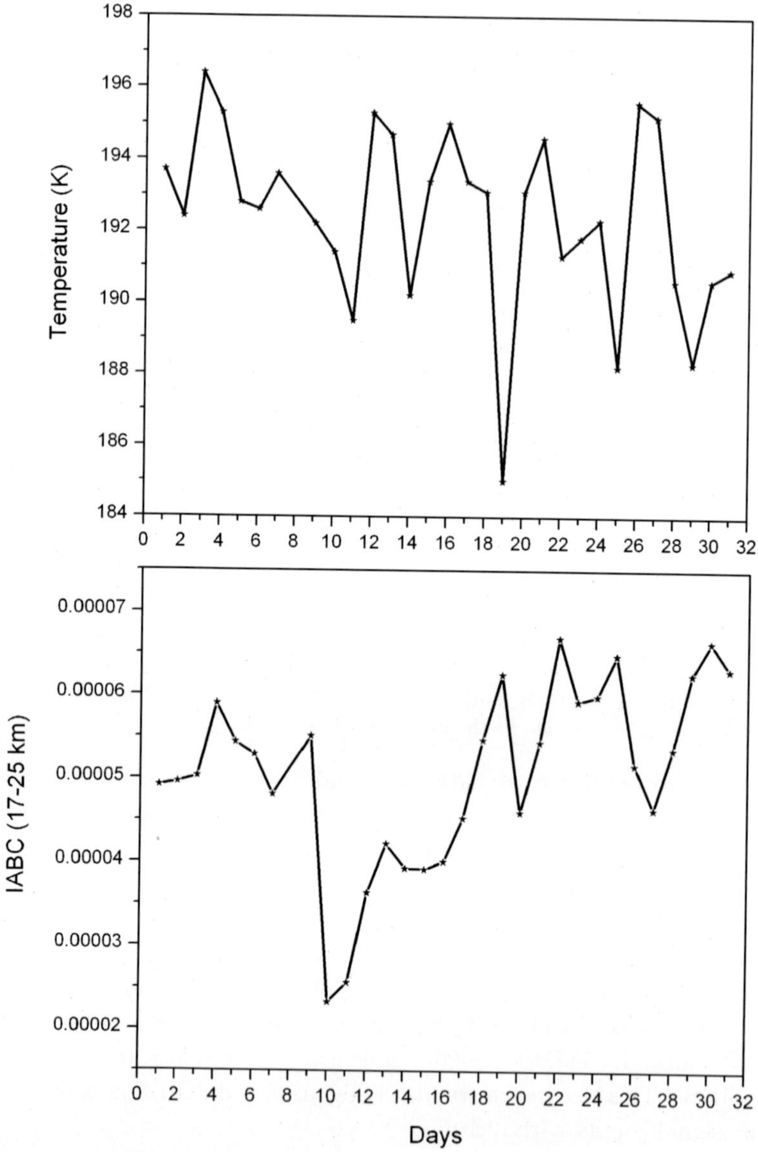

Fig. 6. Comparison of lidar derived IABC with tropopause temperature data (inferred from Radiosonde data).

temperature at a particular stratospheric altitude is positive at lower altitudes and negative at the higher altitude above 22 km as can be expected, though the value of the correlation coefficient itself is low and not significant. The aerosol backscatter or extinction at all altitudes in the stratosphere (17–25 km) is negatively correlated with the tropopause temperature.[11] This indicates that this air is transported up in the stratosphere, the effect of which will be more pronounced at higher altitudes because of increased efficiency of stratospheric microphysical processes at higher altitudes.

3.4. *Stratospheric aerosol LDR and atmospheric temperature*

Polar stratospheric clouds (PSC) are observed with highly anisotropic aerosol particles in ice crystalline form. Little is known about the tropical observations with such characteristics of stratospheric aerosol. The observation shows that the tropical stratospheric aerosol can also exhibit the property of anisotropy by significantly depolarizing the laser returns, when ambient temperatures goes below the freezing point of sulfate aerosol. In Fig. 7, the polarization lidar observations of stratospheric aerosol on

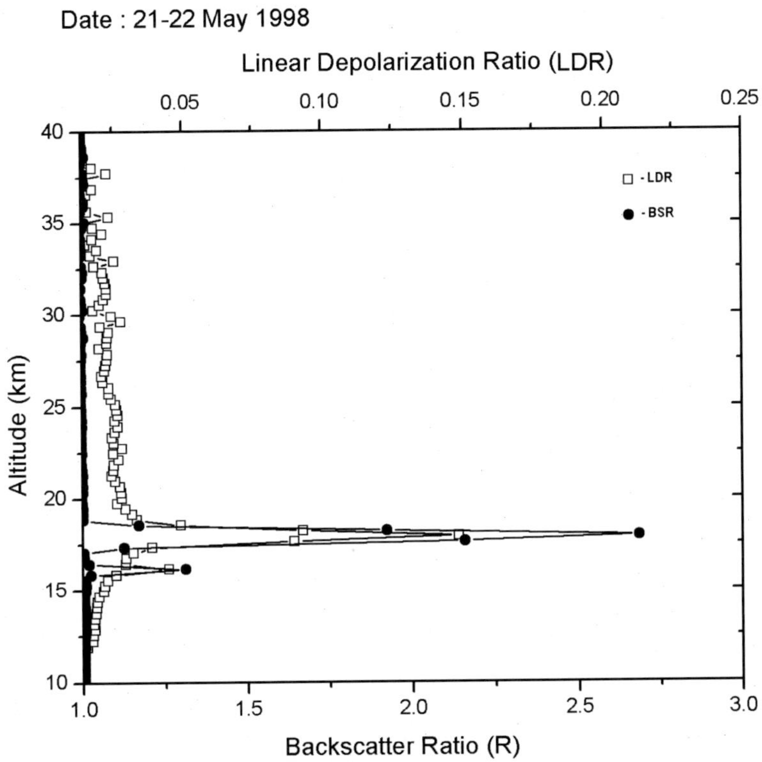

Fig. 7. Lidar observation of enhanced aerosol BSR and LDR on the night of 21–22 May 1998 over Gadanki site.

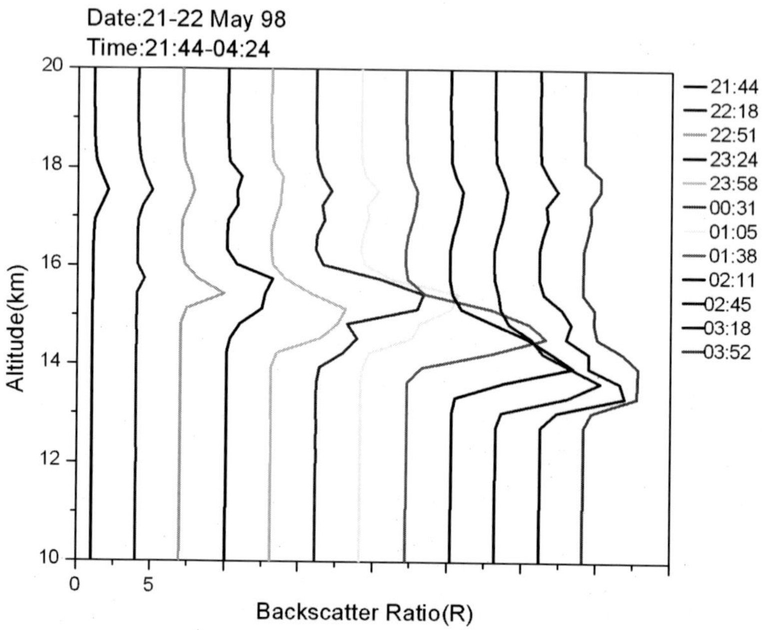

Fig. 8. Spatial and temporal variability of tropical cirrus observed on 21–22 May 1998 over Gadanki site.

the night of 21–22 May 1998 are shown. Figure 8 shows the altitude profile of R, illustrating the presence of cirrus layer on the night of 21–22 May 1998. In this figure R varies as high as 12 between 13 and 16 km which indicates the low altitude clouds and no high altitude clouds were observed.

4. Summary

The measurement of stratospheric aerosols by the lidar system confirms that during the observation period, the tropical upper troposphere and lower stratosphere over Gadanki site appears free from volcanic aerosols and what we observed now is the background variable aerosols. The tropical aerosol backscattering at stratospheric altitudes shows strong negative correlation with tropopause temperature. The integrated aerosol backscatter coefficient at stratospheric altitude range of 17–25 km shows a negative correlation with ambient temperature. The stratospheric aerosol backscatter at levels closer to the tropopause is mainly due to aerosol particle injection from the troposphere whereas at higher levels the contribution due to stratospheric microphysical processes (condensation and nucleation) involving precursor gases transported from the troposphere becomes increasingly important.

Acknowledgments

This work was supported by the University Grants Commission, SERO, Hyderabad in the form of a Minor Research Project sanctioned to the author C. Nageswara Raju, who would like to thank the Deputy Secretary (UGC-SERO) Hyderabad. The author also thanks the Director and the staff of National Atmospheric Research Laboratory, Gadanki for their kind encouragement to this work. The author is also thankful to Sri A. Gangi Reddy, Chairman, S. V. Group of Colleges in Kadapa & Tirupati for his moral support and cooperation.

References

1. A. Lacis, J. Hansen and M. Sato, Climate forcing by stratospheric aerosol, *Geophys. Res. Lett.* **19** (1992) 1607–1610.
2. D. J. Hofmann, T. J. Pepin, and R. G. Pinnick, Stratospheric aerosol measurements. I: Time variations at northern midlatitudes, *J. Atmos. Sci.* **32** (1975) 1446–1456.
3. T. Deshler, G. B. Liley, G. Bodeker, W. A. Matthews and D. J. Hofmann, Stratospheric aerosol following Pinatubo: Comparison of the North and South mid-latitudes using *in-situ* measurements, *Adv. Space Res.* **20** (1997) 2057–2061.
4. Y. B. Kumar, C. N. Raju, and M. Krishnaiah, Indo-Japanese Lidar Observations of the tropical middle atmosphere during 1998 and 1999, *Adv. Atmos. Sci.* **23** (2006) 711–725.
5. Y. B. Kumar, K. Raghunath, V. Siva Kumar, P. B. Rao, A. R. Jain, K. Mizutai, T. Aoki, M. Yasui and T. Itab, Indo-Japanese Lidar system description and data processing, *IRSI Proc.* (1999), pp. 560–567.
6. F. G. Fernald, Analysis of atmospheric lidar observations: Some comments, *Appl. Opt.* **23** (1984) 652–653.
7. T. Takamura, and Y. Sasano, Ratio of aerosol backscatter to extinction coefficients determined from angular scattering measurements for use in atmospheric lidar applications, *Opt. Quant. Electron.* **19** (1987) 293–302.
8. R. Fisher, *Statistical Methods for Research Workers*, 14th ed. (Oliver and Boyd, Edinburg, 1970).
9. A. Jayaraman, Y. B. Acharya, H. Chandra, B. H. Subbaraya, S. Rama chandran, and S. Ramaswamy, Laser radar study of the middle atmosphere over Ahmedabad, *Ind. J. Radio Space Phys.* **25** (1996) 318–327.

10. J. B. Nee, G. B. Wang, P. C. Lee, and S. B. Lin, Lidar studies of particles and temperatures of the atmosphere: First results from National Central University, Japan, *Rad. Sci.* **30** (1995) 1167–1175.
11. B. V. Krishna Murthy, K. Parameswaran, O. Rose and M. Satanarayana, Temperature dependencies of stratospheric aerosol extinction at a tropical station, *J. Atmos. Terr. Physics* **55** (1993) 809–814.